High Performance Computing on Vector Systems 2007

Michael Resch · Sabine Roller · Peter Lammers
Toshiyuki Furui · Martin Galle · Wolfgang Bez
Editors

High Performance Computing on Vector Systems 2007

Springer

Michael Resch
Sabine Roller
Peter Lammers
Höchstleistungsrechenzentrum
Stuttgart (HLRS)
Universität Stuttgart
Nobelstraße 19
70569 Stuttgart, Germany
resch@hlrs.de
roller@hlrs.de
lammers@hlrs.de

Toshiyuki Furui
NEC Corporation
Nisshin-cho 1-10
183-8501 Tokyo, Japan
t-furui@bq.jp.nec.com

Wolfgang Bez
Martin Galle
NEC High Performance Computing
Europe GmbH
Prinzenallee 11
40459 Düsseldorf, Germany
wbez@hpce.nec.com
mgalle@hpce.nec.com

Front cover figure: Impression of the projected tidal current power plant to be built in the South Korean province of Wando. Picture due to RENETEC, Jongseon Park, in cooperation with Institute of Fluid Mechanics and Hydraulic Machinery, University of Stuttgart

ISBN 978-3-642-09381-4 e-ISBN 978-3-540-74384-2

DOI 10.1007/978-3-540-74384-2

Mathematics Subject Classification (2000): 68Wxx, 68W10 , 68U20, 76-XX, 86A05, 86A10, 70Fxx

Cover design: WMX Design GmbH, Heidelberg

Printed on acid-free paper

9 8 7 6 5 4 3 2 1

springer.com

Preface

In 2004 the High Performance Computing Center of the University of Stuttgart and NEC established the TERAFLOP Workbench collaboration. The TERAFLOP Workbench is a research & Service Project for which the following targets have been defined:

- Make new science and engineering possible that requires TFLOP/s sustained application performance
- Support the HLRS user community to achieve capability science by improving existing codes
- Integrate different system architectures for simulation, pre- and post-processing, visualisation into a computational engineering workbench
- Assess and demonstrate system capabilities for industry relevant applications

In the TERAFLOP Workbench significant hardware and human resources have been made available by both partners. The hardware provided within the TERAFLOP project consists of 8 nodes NEC SX-8 Vector Computers and a cluster of 200 Intel Xeon nodes. The complete NEC installation at HLRS comprises 72 nodes SX-8, two TX-7 (i.e. 32-way Itanium based SMP systems) which are also used as front end for the SX nodes and the previously described Xeon cluster.

Six computer experts, who are dedicated to advanced application support and user services, are funded over the complete project runtime. The support is carried out on the basis of small project groups working on specific applications. These groups usually consist of application experts, in most cases members of the development team, and TERAFLOP workbench representatives. This setup to combines detailed application know-how and physical background with computer science and engineering knowledge and sound numerical mathematics expertise. The combination of these capabilities forms the basis for leading edge research and computational science.

Following the above formulated targets, the cooperation was successful in achieving sustained application performance of more than 1 TFLOP/s for more than 10 applications so far. The best performing application is the hydrodynamics code BEST, which is based on the solution of the Lattice Boltzmann equations. This application achieves a performance of 5.7 TFLOP/s on the 72 nodes SX-8. Also other hydrodynamics as well as oceanography and climatology applications are running highly efficient on the SX-8 architecture. The enhancement of applications and their adaptation to the SX-8 Vector architecture within the collaboration will continue.

On the other hand, the Teraflop Workbench project works on supporting future applications, looking at the requirements users ask for. In that context, we see an increasing interest in Coupled Applications, in which different codes are interacting to simulate complex systems of multiple physical regimes. Examples for such couplings are Fluid-Structure or Ocean-Atmosphere interactions. The codes in these coupled applications may have completely different requirements concerning the computer architecture which often results in the situation that they are running most efficient on different platforms. The efficient execution of coupled application requires a close integration of the different platforms. The platform integration and the support for coupled applications will become another important share in the TERAFLOP Workbench collaboration.

Within the framework of the TERAFLOP Workbench collaboration, semiannual workshops are carried out in which researchers and computer experts come together to exchange their experiences. The workshop series started in 2004 with the 1st TERAFLOP Workshop in Stuttgart. In autumn 2005, the TERAFLOP Workshop Japan session was established with the 3rd TERAFLOP Workshop in Tokyo.

The following book presents contributions from the 6th TERAFLOP Workshop which was hosted by Tohoku University in Sendai, Japan in autumn 2006 and the 7th Workshop in Stuttgart which was held in spring 2007 in Stuttgart. Focus is layed on current applications and future requirements, as well as developments of next generation hardware architectures and installations.

Starting with a section on geophysics and climate simulations, the suitability and necessity of vector systems is justified showing sustained teraflop performance. Earthquake simulations based on the Spectral-Element Method demonstrate that the systhetic seismic waves computed by this numerical technique match with the observed seismic waves accurately. Further papers address cloud-resolving simulation of tropical cyclones, or the question: What is the impact of small-scale phenomena on the large-scale ocean and climate modeling? Ensemble climate model simulation discribed in the closing paper in this section enable scientists to better distinguish the forced signal due to the increase of greenhouse gases from internal climat variability.

A section on computational fluid dynamics (CFD) starts with a paper discussing the current capability of CFD and the maturity to reduce wind tunnel testings. Further papers in this section show simulations in applied fields as aeronautics and flows in gas and steam turbines, as well as basic research and detailed performance analysis.

The following section considers multi-scale and multi-physics simulations based on CFD. Current applications in aero-acoustics and the coupling of Large-Eddy Simulation (LES) with acoustic perturbation equations (APE) start the section, followed by fluid-structure interaction (FSI) in such different aereas as tidal current turbines or the respiratory systems. The section is closed by a paper addressing the algorithmic and imlementation issues associated with FSI simulations on vector architecture. These examples show to us the tendency to coupled applications and the requirements coming up with future simulation techniques.

The section on chemistry and astrophysics combines simulation of premixed swirling flames and supernova simulations. The common basis for both applications is the combination of a hydrodynamic module with processes as chemical kinetics or multi-floavour, multi-frequencey neutrino transport based on the Boltzmann transport equation, respectively.

A section on material science closes the applications part. Green chemistry from supercomputers considers Car-Parrinello simulations of ionic liquids. Micromagnetic simulations of magnetic recording media allow new head and media designs to be evaluated and optimized prior to fabrications.

These sections show the wide range of application areas performed on current vector systems. The closing section on Future High Performance Systems consider the potential of on-chip memory systems for future vector architectures. A technical note describing the TSUBAME installation at Tokyo Institute of Technology (TiTech) closes the book.

The papers presented in this book lay out the wide range of fields in which sustained performance can be achieved if engineering knowledge, numerical mathematics and computer science skills are brought together. With the advent of hybrid systems, the Teraflop workbench project will continue the support of leading edge computations for future applications.

The editors would like to thank all authors and Springer for making this publication possible and would like to express their hope that the entire high performance computing community will benefit from it.

Stuttgart, Juli 2007

M. Resch, W. Bez
S. Roller, M. Galle
P. Lammers, T. Furui

Contents

Applications I Geophysics and Climate Simulations

Applications II Computational Fluid Dynamics

Future High Performance Systems

List of Contributors

Applications I

Geophysics and Climate Simulations

Sustained Performance of 10+ Teraflop/s in Simulation on Seismic Waves Using 507 Nodes of the Earth Simulator

Seiji Tsuboi

Institute for Research on Earth Evolution, JAMSTEC tsuboi@jamstec.go.jp

Summary. Earthquakes are very large scale ruptures inside the Earth and generate elastic waves, known as seismic waves, which propagate inside the Earth. We use a Spectral-Element Method implemented on the Earth Simulator in Japan to calculate seismic waves generated by recent large earthquakes. The spectral-element method is based on a weak formulation of the equations of motion and has both the flexibility of a finite-element method and the accuracy of a pseudospectral method. We perform numerical simulation of seismic wave propagation for a fully three-dimensional Earth model, which incorporates realistic 3D variations of Earth's internal properties. The simulations are performed on 4056 processors, which require 507 out of 640 nodes of the Earth Simulator. We use a mesh with 206 million spectral-elements, for a total of 13.8 billion global integration grid points (i.e., almost 37 billion degrees of freedom). We show examples of simulations and demonstrate that the synthetic seismic waves computed by this numerical technique match with the observed seismic waves accurately.

1 Introduction

The Earth is an active planet, which exhibits thermal convection of solid mantle and resultant plate dynamics at the surface. As a result of continuous plate motion, we have seismic activities and sometimes we experience huge earthquakes, which causes devastating damage to the human society. In order to know the rupture process during large earthquakes, we need to have accurate modeling of seismic wave propagation in fully three-dimensional (3-D) Earth models, which is of considerable interest in seismology. However, significant deviations of Earth's internal structure from spherical symmetry, such as the 3-D seismic-wave velocity structure inside the solid mantle and laterally heterogeneous crust at the surface of the Earth, have made applications of analytical approaches to this problem a formidable task. The numerical modeling of seismic-wave propagation in 3-D structures has been significantly advanced in the last few years due to the introduction of the Spectral-Element Method (SEM), which is a high-degree version of the finite-element method.

The 3-D SEM was first used in seis-mology for local and regional simulations ([Ko97]-[Fa97]), and more recently adapted to wave propagation at the scale of the full Earth ([Ch00]-[Ko02]) .

In addition, massively parallel super computer has started its operation in 2002 at Japan Agency for Marine-Earth Science and Technology (JAMSTEC). The machine is called the Earth Simulator and dedicated specifically to basic research in Earth sciences, such as climate modeling (http://www.es.jamstec.go.jp). The Earth Simulator consisted of 640 processor nodes with each equipped with 8 vector processors. Each vector processor has its peak performance of 8 GFLOPS and main memory is 2 gigabytes per processor. In total, the peak performance of the Earth Simulator is about 40 TFLOPS and maximum memory size is about 10 terabytes. In 2002, the Earth Simulator has scored 36 TFLOPS as its sustained performance and has been ranked as No.1 of TOP500.

Here we show that our implementation of the SEM on the Earth Simulator in Japan allows us to calculate theoretical seismic waves which are accurate up to 3.5 seconds and longer for fully 3-D Earth models. We include the full complexity of the 3-D Earth in our simulations, i.e., a 3-D seismic wave velocity [Ri99] and density structure, a 3-D crustal model [Ba00], ellipticity as well as topography and bathymetry. Because dominant frequency of body waves, which are one of the seismic waves that travel inside the Earth, is 1 Hz, it is desirable to have synthetic seismograms with the accuracy of 1Hz. However, synthetic waveforms at this resolution (periods of 3.5 seconds and longer) also allow us to perform direct comparisons between observed and synthetic seismograms with various aspects, which has never been accomplished before. Conventional seismological algorithms, such as normal-mode summation techniques that calculate quasi-analytical synthetic seismograms for one-dimensional (1-D) spherically symmetric Earth models [Dz81], are typically accurate down to 8 seconds [Da98]. In other words, the SEM on the Earth Simulator allows us to simulate global seismic wave propagation in fully 3-D Earth models at periods shorter than current seismological practice for simpler 1-D spherically symmetric models. The results of our simulation show that the synthetic seismograms calculated for fully 3-D Earth models by using the Earth Simulator and the SEM agree well with the observed seismograms, which enables us to investigate the earthquake rupture history and the Earth's internal structure in much higher resolution than before.

2 Numerical Technique

We use the spectral-element method (SEM) developed by Komatitsch and Tromp [Ko02a, Ko02b] to simulate global seismic wave propagation throughout a 3-D Earth model, which includes a 3-D seismic velocity and density structure, a 3-D crustal model, ellipticity as well as topography and bathymetry. The SEM first divides the Earth into six chunks. Each of the six chunks is divided into slices. Each slice is allocated to one CPU of the Earth Simulator. Communication between each CPU is done by MPI. Before the system can be marched forward in time, the contributions from all the elements that share a common global grid point need to be summed. Since the global mass matrix is diagonal, time discretization of the second-order ordinary differential equation is achieved based upon a classical explicit second-order finite-difference scheme.

The number of nodes we used for this simulation is 4056 processors, i.e., 507 nodes out of 640 of the Earth Simulator. This means that each chunk is subdivided into 26×26 slices ($6 \times 26 \times 26 = 4056$). Each slice is allocated to one processor of the Earth Simulator and subdivided with a mesh of 48×48 spectral-elements at the surface of each slice. Within each surface element we use $5 \times 5 = 25$ Gauss-Lobatto-Legendre (GLL) grid points to interpolate the wave field [Ko98, Ko99], which translates into an average grid spacing of 2.0 km (i.e., 0.018 degrees) at the surface. The total number of spectral elements in this mesh is 206 million, which cor-responds to a total of 13.8 billion global grid points, since each spectral element contains $5 \times 5 \times 5 = 125$ grid points, but with points on its faces shared by neighboring elements. This in turn corresponds to 36.6 billion degrees of freedom (the total number of degrees of freedom is slightly less than 3 times the number of grid points because we solve for the three components of displacement everywhere in the mesh, except in the liquid outer core of the Earth where we solve for a scalar potential). Using this mesh, we can calculate synthetic seismograms that are accurate down to seismic periods of 3.5 seconds. Total performance of the code, measured using the MPI Program Runtime Performance Information was 10 teraflops, which is about one third of the expected peak performance for this number of nodes (507 nodes $\times$ 64gigaflops = 32 teraflops). Figure 1 shows a global view of the spectral-element mesh at the surface of the Earth. Before we could use 507 nodes of the Earth Simulator for this simulation, we could have successfully used 243 nodes to calculate synthetic seismograms. Using 243 nodes (1944 CPUs), we can subdivide the six chunks into 1944 slices ($1944 = 6 \times 18 \times 18$). Each slice is then subdivided into 48 elements in one direction. Because each element has 5 Gauss-Lobatto Legendre integration points, the average grid spacing at the surface of the Earth is about 2.9 km. The number of grid points in total amounts to about 5.5 billion. Using this mesh, it is expected that we can calculate synthetic seismograms accurate up to 5 sec all over the globe. For the 243 nodes case, the total performance we achieved was about 5 teraflops, which also is about one third of the peak performance. The fact that when we double the number of nodes from 243 to 507 the total performance also doubles from 5 teraflops to 10 teraflops shows that our SEM code exhibits an excellent scaling relation with respect to performance. The details of our computation with 243 nodes of the Earth Simulator were described in Tsuboi et al (2003) [Ts03] and Komatitsch et al (2003) [Ko03a], which was awarded 2003 Gordon Bell prize for peak performance in SC2003.

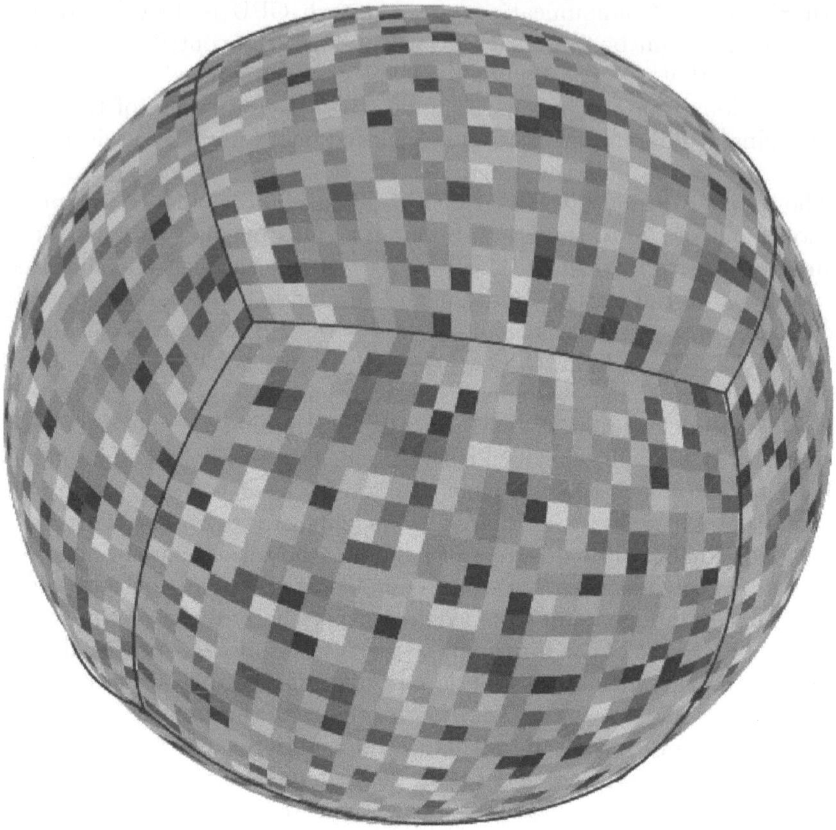

Fig. 1. The SEM uses a mesh of hexahedral finite elements on which the wave field is interpolated by high-degree Lagrange polynomials on Gauss-Lobatto-Legendre (GLL) integration points. This figure shows a global view of the mesh at the surface, illustrating that each of the six sides of the so-called "cubed sphere" mesh is divided into 26 $\times$ 26 slices, shown here with different colors, for a total of 4056 slices (i.e., one slice per processor).

3 Examples of 2004 Great Sumatra Earthquake

On December 26, 2004, one of the great earthquakes ever recorded by modern seismographic instruments has occurred along the coast of Sumatra Island, Indonesia. The magnitude of this huge earthquake was estimated to be 9.1 -9.3, whereas the greatest earthquake ever recorded was 1960 Chilean earthquake with the magnitude 9.5. Since this event has caused devastating tsunami hazard around the Indian Ocean, it is important to know how this event has started its rupture and propagated along the faults because the excitation mechanism of tsunami is closely related to the earthquake source mechanisms. It is now estimated that the earthquake started its rupture at the west of northern part of Sumatra Island and propagated in a northwestern direction up to Andaman Islands. Total length of the earthquake fault is estimated to be more than 1000 km and the rupture duration lasts for more than 500 sec. To simulate synthetic seismograms for this earthquake, we represent the earthquake source by more than 800 point sources distributed both in space and time, which are obtained by seismic wave analysis. In Fig. 2, we show snapshots of seismic wave propagation along the surface of the Earth. Because the rupture along the fault propagated in a northwest direction, the seismic waves radiated in this direction are strongly amplified. This is referred

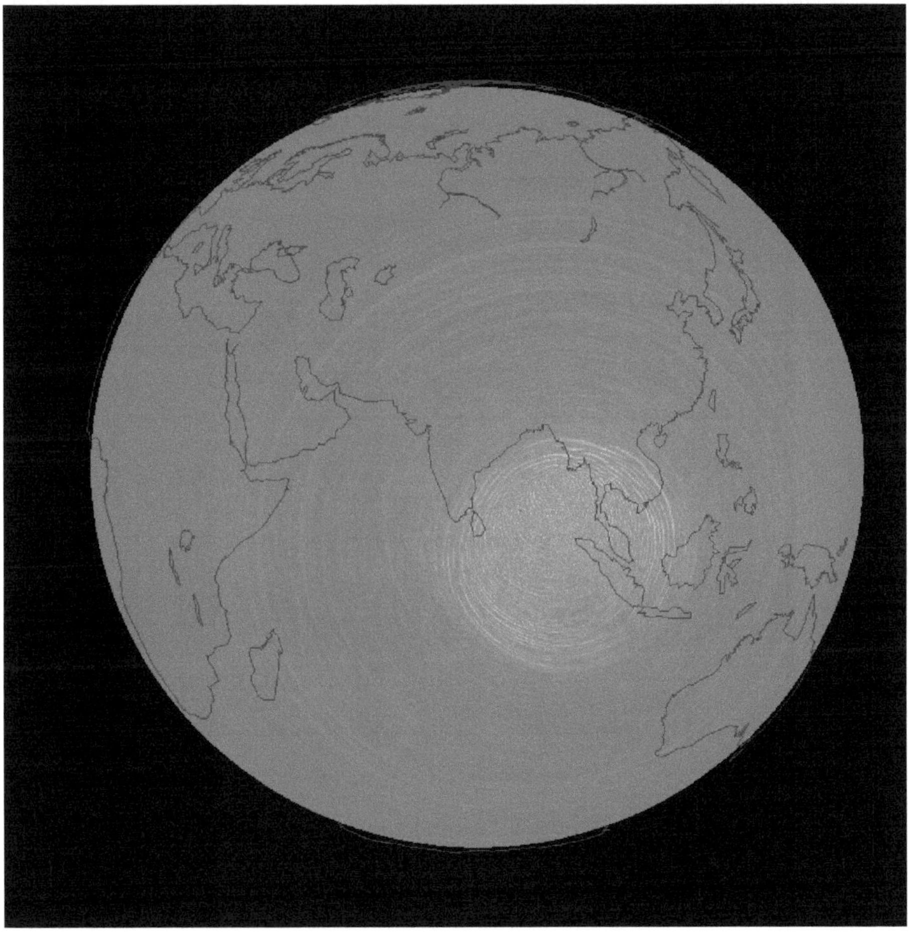

Fig. 2. Snapshots of the propagation of seismic waves excited by the December 26, 2004 Sumatra earthquake. Total displacement at the surface of the Earth is plotted at 10 min after the origin time of the event.

as the directivity caused by the earthquake source mechanisms. Figure 2 illustrate that the amplitude of the seismic waves becomes large in the northwest direction and shows that this directivity is modeled well. Because there are more than 200 seismographic observatories, which are equipped with broadband seismometers all over the globe, we can directly compare the synthetic seismograms calculated with the Earth Simulator and the SEM with the observed seismograms. Figure 3 shows comparisons of observed seismograms and synthetic seismograms for some broadband seismograph stations.

The results demonstrate that the agreement between synthetic and observed seismograms is generally excellent and illustrate that the earthquake rupture model that we have used in this simulation is accurate enough to model seismic wave propagation on a global scale. Because the rupture duration of this event is more than 500 sec, the first arrival P waveform overlapped with the surface reflected wave of P-wave, which is called PP wave. Although this effect may obscure the analysis of earthquake source mechanism, it has been shown that the synthetic seismograms computed with Spectral-Element Method on the Earth Simulator can fully take these effects into account and are quite useful to study source mechanisms of this complicated earthquake.

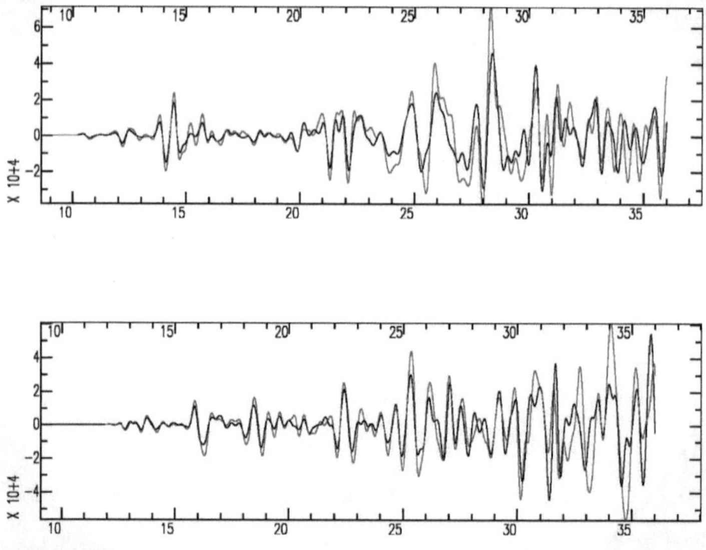

Fig. 3. Comparison of synthetic seismograms (red) and observed seismograms(black) for December 26, 2004, Sumatra earthquake. 1 hour ground vertical velocity seismograms are shown. Seismograms are lowpass filtered at 0.02Hz. Top figure shows seismogram recorded at Tucson, Arizona, USA and bottom figure shows those recorded at Puerto Ayora in Galapagos Islands.

4 Application to estimate the Earth's internal structure

The Earth's internal structure is another target that we can study by using our synthetic seismograms calculated for fully 3-D Earth model. We describe the examples of Tono et al (2005) [To05]. They used records of 500 tiltmeters of the Hi-net, in addition to 60 broadband seismometers of the F-net, operated by the National Research Institute for Earth Science and Disaster Prevention of Japan (NIED). They analyzed pairs of sScS waves, which means that the S-wave traveled upward from the hypocenter reflected at the surface and reflected again at the core-mantle boundary, and its reverberation from the 410- or 660-km reflectors (sScSSdS where d=410 or 660 km) for the deep earthquake of the Russia-N.E. China border (PDE; 2002:06:28; 17:19:30.30; 43.75N; 130.67E; 566 km depth; 6.7 Mb). The two horizontal components are rotated to obtain the transverse component. They have found that these records show clearly the near-vertical reflections from the 410- and 660-km seismic velocity discontinuities inside the Earth as post-cursors of sScS phase. By reading the travel time difference between sScS and sScSSdS, they concluded that this differential travel time anomaly can be attributed to the depth anomaly of the reflection point, because it is little affected by the uncertainties associated with the hypocentral determination, structural complexities near the source and receiver and long-wavelength mantle heterogeneity. The differential travel time anomaly is obtained by measuring the arrival time anomaly of sScS and that of sScSSdS separately and then by taking their difference. The arrival time anomaly of sScS (or sScSSdS) is measured by cross-correlating the observed sScS (or sScSSdS) with the corresponding synthetic waveform computed by SEM on the Earth Simulator. They plot the measured values of the two-way near-vertical travel time anomaly at the corresponding surface bounce points located beneath the Japan Sea. The results show that the 660-km boundary is depressed at a constant level of 15 km along the bottom of the horizontally extending aseismic slab under southwestern Japan. The transition from the normal to the depressed level occurs sharply, where the 660-km boundary intersects the bottom of the obliquely subducting slab. This observation should give important imprecations to geodynamic activities inside the Earth.

Another topic is the structure of the Earth's inner most core. The Earth has solid inner core inside the fluid core with the radius of about 1200 km. It is proposed that the inner core has anisotropic structure, which means that the seismic velocity is faster in one direction than the other, and used to infer inner core differential rotation [Zh05]. Because the Earth's magnetic field is originated by convective fluid motion inside the fluid core, the evolution of the inner core should have important effect to the evolution of the Earth's magnetic field.

Figure 4 illustrates definitions of typical seismic waves which travel through the Earth's core. The seismic wave, labeled as PKIKP, penetrates inside the inner core and its propagation time from the earthquake hypocenter to the

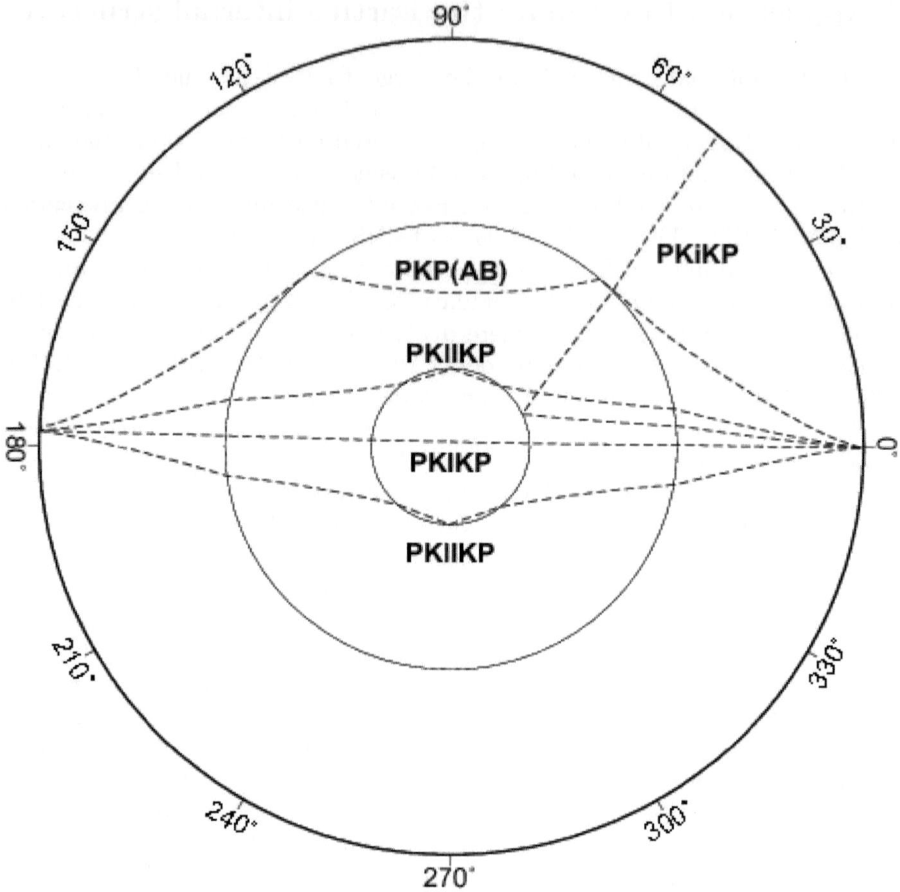

Fig. 4. Raypaths and its naming conventions of seismic waves, which travel inside the Earth's core.

seismic station (that is travel time) is used to infer the seismic velocity structure inside the inner core. Especially the dependence of PKIKP travel time to the direction is useful to estimate anisotropic structure of the inner core. We calculate synthetic seismograms for those PKIKP and PKP(AB) waves and evaluate the effect of inner core anisotropy to these waves. When we construct global mesh in SEM computation, we put one small slice at the center of the Earth. Because of this, we do not have any singularity at the center of the Earth, which makes our synthetic seismograms very accurate and unique. We calculate synthetic seismograms by using the Earth Simulator and SEM for deep earthquake on April 8, 1999, at E. USSR-N.E. CHINA Border region (43.66N 130.47E depth 575.4km Mw7.1). We calculate synthetic seismograms for both isotropic inner core model and anisotropic inner core model and compare with the observed seismograms. Figure 5 summarizes comparisons of

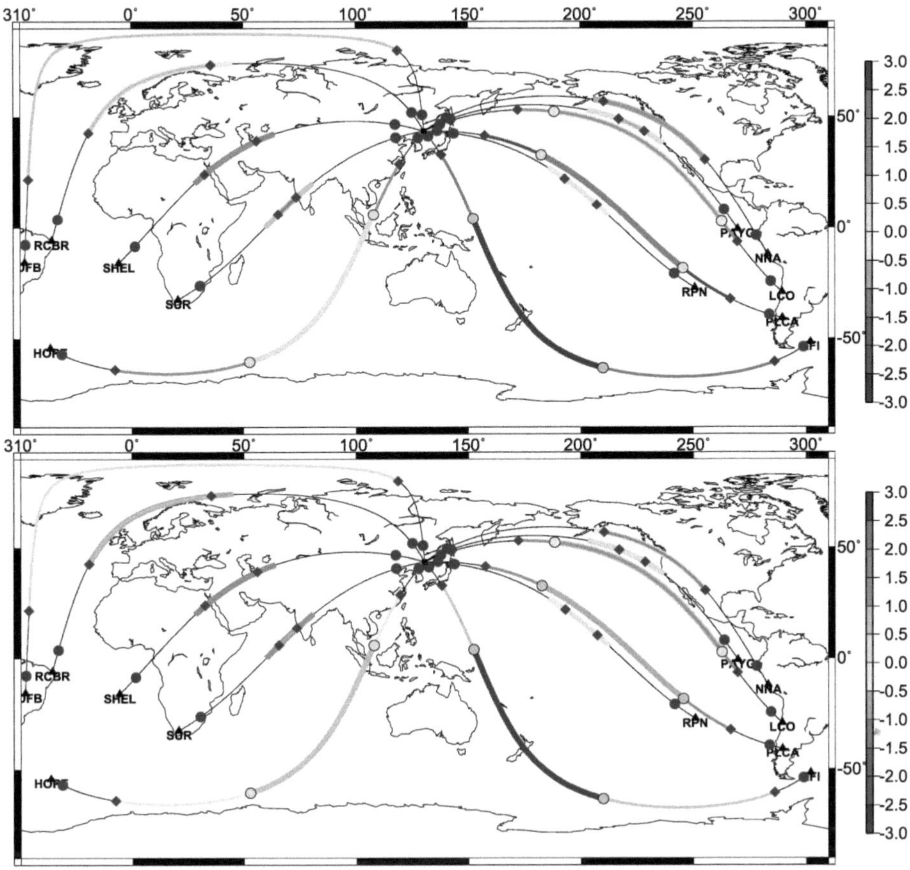

Fig. 5. Great circle paths to the broadband seismograph stations from the earthquake. Open circles show crossing points of Pdiff paths along the core mantle boundary (CMB). Red circles show crossing point at CMB for PKP(AB). Blue circles show crossing point at CMB for PKIKP. Blue squares show crossing point at ICB for PKIKP.Travel time differences of (synthetics)-(observed) are overlaid along the great circle paths with the color scale shown in the right of the figures. Comparison for isotropic inner core model (top) and anisotropic inner core model (bottom).

synthetics and observation. Travel time differences of (synthetics)-(observed) are overlaid along the great circle paths with the color scale shown in the right of the figures. The results show: (1) Travel time differences of PKIKP phases are decreased by introducing anisotropic inner core. (2) For some stations, there still left significant differences in travel time differences for PKIKP. (3) Observed Pdiff phases, which are diffracted wave along the core mantle boundary, are slower than the synthetics, which shows that we need to introduce slow velocity at CMB.

These results illustrate that the current inner core anisotropic model does improve the observation but it also should be modified to get much better agreement. They also demonstrate that there exist some anomalous structure along some portion of the core mantle boundary. This kind of anomalous structure should be incorporated in the Earth model to explain observed travel time anomaly of Pdiff waves.

5 Discussion

We have shown that we now calculate synthetic seismograms for realistic 3D Earth model with the accuracy of 3.5 sec by using the Earth Simulator and SEM. 3.5 second seismic wave is sufficiently short enough to explain various characteristics of seismic waves which propagate inside aspherical Earth. However, it also is true that we need to have 1Hz accuracy to explain body wave travel time anomaly. We will consider if it will be possible to calculate 1Hz seismograms in near future using the next generation Earth Simulator. We could calculate seismograms of 3.5 sec accuracy with 507 nodes (4056 CPUs) of the Earth Simulator. The key to how we increase the accuracy is the size of the mesh used in SEM. If we reduce the size of one slice by half, the required memory will become quadruple and the accuracy is increased by $\sqrt{2}$. Thus to have 1Hz accuracy, we should reduce the size of mesh at least one fourth of 3.5 sec (507 nodes) case. If we assume that the size of memory available per each CPU is the same as the current Earth Simulator, we need to have at least $16 \times 507 = 8112$ nodes (64,896 CPUs). If we can use 4 times larger memory per CPU, then number of CPU becomes 16,224, which is a realistic value. We have examined if we will be able to have expected performance for possible candidate of next generation machine. We have NEC SX-8R at JAMSTEC, of which peak performance of each CPU is about 4 times faster than that of the Earth Simulator. We have compiled our SEM program on SX-8R as it is and measured the performance. The result shows that the performance is less than two times faster than the Earth Simulator. Because we have not optimized our code so that it fits the architecture of SX-8R, there is still a possibility that we may have good performance. However, we have found that the reason why we did not get good performance is because of the memory access speed. As the memory used in SX-8R is not as fast as the Earth Simulator, bank conflict time becomes bottleneck of the performance. This result illustrates that it may become feasible to calculate 1Hz synthetic seismograms on the next generation machine but it is necessary to have good balance between CPU speed and memory size to get excellent performance.

Acknowledgments

The author used the program package SPECFEM3D developed by Jeroen Tromp and Dimitri Komatitsch at Caltech to perform Spectral-Element method computation. All of the computation shown in this paper was done by using the Earth Simulator operated by the Earth Simulator Center of JAMSTEC. The rupture model of 2004 Sumatra earthquake was provided by Chen Ji of University of California Santa Barbara. Figures 3 through 5 are prepared by Dr. Yoko Tono of JAMSTEC. Implementation of SEM program on SX-8R was done by Dr. Ken'ichi Itakura of JAMSTEC.

References

[Ko97] Komatitsch, D.: Spectral and spectral-element methods for the 2D and 3D elasto-dynamics equations in heterogeneous media, PhD thesis, Institut de Physique du Globe, Paris (1997)

[Fa97] Faccioli, E., F. Maggio, R. Paolucci, A. Quarteroni,: 2D and 3D elastic wave propagation by a pseudo-spectral domain decomposition method. J. Seismol., **1**, 237–251 (1997)

[Se98] Seriani, G.: 3-D large-scale wave propagation modeling by a spectral element method on a Cray T3E multiprocessor. Comput. Methods Appl. Mech. Engrg., **164**, 235–247 (1998)

[Ch00] Chaljub, E.: Numerical modelling of the propagation of seismic waves in spherical geometry: applications to global seismology. PhD thesis, Universit Paris VII Denis Diderot, Paris (2000)

[Ko02a] Komatitsch, D., J. Tromp: Spectral-element simulations of global seismic wave propagation-I. Validation. Geophys. J. Int. **149**, 390–412 (2002)

[Ko02b] Komatitsch, D, J. Tromp: Spectral-element simulations of global seismic wave propagation-II. 3-D models, oceans, rotation, and self-gravitation. Geophys. J. Int. **150**, 303–318 (2002)

[Ko02] Komatitsch, D., J. Ritsema, J. Tromp: The spectral-element method, Beowulf computing, and global seismology. Science, **298**, 1737–1742 (2002)

[Ri99] Ritsema, J., H. J. Van Heijst, J. H. Woodhouse: Complex shear velocity struc-ture imaged beneath Africa and Iceland. Science **286**, 1925–1928 (1999)

[Ba00] Bassin, C., G. Laske, G. Masters: The current limits of resolution for surface wave tomography in North America. EOS Trans. AGU. **81**: Fall Meet. Suppl., Abstract S12A-03 (2000)

[Dz81] Dziewonski, A. M., D. L. Anderson: Preliminary reference Earth model. Phys. Earth Planet. Inter. **25**, 297–356 (1981)

[Da98] Dahlen, F. A., J. Tromp: Theoretical Global Seismology. Princeton University Press, Princeton (1998)

[Ko98] Komatitsch, D., J. P. Vilotte: The spectral-element method: an efficient tool to simulate the seismic response of 2D and 3D geological structures. Bull. Seismol. Soc. Am. **88**, 368–392 (1998)

[Ko99] Komatitsch, D., J. Tromp: Introduction to the spectral-element method for 3-D seismic wave propagation. Geophys. J. Int. **139**, 806–822 (1999)

[Ts03] Tsuboi, S., D. Komatitsch, C. Ji, J. Tromp: Broadband modeling of the 2003 Denali fault earthquake on the Earth Simulator, Phys. Earth Planet. Int., **139**, 305–312 (2003)

[Ko03a] Komatitsch, D., S. Tsuboi, C. Ji, J. Tromp: A 14.6 billion degrees of freedom, 5 teraflops, 2.5 terabyte earthquake simulation on the Earth Simulator, Proceedings of the ACM/IEEE SC2003 confenrence, published on CD-ROM, (2003)

[To05] Tono, Y., T. Kunugi, Y. Fukao, S. Tsuboi, K. Kanjo, K. Kasahara: Mapping the 410- and 660-km discontinuities beneath the Japanese Islands. J. Geophys. Res., **110**, B03307, doi:10.1029/2004JB003266 (2005)

[Zh05] Zhang, J., X. Song, Y. Li, P. G. Richards, X. Sun, F. Waldhauser: Inner core differential motion confirmed by earthquake waveform doublets. Science, **309**, 1357–1360 (2005)

Cloud-Resolving Simulation of Tropical Cyclones

Toshiki Iwasaki[1] and Masahiro Sawada[1]

Department of Geophysics, Graduate School of Science, Tohoku University
iwasaki@wind.geophys.tohoku.ac.jp

1 Numerical simulation of tropical cyclones

Many casualties have been recorded due to typhoons (tropical cyclones in the western North Pacific region) for a long time in Japan. In 1959, T15, VERA, which is known as the "Isewan Typhoon" in Japan, landfalled the Kii Peninsula, bringing heavy precipitation, generated the storm surge of 3.5 m on the coast of Ise Bay, flooding over Nobi Plain very widely and killing more than 5000 people. Since this event, many efforts have been made to improve social infrastructures against typhoon, such as seawall and riverbank, and to establish its observation and prediction system. These efforts succeeded in significantly reducing casualties. Even now, however, typhoons are among the most hazardous meteorological phenomena in Japan. In the North America, the coastal regions suffer from tropical cyclones, hurricanes as well. Accurate forecasts of tropical cyclones are of great concern in societies all over the world.

The Japan Meteorological Agency (JMA) issues typhoon track prediction basis on numerical weather prediction (NWP). In some sense, it is more difficult to simulate tropical cyclones than extratropical cyclones. Numerical models for typhoon prediction must cover large domains with high resolutions, because tropical cyclones are very sharp near the centers but move around over very wide areas. Thus, many years ago, JMA began conducting typhoon predictions using a moving two-way multi-nested model, whose highest resolution region can be moved to follow the typhoon center, in order to save computational resources (Ookochi, 1978,[4]). The resolution gap of the multi-nesting, however, has limited forecast performance because of differences in characteristics of simulations. Another reason for the difficulty in the typhoon prediction is related to the energy source of typhoon, that is, released condensation heating of water vapor. Iwasaki et al. (1989) developed a numerical model with uniform resolution over the entire domain, a parameterization scheme for deep cumulus convections and a bogussing technique for initial vortices. It was confirmed that implementing cumulus parameterization considerably improves the performance of typhoon track forecasts.

The global warming is an important issue which may affect the future climatology of tropical cyclones. This issue has been thoroughly discussed, since Emanuel (1987) [1] suggested the possibility of a "super typhoon". In his hypothesis, increased sea surface temperature (SST) moist statically destabilizes the atmosphere and increases the maximum intensity of tropical cyclones. Many general circulation model (GCM) experiments have been conducted to analyze tropical cyclone climatology under the influence of global warming. Most experiments have shown that the global warming statically stabilizes the atmosphere and considerably reduces the number of tropical cyclones (e.g., Sugi et al, 2002,[7]). Recently, however, Yoshimura et al. (2006) [8] found that the number of strong tropical cyclones will increase due to the global warming, although that of weak tropical cyclones will decrease. This seems to be consistent with the hypothesis by Emanuel (1987) [1]. Intensive studies are underway to reduce the uncertainty of predicting tropical cyclone climatology under the influence of global warming.

Numerical models are key components for typhoon track and intensity prediction track and for forecasting cyclone climatology under the influence of global warming. NWP models and GCMs currently used do not have enough horizontal resolutions to simulate deep cumulus clouds. Thus, they implement deep cumulus parameterization schemes, which express vertical profiles of condensation heating and moisture sink, considering their undulations within grid boxes. They have a lot of empirical parameters which are determined from observations and/or theoretical studies. As a result, forecast errors and uncertainties of future climate are attributed to cumulus parameterization schemes in the model. In particular, the deep cumulus convection schemes can not explicitly handle cloud microphysics. In order to eliminate the uncertainty of deep cumulus parameterization schemes, we should use high resolution cloud-resolving models with a horizontal grid spacing of equal to or less than 2km, which incorporate appropriate cloud microphysics schemes. Of course, cloud microphysics has many unknown parameters to be studied. To survey parameters efficiently, we should know their influences on tropical cyclones in advance.

In this note, we demonstrate effects of ice-phase processes on the development of an idealized tropical cyclone. In deep cumulus convections, latent heat is released (absorbed), when liquid water is converted into (from) snow, respectively. Snow flakes fall at much slower speed than liquid rain drops. Such latent heating and different falling speeds affect organized large-scale features of tropical cyclones.

2 Cloud-resolving models for simulations of tropical cyclones

Current NWP models generally assume the hydrostatic balance in their dynamical cores for economical reasons. The assumption of hydrostatic balance is valid only for atmospheric motions, in which their horizontal scales are larger than their vertical scales. The energy source of tropical cyclones is the latent heating generated in the deep cumulus convections. Horizontal and vertical scales of deep cumulus clouds are about 10km so that they do not satisfy the hydrostatic balances which is usually assumed in coarse resolution models. The Boussinesq approximation, which is sometime assumed to economically simulate small-scale phenomena, is not applicable to deep atmosphere. Thus, the cloud-resolving models adopt fully compressible nonhydrostatic dynamical cores. Their grid spacings need to be much smaller than the scale of clouds.

Cloud-resolving models explicitly express complicated cloud microphysics as shown in Fig. 1. There are many water condensates, such as water vapor, cloud water, rainwater, cloud ice, snowCand graupel, whose mixing ratios and effective sizes must be treated as prognostic variables in the model. Terminal velocities of liquid and solid substances are important parameters for their densities. They can be transformed into each other as indicated by arrows, and their change rates are important empirical parameters. The changes among gas, liquid and solid phases accompany the latent heat absorption from the atmosphere or its release to the atmosphere. Total content of water substances provides the water loading and changes the buoyancy of the atmosphere. Optical properties are different from each other and are used for computing atmospheric radiations.

3 An experiment on effects of ice phase processes on an idealized tropical cyclone

3.1 Experimental design

An important question is how the cloud microphysics affects organizations of tropical cyclone. Here, the ice phase processes, which can not be explicitly expressed in deep cumulus parameterization scheme used in the conventional numerical weather prediction and climate models, are of great interest. We describe this problem based on the idealized experiment by Sawada and Iwasaki (2007) [6].

The dynamical framework of the cloud-resolving model is the NHM developed at JMA (Saito et al. 2006, [5]). We set the computational domain of 1200km by 1200km covered with a 2km horizontal grid and 42 vertical layers up to 25km. The domain is surrounded by an open lateral boundary condition. Cloud microphysics is expressed in terms of double-moment bulk

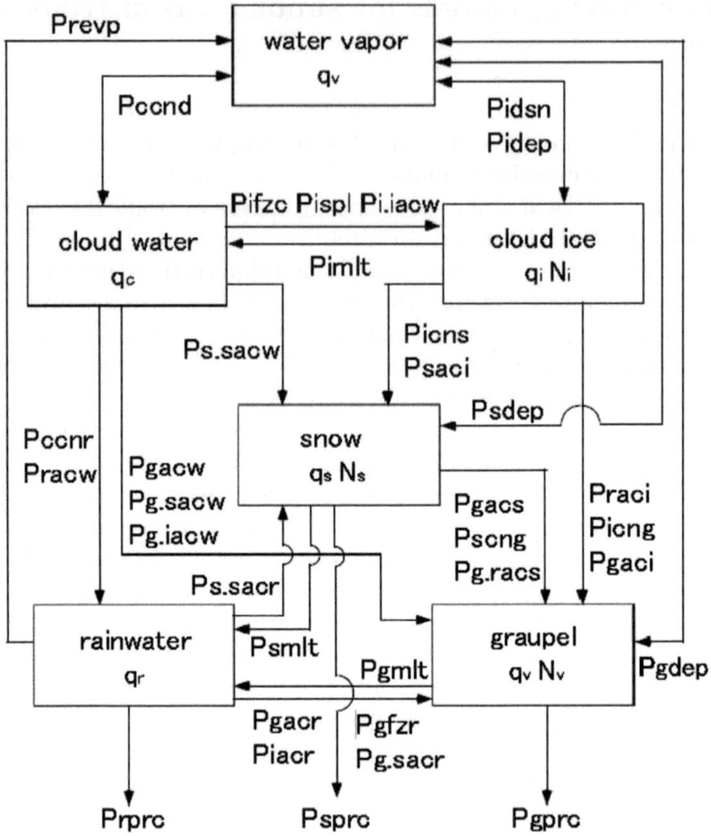

Fig. 1. Cloud microphysical processes. Boxes indicate water condensate and arrows do transformation or fallout processes. After Murakami (1990) [3]

method by Murakami (1990) [3]. Further details on the physics parameterization schemes used in this model can be found in Saito et al. (2006) [5] and Sawada and Iwasaki (2007) [6].

Real situations are too complicated to consider effects of cloud microphysics in detail, because of their inhomogeneous environments. In addition, the computational domain of a real situation needs to be much larger than the above, because tropical cyclones move around. Therefore, instead of real situations, we examine TC development under an idealized calm condition and a constant Coriolis parameter at 10N. The horizontally uniform environment is derived from averaging temperature and humidity over the active regions of tropical cyclogenesis, the subtropical Western North Pacific (EQ to 25N, 120E to 160E) in August for five years (1998-2002). As an initial condition, an axially symmetric weak vortex (Nasuno and Yamasaki 1997) is superposed on the environment. From the initial condition, the cloud-resolving model is integrated for 120 hours with and without ice phase processes, and the results are compared to see their impacts. Hereafter, integrations with and without ice phase processes are called as control run and warm rain run, respectively.

3.2 Results

The development and structure of a tropical cyclone are very different between control and warm rain runs. Figure 2 depicts time sequences of central mean sea level pressure (MSLP), maximum azimuthally averaged tangential wind, area-averaged precipitation rate and area-averaged kinetic energy. The warm rain run exhibits greater precipitation and increases in maximum wind and depth of the central pressure more rapidly than the control run. It indicates that ice phase processes interfere with the organization of tropical cyclones, which is called as Conditional Instability of the Second Kind (CISK). As far as the maximum wind and pressure depth are concerned, the control run catches up to the warm rain run at about day 3 of integration and achieves the same levels. However, the kinetic energy of warm rain run is still increasing at day 5 of integration and becomes much larger than the control run. The warm rain run continues to extend its strong wind area.

Figure 3 presents the horizontal structure distributions of vertically integrated total water condensates for the control and warm-rain experiments after 5 days of integration. Although the two runs have circular eyewall clouds of the total water condensate maximum, their radii differ significantly. The radius of the eyewall is about 30km for the control run, and about 60km for the warm-rain run. Figure 4 presents radius-height distributions of azimuthally averaged tangential wind and water condensates. In the control run, the total water content has double peaks near the ground and above the melting level, where the latter is due to the slow vertical velocity of snow. The low-level tangential wind is maximal at around 30km in the control run and around 50km in the warm rain run. Thus, ice phase processes considerably shrink the tropical cyclone in the horizontal dimension even in at their equilibrium states. The impact on the radius is very important for actual typhoon track predictions, because the typhoon tracks are very sensitive to their simulated sizes (Iwasaki et al., 1987 [2]).

Figure 5 plots the vertical profile of 12-hourly and azimuthally averaged diabatic heating for a mature tropical cyclone at day 5 of integration. In the control run with ice phase processes, the total diabatic heating rates are of three categories. Figures 5a to c present the sum of condensational heating and evaporative cooling rates (cnd+evp), the sum of freezing heating and melting cooling rates (frz+mlt), and the sum of deposition heating and sublimation cooling rates (dep+slm) respectively. Strong updrafts are induced near the eyewall by condensation heating below the melting layer and depositional heating and in the upper troposphere respectively. In reverse, the updrafts induce the large condensation heating and depositional heating (see Figs. 5a, c). Also, the updrafts induce the small freezing heating above the melting layer

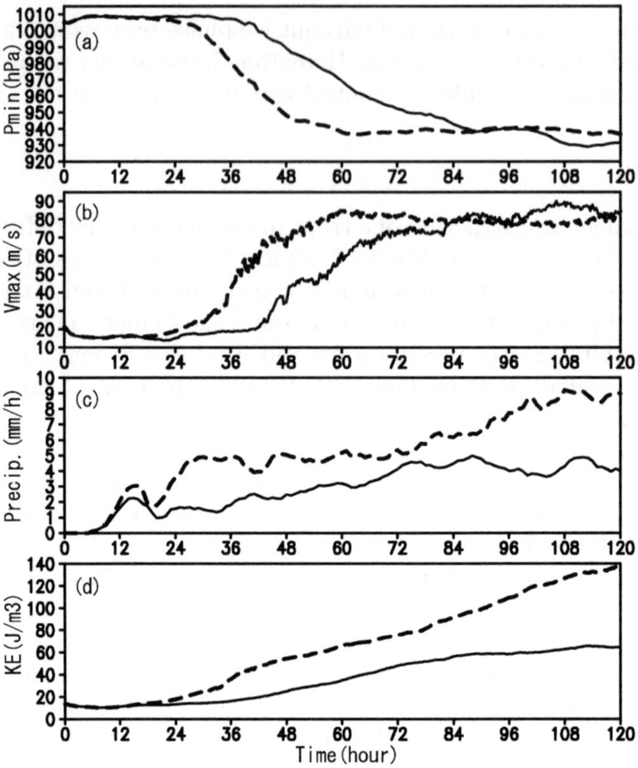

Fig. 2. Impacts of ice phase processes on the time evolution of an idealized tropical cyclone. The panels depict (a) minimum sea-level pressure, (b) maximum azimuthally averaged tangential wind, (c) area-averaged precipitation rate within a radius of 200km from the center of the TC, and (d) area-averaged kinetic energy within a radius of 300km fron the center of the TC; the solid and dashed lines indicate the control experiment including ice phase processes and the warm-rain experiment, respectively. After Sawada and Iwasaki (2007) [6].

(Fig. 5b). Melting and sublimation cooling spread below and above the melting layer (4-7km), respectively (Figs. 5b, c). Graupel cools the atmosphere four times larger than snow near the melting layer. Figure 5d illustrates cross sections of the sum of above all diabatic heating related to phase transition of water. Significant cooling occurs at the outside of the eyewall near the melting layer, which reduces the size of tropical cyclone. Figure 5e shows diabatic heating of warm rain run, which consist only of condensation and evaporation. In the lower troposphere, there is evaporative cooling from rain drop. Comparing between Figs. 5d and 5e, we see that ice phase processes produce the radial differential heating in the middle troposphere, which reduces the typhoon size. The detailed mechanisms are discussed in Sawada and Iwasaki (2007) [6].

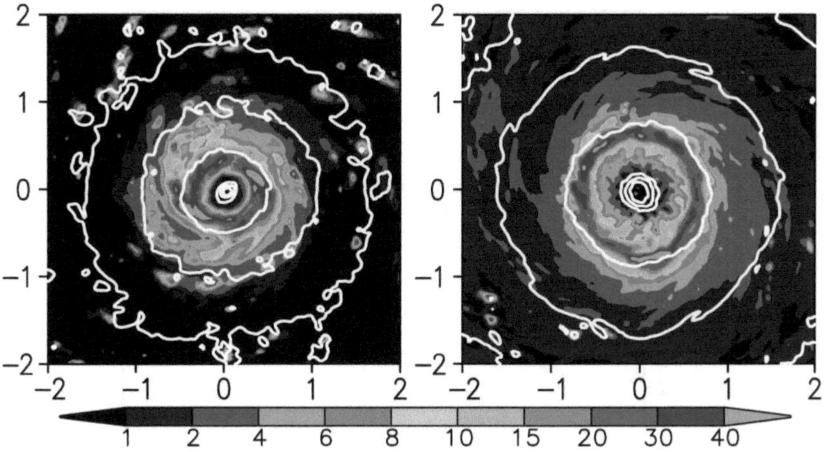

Fig. 3. Horizontal distributions of vertically integrated total water condensates (kg m^{-2}) of the control (left) and warm-rain (right) experiments over a domain of 600 x 600 km^2 at T=120hs.

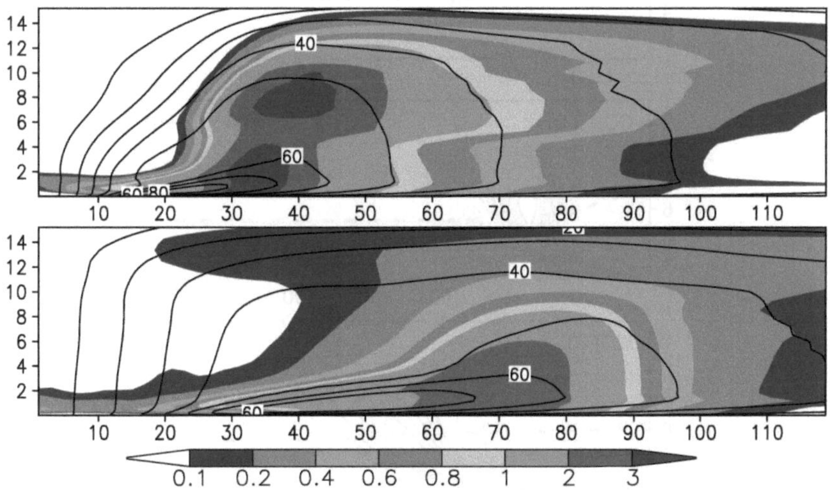

Fig. 4. Radial-height cross sections of tangential wind speed with a contour interval of 10 ms^{-1} and water condensates with coulars (g/kg) at the mature stage (T=108-120hs) in the control (upper panel) and in the warm rain (lower panel).

The cloud-resolving simulation indicates that the ice phase processes delay the organization of a tropical cyclone and reduce its size. This is hardly expressed in coarse resolution models with deep cumulus parameterization schemes.

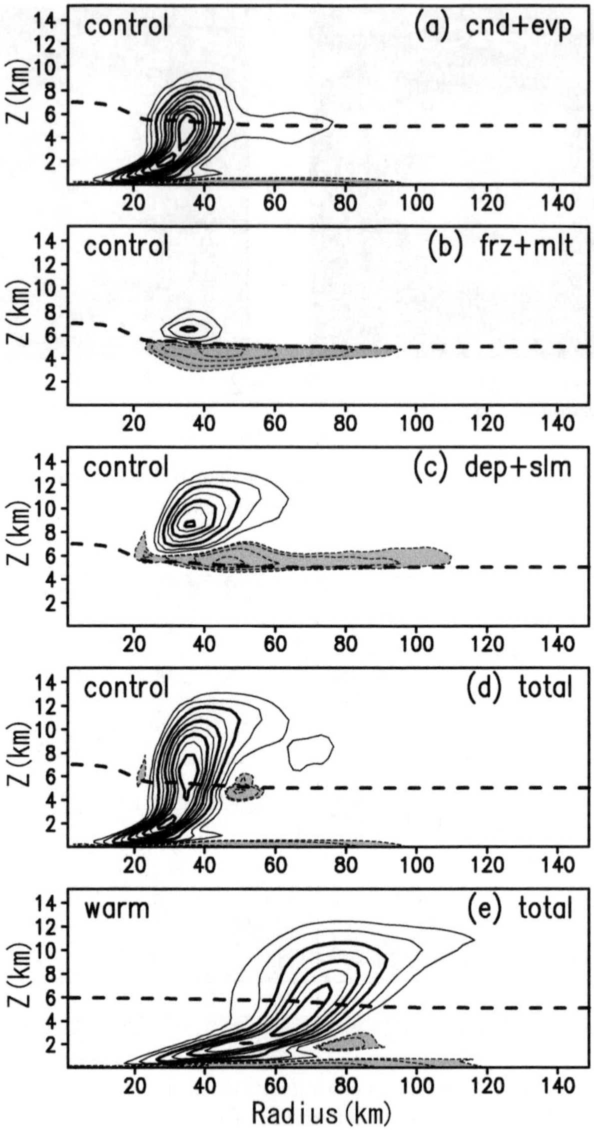

Fig. 5. Radius-height cross sections of diabatic heating for mature tropical cyclone (T=108-120hs) in the (a)-(d) control and (e) warm-rain experiments. (a), (b), (c), (d) and (e) show the sum of condensational heating and evaporative cooling rates (cnd+evp), the sum of freezing heating and melting cooling rates (frz+mlt), the sum of deposition heating and sublimation cooling rates (dep+slm), and the total diabatic heating rates due to phase change (total), respectively. Contour values are -10, -5, -2, -1, 5, 10, 20, 30, 40, 50, 60, 70, 80K/h. Shaded areas indicate regions of less than -1K/h. The dashed line denotes the melting layer (T=0C). After Sawada and Iwasaki (2007) [6].

4 Future Perspective of Numerical Simulation of Tropical Cyclones

Prediction and climate change of tropical cyclone are of great interest to the society. In conventional numerical weather prediction models and climate models, deep cumulus parameterization schemes obscure the nature of tropical cyclones. Accurate prediction and climate simulation require high-resolution cloud-resolving models without deep cumulus parameterization schemes, as indicated in the above experiment.

Regional models have previously been used for typhoon prediction. In the future, however, typhoon predictions both of track and intensity will be performed by global models, because we should expand computational domains to extend forecast periods. The current operational global model at JMA has a grid distance of about 20km. For the operational use of global cloud-resolving models with a grid spacing of 1km, we need about 10000 times more computational resources than current super computing systems. Computational resources must be assigned not only to the increase in the resolution but also to data assimilation and ensemble forecasts, to cope with difficulties associated with the predictability. The predictability depends strongly on accuracy of the initial conditions provided through the four-dimensional data assimilation system. The predictability can be directly estimated from probability distribution function (PDF) by ensemble forecasts.

Climate change is studied by using climate system models, whose grid spacings are about 100km. In order to study climate change based on cloud-resolving global models, we need computational resources exceeding 1000000 times that of current super computer systems.

We have many subjects requiring intensive use of cloud-resolving global models. At present most of dynamical cores of the global atmosphere are based on the spherical harmonics. Spectral conversion, however, becomes inefficient with increasing horizontal resolution. Considering such a situation, many dynamical cores without spectral conversions are being studied and proposed. It is very important to develop physics parameterization schemes, suitable for cloud-resolving models. Cloud microphysics has many uncertain parameters that need to be intensively validated with observations. Also, cloud microphysical parameters are affected by other physical processes, such as cloud-radiation interactions and the planetary boundary layer.

References

1. Emanuel, K. A.: The dependence of hurricane intensity on climate. Nature, **326**, 483–485 (1987)
2. Iwasaki, T., Nakano, H., Sugi, H.: The performance of a typhoon track prediction model with cumulus parameterization. J. Meteor. Soc. Japan, **65**, 555–570 (1987)

3. Murakami, M.: Numerical modeling of dynamical and microphysical evolution of an isolated convective cloud - The 19 July 1981 CCOPE cloud. J. Meteor. Soc. Japan, **68**, 107–128 (1990)
4. Ookochi, Y.: Preliminary test of typhoon forecast with a moving multi-nested grid (MNG). J. Meteor. Soc. Japan, **56**, 571–583 (1978)
5. Saito, K., Fujita, T., Yamada, Y., Ishida, J., Kumagai, Y., Aranami, K., Ohmori, S., Nagasawa, R., Kumagai, S., Muroi, C., Kato, T., Eito, H., Yamazaki, Y.: The operational JMA nonhydrostatic mesoscale model. Mon. Wea. Rev., **134**, 1266–1298 (2006)
6. Sawada, M., Iwasaki, T.: Impacts of ice phase processes on tropical cyclone development. J. Meteor. Soc. Japan, **85**, (in press)
7. Sugi, M., Noda, A., Sato, N.: Influence of the Global Warming on Tropical Cyclone Climatology: An Experiment with the JMA Global Model. J. Meteor. Soc. Japan, **80**, 249–272 (2002)
8. Yoshimura, J., Sugi, M., Noda, A.: Influence of Greenhouse Warming on Tropical Cyclone Frequency. J. Meteor. Soc. Japan, **84**, 405–428 (2006)
9. Miller, B.M., Runggaldier, W.J.: Kalman filtering for linear systems with coefficients driven by a hidden Markov jump process. Syst. Control Lett., **31**, 93–102 (1997)

OPA9 – French Experiments on the Earth Simulator and Teraflop Workbench Tunings

S. Masson[1], M.-A. Foujols[1], P. Klein[2], G. Madec[1], L. Hua[2], M. Levy[1], H. Sasaki[3], K. Takahashi[3], and F. Svensson[4]

[1] Institut Pierre Simon Laplace (IPSL), Paris, France
[2] French Research Institute for Exploitation of the Sea (IFREMER), Brest, France
[3] Earth Simulator Center (ESC), Yokohama, Japan
[4] NEC High Performance Computing Europe (HPCE), Stuttgart, Germany

1 Introduction

Japanese and French oceanographers built close collaborations since numerous years but the arrival of the Earth Simulator (highly parallel vector supercomputer system, 5120 vector processors, 40 Teraflops of peak performance) reinforced and speeded-up this cooperation. The Achievement of this exceptional computer motivated the creation of a 4-year (2001-2005) postdoc position for a French researcher in Japan, followed by 2 new post-doc positions since 2006. This active Franco-Japanese collaboration already lead to 16 publications. In addition, the signature of a Memorandum of Understanding (MoU) between the Earth Simulator Center, the French National Scientific Research Center (CNRS) and French Research Institute for Exploitation of the Sea (l'IFREMER) formalizes this scientific collaboration and guarantees access to the ES until the end of 2008.

Within this frame, four French teams are currently using the ES to explore a common interest: What is the impact of the small-scale phenomena on the large-scale ocean and climate modeling?

Figure 1 illustrates the large variety of scales that are, for example, observed in the ocean. Left panel presents the Gulf Stream as a narrow band looking like a river within the Atlantic Ocean. The middle panel reveals that the Gulf Stream path is in fact a characterized by numerous clockwise and anticlockwise eddies. When looking even closer, we observe that these eddies are made of filaments delimiting waters with different physical characteristics (right panel). Today, because of computational cost, the very large majority of climate simulations (for example most IPCC experiments) "see" the world as shown on the left panel. Smaller phenomena are then parameterized or sometimes even ignored. The computing power of the ES offers us the unique opportunity to explicitly take into account some of these scale phenomena and quantify their impact on the large-scale climate. This project represents

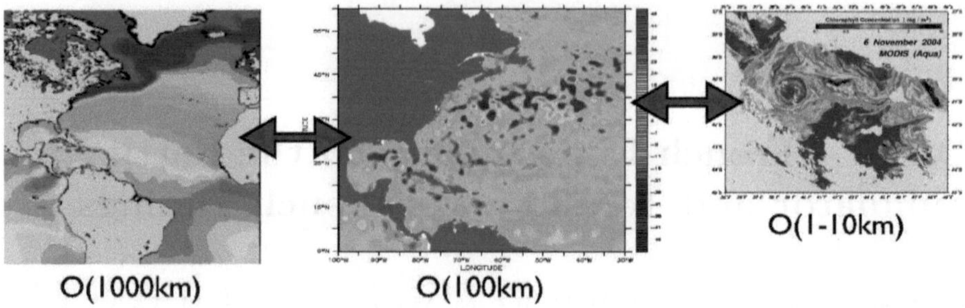

Fig. 1. Satellite observation of the sea surface temperature (left panel, Gulf Stream visible in blue), the sea level anomalies (middle panel, revealing eddies along the Gulf Stream path) and chlorophyll concentration (right panel, enlightening filaments within eddies).

a technical and scientific challenge. The size of the simulated problem is 100 to 1500 times bigger than existing work. New small-scale mechanisms with potential impacts on large-scale circulation and climate dynamics are explored. At the end, this work will help us to progress in our understanding of the climate and thus improve parameterizations used in climate change simulation for example.

The next four sections give a brief overview of the technical and scientific aspects of the four parts of the MoU project that started at the end of 2004.

In Europe the OPA9 application is being investigated and improved in the Teraflop Workbench project at Hchstleistungsrechenzentrum Stuttgart (HLRS), University of Stuttgart, Germany. The last section describes the work that have been done and the performance improvment for OPA9 using a dataset provided by the project partner, Institute für Meereskunde IFM GEOMAR of the University of Kiel, Germany.

2 Vertical pumping associated with eddies

The goal of this study is to explore the impact of very small-scale phenomena (order of 1 km) on vertical and horizontal mixing of the upper part of the ocean that plays a key role in air-sea exchange and thus climate regulation. These phenomena are created by the interactions between mesoscale eddies (50-100 km) that are, for example, observed in the Antarctic Circumpolar Current known for its very high eddy activity. In this process study, we therefore selected this region and modeled this circular current by a simple periodic canal. A set of 3-year long experiments are performed with the horizontal and vertical resolution increasing step by step until 1 km $\times$ 1 km $\times$ 300 levels (or $3000\times2000\times300$ points). Major results show that, first, the increase of resolution is associated with an explosion on the number of eddies. Second, when reaching the highest resolutions, these eddies are encircled with very energetic filaments where very high vertical velocities (upward or downward according to the rotating direction of the eddy) are observed, see Fig. 2. Our first results conclude that being able to explicitly represent these very fine filaments warms the upper part of the ocean in areas such as the Antarctic Circumpolar Current. This could therefore have a significant impact on the earth climate that is at first driven by the heat redistribution from equatorial regions toward the higher latitudes. The biggest simulations we perform in this study use 512 vector processors (or about 10% of the ES). Future experiments should reach 1024 processors. Performances in production mode are 1.6 Teraflops corresponding to 40% of the peak performance that is excellent.

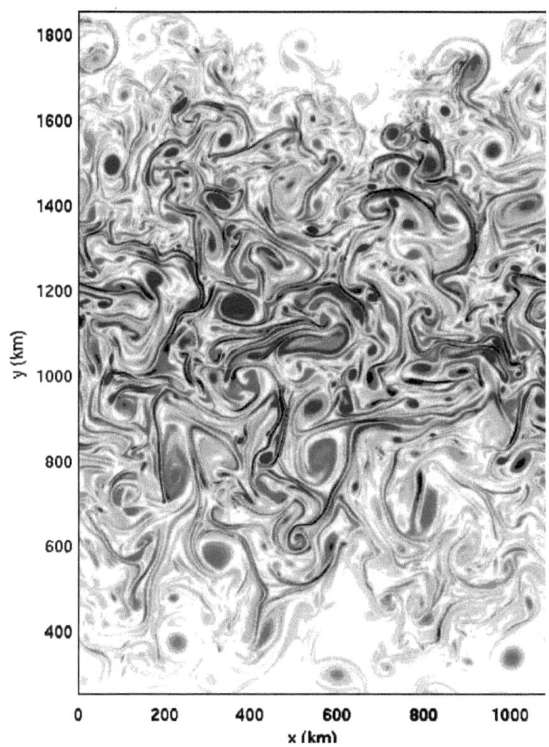

Fig. 2. Oceanic vertical velocity at 100 m. Blue/red denotes upward/downward currents.

3 Oceanic CO_2 pumping and eddies

The capacity of the ocean to pump or reject CO2 is a key point to quantify the climatic response to the atmospheric CO2 increase. Through biochemistry processes involving life and death of phyto and zooplankton, oceans reject CO2 to the atmosphere in the equatorial regions where upwelling are observed, whereas at mid and high latitudes, oceans pump and store CO2. It is thus primordial to explore the processes that keep balance between oceanic CO2 reject and pumping and understand how this equilibrium could be affected in the global warming context: Will the ocean be able to store more CO2 or not? The efficiency of CO2 pumping at mid latitude is linked to ocean dynamics and particularly the meso-scale eddies. This second study will thus aim to explore impacts of small-scale ocean dynamics on biochemistry activity with a focus on the carbon cycle and fluxes at the air-sea interface. Our area of interest is this time much larger and represents the western part of the North Atlantic (see Fig. 3) that is one of the most important regions for oceanic CO2 absorption. Our ocean model is also coupled to a biochemistry model

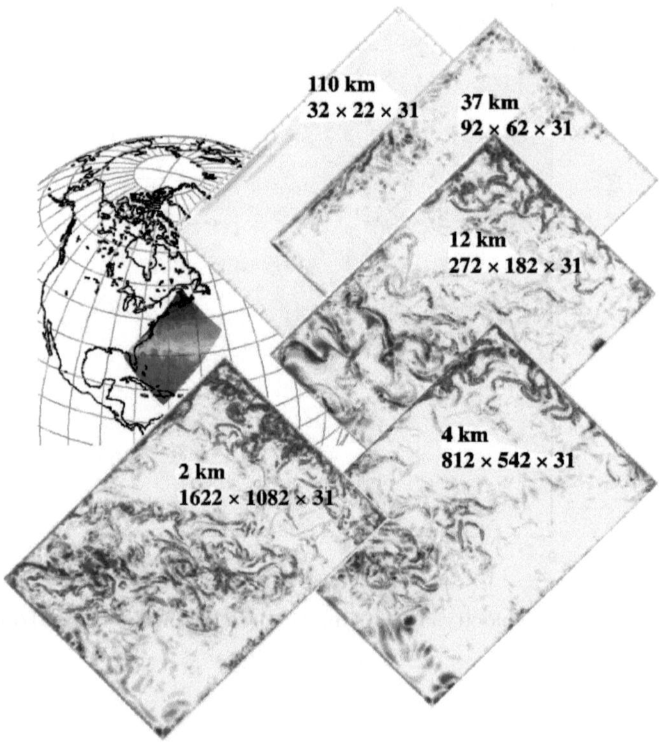

Fig. 3. Localization of the model domain. Oceanic vertical velocity at 100 m for different model resolutions. Blue/red denotes upward/downward currents.

during experiments of one to two hundred years with resolution increasing until 2 km × 2 km × 30 levels (or 1600×1100×30 points). The length of our experiments is constrained by the need to reach equilibrium for deep-water characteristics on a very large domain. Today, our firsts results concern only the ocean dynamics as the biochemistry will be explored in a second step. As in the first part, the highest resolutions are associated to an explosion of energetic eddies and filaments Atlantic (see Fig. 3). Strong upward and downward motions are observed within the fine filaments. It modifies the formation of sub-surface waters. These waters are rich in nutriment and play a key role in the carbon cycle in the northern Atlantic. In this work, the larges simulations with the ocean dynamic only use 423 vector processors (8.2% of the ES) and reach 875 Gigaflops (or 25% of the peak performance).

4 Dynamics of deep equatorial transport and mixing

Recent observations shown a piling up of eastward and westward zonal currents in the equatorial Atlantic and Pacific from the surface to depths exceeding 2000 m. These currents are located along the path of the so called "global conveyor belt" (left panel of Fig. 4), a global oceanic circulation, that regulates the whole oceans at long term. The vertical shear associated with this alternation of zonal jets could favor the equatorial mixing in the deep ocean. This would impact (1) the cross-equatorial transport transport and (2) the swallowing of the deep and cold waters along the global conveyor belt path. This work uses an idealized representation of an equatorial basin with a size comparable to the Atlantic or the Pacific. The biggest configuration tested has 300 vertical levels and a horizontal resolution of 4 km. Model is integrated during 30 years to reach the equilibrium of the equatorial dynamics at depth. This very high resolution (more than 300 times the usual resolutions), accessible thanks to the ES computational power, is crucial to obtain our results. First for the first time the deep equatorial jets are correctly represented in a 3D numerical model. Analyze of our simulations explained the mechanisms that

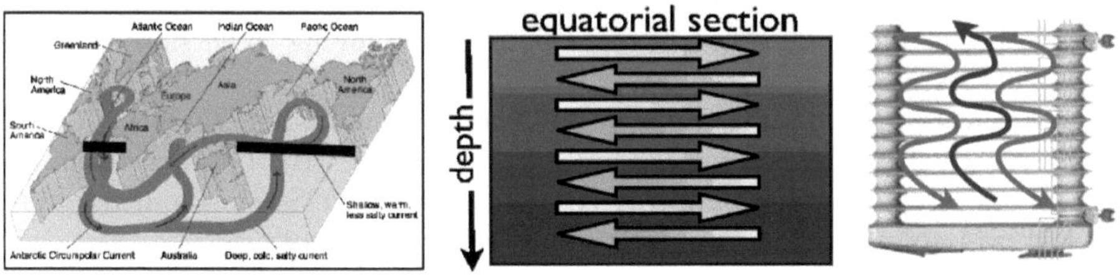

Fig. 4. Schematic representation of the global conveyor belt (left), the piling up of equatorial jets (middle) and their potential impact on the vertical mixing (acting like radiator blades, right)

drives theses jets and their characteristics differences between the Atlantic and the Pacific. Further studies are ongoing to now explore their impact on the global ocean As for the first part, these simulations reaches almost 40% of the ES peak performance.

5 First meters of the ocean and tropical climate variability

The tropical climate variability (El Nio, Indian monsoon) partly relies on the response of the ocean to the high frequency forcing from the atmosphere. The thermal inertia of the full atmospheric column is comparable to the thermal inertia of a few meter of the ocean water. An ocean model with a vertical resolution reaching 1 meter in the surface layers is thus needed to explicitly resolve the high frequency air-sea interactions like the diurnal cycle. In this last part, the computational power of the ES allows us to realize unprecedented simulations: a global ocean-atmosphere coupled simulation with a very high oceanic vertical resolution. These simulations are much more complex than the previous one because they assemble a global ocean/sea-ice model (720×510×300 points) to a global atmospheric model (320×160×31 points) through the use of an intermediate "coupling" model. Until today, we dedicated most of our work to the technical and physical set-up of this unique model. Our first results show an impact of the diurnal cycle on the sea surface temperature variability with a potential effect on the intraseasonal variability of the monsoon. Further work is ongoing to explore this issue.

The ocean model use 255 vector processors (5% of the ES) and reach 33of the peak performance, an excellent number for this kind of simulation involving a large quantity of inputs and outputs.

6 Application tunings for vector computers under the Teraflop Workbench

The project in Teraflop Workbench to improve OPA9 was launched in June 2005, working with the Institute für Meereskunde IFM GEOMAR of the University of Kiel, Germany. Several steps were made moving to version 1.09 and a 1/4 degrees model in March 2006.

There were several OPA tunings implemented and tested. The calculation of the sea ice rheology was revised and improved. The MPI boarder exchange was completely rewritten to improve the communication. Several routines were worked over to improve vectorization and usage of memory.

6.1 Sea ice rheology

The sea ice rheology calculation was rewritten in several steps. As ice in the worlds oceans are limited to the poles the new routine first scans what parts will result in calculations and select those array limits to work on. As ice changes the limits are adjusted in each iteration, and to be sure to handle growths of ice the band where the ice is calculated is set larger than where ice have been detected. The scanning of the ice arrays increase the the runtime but for the tested domain decompositions the amount of ice on a CPU were always less than 50% or even less than 10% for some. This reduction of data does reduce the runtime so much that the small increase for scanning the data is negligible.

The second step taken was to merge the many loops, especially inside the relaxation iteration part where most time where spent. At the end there were one major loop in the relaxation after the scanning of the bands and at the end the convergence test. By merging the loops, the access to the input arrays u_ice and v_ice was limited, and several temporary array could be replaced by temporary variables. In the original version the different arrays addressed multiple times in more complex structures and also by using temporary variables for these kind of calculations the compiler could do a better job in scheduling the calculations.

6.2 MPI boarder exchange

The MPI boarder exchange went through several steps of improvements. The original code consisted of five steps and two branches depending on if the northern area of the globe were covered by one MPI thread or by several. The first step was to exchange the boarder nodes of the arrays with the east-west neighbors. Then secondly to do a north-south exchange. The third steps treats the four most north lines of the dataset. If this array part is owned by one MPI thread the calculation is made in the subroutine, and if it is distributed over more routines a MPI_Gather getting all the data to the zero rank MPI thread of the northern communicator, and there a subroutine was called to do the same kind of transformation that was made in-line in the one MPI thread case. After that the data was distributed with MPI_Scatter to the waiting MPI threads. The fourth step was to make another east-west communication to make sure that the top four lines were properly exchanged after the northern communication. All these exchanges were configurable to use MPI_Send, MPI_Isend and MPI_Bsend with different version of MPI_Recv. The MPI_Recv were using MPI_ANY_SOURCE.

The first step taken was to combine the two different branches of treating the northern part. They were basically the same and only had small differences in variable naming. By merging this part into a subroutine, it was possible to just arrange the input variables differently depending on if one or more MPI threads were responsible for the northern part. The communication pattern was also changed (see Fig. 5) to use MPI_Allgather so that each MPI thread in the northern communicator have access to the data, and each calculate the region. By doing this the MPI_Scatter with its second synchronization point

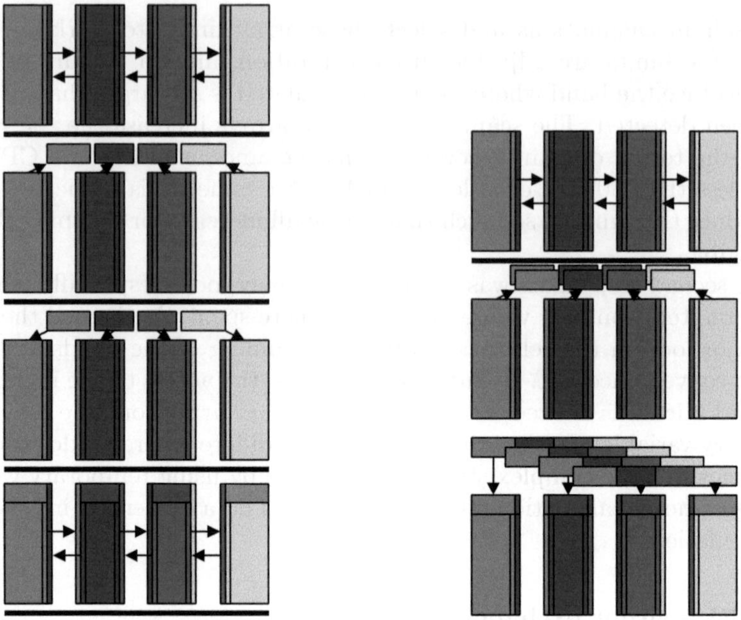

Fig. 5. The new communication contains less synchronization and less communication than the old version.

can be avoided and also the last East-West exchange as this data already are available at all MPI threads.

It is an important tuning as the boarder exchange is used by many routines, for example the solver of the transport divergence system using a diagonal preconditioned conjugate gradient method. One boarder exchange per iteration is made in the solver.

The selection of exchange models were before made with CASE and CHAR constants, this was changed to IF statements with INTEGER constants to improve constant folding done by the compiler, and to facilitate inlineing.

6.3 Loop tunings

Several routines were reworked to improve memory access and loop layouts. Simple tunings like precalclulating some masks that remain constant during the simulation (like the land/water mask). Min/max calculation without having to store the intermediate results in temporary arrays, to reduce the memory access.

6.4 Results

The original version that was used as a baseline was the 1/4 degree model (1442×1021×46 points) that was delivered from IFM GEOMAR in Kiel in

March 2006. The some issues with reading the input deck of such high resolution models were fixed and a baseline run was made with that version. It is here referenced as the Original version. To put it in relation to the OPA production versions it is named OPA 9.0 , LOCEAN-IPSL (2005) in opa.F90 with the CVS tag `$Header: /home/opalod/NEMOCVSROOT/NEMO/OPA_SRC/opa.F90,v 1.19 2005/12/28 09:25:03 opalod Exp $` The model settings that were used is the production settings from IFM GEOMAR in Kiel.

The OPA9 version called 5:th TFW is the state of the OPA9 tunings before the 5:th Teraflop Workshop at Tohoku University in Sendai, Japan. (November 20th and 21st 2006) This contains the sea ice rheology tunings and the MPI tunings for 2D arrays. The results named 6:th TFW are the results measured before the 6:th Teraflop Workshop at the University of Stuttgart, Germany (March 26th and 27th 2007).

All tunings brouht an improvement of 17.2% in runtime using 16 SX-8 CPUs and 14.1% in runtime using 64 SX-8 CPUs compared to the original version.

7 Conclusion

For its fifth anniversary, the ES remains a unique supercomputer with exceptional performances when considering real application in climate research filed. The MoU between ESC, CNRS and IFREMER allowed four French teams to benefit of the ES computational power to investigate challenging and unexplored scientific questions on climate modeling. The size of the studied problems is at least two orders of magnitude larger than the usual simulations done on French computer resources. Accessing the ES allowed us to remain at the top of the world climate research. However, we deplore that such computational facilities do not exist in Europe. We are afraid that within a few

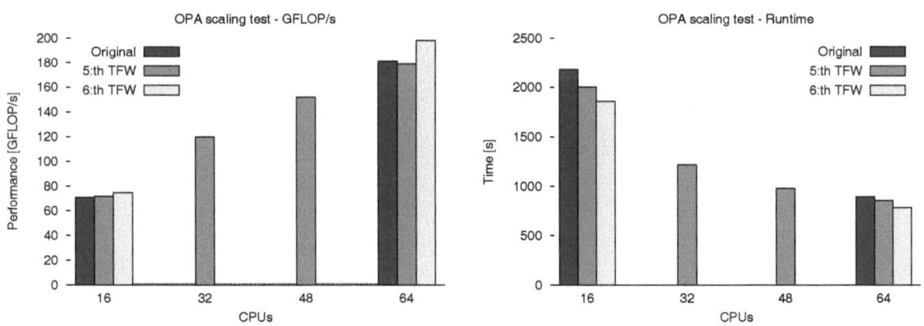

Fig. 6. Performance and time plot - Scaling results made with OPA initial version, before 5:th Teraflop workbench and before 6:th Teraflop workbench - 1200 simulation cycles without initialization

years European climate research will decline in comparison with work done in US or Japan.

The work in the Teraflop Workbench is to enable this kind of research on new larger models, being able to test limits of models and improve the application.

TERAFLOP Computing and Ensemble Climate Model Simulations

Henk A. Dijkstra

Institute for Marine and Atmospheric Research Utrecht (IMAU), Utrecht University, P.O. Box 80005, NL-3508 TA Utrecht, The Netherlands
dijkstra@phys.uu.nl

Summary. A short description is given of ensemble climate model simulations which have been carried out since 2003 by the Dutch climate research community on Teraflop computing facilities. In all these simulations, we were interested in the behavior of the climate system over the period 2000-2100 due to a specified increase in greenhouse gases. The relatively large size of the ensembles has enabled us to better distinguish the forced signal (due to the increase of greenhouse gases) from internal climate variability, look at changes in patterns of climate variability and at changes in the statistics of extreme events.

1 Introduction

The atmospheric concentrations of CO_2, CH_4 and other so-called greenhouse-gases (GHG) have increased rapidly since the beginning of the industrial revolution, leading to an increase of radiative forcing of 2.4 W/m^2 up to the year 2000 compared to pre-industrial times [1]. Simultaneously, long term climate trends are observed everywhere on Earth. Among others, the global mean surface temperature has increased by 0.6 ± 0.2 °C over the 20th century, there has been a widespread retreat of non-polar glaciers, and patterns of pressure and precipitation have changed [2].

Although one may be tempted to attribute the climate trends directly to changes in the radiative forcing, the causal chain is unfortunately not so easy. The climate system displays strong internal climate variability on a number of time scales. Hence, even when one would be able to keep the total radiative forcing constant, substantial natural variability would still be observed. In many cases, this variability expresses itself in certain patterns with names such as the North Atlantic Oscillation, El Niño/Southern Oscillation, the Pacific Decadal Oscillation and the Atlantic Multidecadal Oscillation. The relevant time scales of variability of these patterns are in many cases comparable to the trends mentioned above and the observational record is too short to accurately establish their amplitude and phase.

To determine the causal chain between the increase in radiative forcing and climate change observed, climate models are essential. Over the last decade, climate models have grown in complexity at a fast pace due to increased detail of description of the climate system and increased spatial resolution. One of the standard numerical simulations with a climate model is the response to a doubling in CO_2 over a period of about 100 years. Current climate models, predict an increase in global averaged surface temperature within the range from 1°C to 4°C [2]. Often just one or a few transient coupled climate simulations are performed for a given emission scenario due to the high computational demand of a single simulation. This allows an assessment of the mean climate change but, because of the strong internal variability, it is difficult to attribute certain trends in model response to increased radiative forcing. To distinguish the forced signal from internal variability, a large ensemble of climate simulations is necessary.

The Dutch climate community, grouped into the Center for Climate research (CKO), has played a relatively minor role in running climate models compared to that of the large climate research centers elsewhere (Hadley Center in the UK, DRKZ in Germany and centers in the USA such as NCAR and GFDL). Since 2003, however, there have been two relatively large projects on Teraflop machines where the CKO group (Utrecht University and the Royal Dutch Meteorological Institute) has been able to perform relatively large ensemble simulations with state-of-the-art climate models.

2 The Dutch Computing Challenge Project

The Dutch Computing Challenge project was initiated in 2003 by the CKO group and run on an Origin3800 system (1024 x 500 Mhz processors) situated at the Academic Computation Centre (SARA) in Amsterdam. For a period of 2 months (June July 2003, an extremely hot Summer by the way) SARA reserved 256 processors of this machine for the project. The National Center for Atmospheric Research (NCAR) Community Climate System Model (CCSM) was used for simulations over the period 1940-2080 using the SRES A1b (Business as Usual) scenario of greenhouse emissions. Initial conditions of 62 ensemble members were slightly different in 1940, the model was run under known forcing (aerosols, solar and greenhouse) until 2000 and from 2000 under the SRES A1b scenario. An extensive technical report is available through http://www.knmi.nl/onderzk/CKO/Challenge/ which serves as a website for the project.

One of the main aims of the project was to study the changes in the probability density distribution of Western European climate characteristics such as temperature and precipitation. For example, in addition to a change in the mean August temperature, we also found changes in the probability of extreme warmth in August. This is illustrated in Fig. 1 which shows the probability distribution of daily mean temperatures in a grid cell that partially

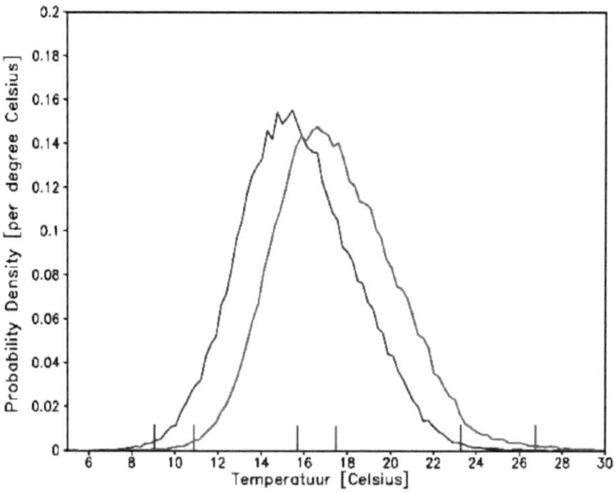

Fig. 1. Probability density of daily mean temperatures in August in a grid cell located at 7.5°E and 50°N. In blue for the period 1961-1990, in red for 2051-2080. The vertical lines indicate the temperature of the 1 in 10 year cold extreme (left ones), the mean (middle ones) and the 1 in 10 year warm extreme (right ones), results from [3].

overlaps the Netherlands. On average, August daily mean temperatures warm about 1.4 degrees. However, the temperature of the warm extreme that is exceeded on average once every 10 years increases by a factor two. Further analyses have suggested that this is due to an increased probability of dry soil conditions in which the cooling effect of the evaporation of soil moisture diminishes. Another contributing factor is a change in the circulation with more often winds blowing from the southeast.

Although there were many technical problems during execution of the runs, these were relatively easy to overcome. From the project, about 8 Tb of data resulted which were put on tape at SARA for further analysis. The biggest problem in this project has been the generation of scientific results. As one of the project leaders of the project (the other being Frank Selten from KNMI), I have been rather naive in thinking that if the results were there, everyone would immediately start to analyze them. At the KNMI this indeed happened, but at the UU many people were involved in so many other projects and nobody except myself was eventually involved in resulting publications.

Nevertheless, as of April 2007 many interesting publications have resulted from the project; an overview of the first results appeared in [3]. [4], it was demonstrated that the observed trend in the North Atlantic Oscillation (associated with variations of the strength of the northern hemispheric midlatitude jetstream) over the last decades can be attributed to natural variability. There is, however, a change in the winter circulation in the northern hemisphere due to the increase of greenhouse gases which has its origin in precipitation changes in the Tropical Pacific. In [5], it was shown that the El Niño/Southern Oscillation does not change under an increase of greenhouse gases. Further analysis of the data showed, however, that deficiencies in the climate model, i.e. a small meridional scale of the equatorial wind response, were the cause of this negative result. In [6] it is shown that fluctuations in ocean surface temperatures lead to an increase in Sahel rainfall in response to anthropogenic warming. In [7, 8], it is found that anthropogenically forced changes in the thermohaline ocean circulation and its internal variability are distinguishable. Forced changes are found at intermediate levels over the entire Atlantic basin. The internal variability is confined to the North Atlantic region. In [9], the changes in extremes in European daily mean temperatures were analyzed in more detail. The ensemble model data were also used for development of a new detection technique of climate change by [10].

Likely the data have been used in several other publications but we had no policy that manuscripts would pass along project leaders, while data were open to everyone. Even the people directly involved sometimes did not care to provide the correct acknowledgment of project funding and management. For people involved in arranging funding for the next generation of supercomputing systems this has been quite frustrating as output from the older systems cannot be fully justified and it is eventually the resulting science which will convince funding agencies.

During the project, there have been many interactions with the press and several articles have appeared in the national newspapers. In addition, results have been presented on several national and international meetings. On October 15, 2004, many of the results were presented at a dedicated meeting at SARA and in the evening some of these results were highlighted on the national news. With SARA assistance, an animation of a superstorm has been made within the CAVE at SARA. During the meeting on October 15, many of the participants took the opportunity to see this animation. Finally, several articles have been written for popular magazines (in Dutch).

3 The ESSENCE Project

Early 2005, the first call of the Distributed European Infrastructure for Supercomputer Applications (DEISA) Extreme Computing Initiative was launched and the CKO was successful with the ESSENCE (Ensemble SimulationS of Extreme events under Nonlinear Climate changE) project. For this project, we were provided 170,000 CPU hours on the NEC-SX8 at the High Performance Computing Stuttgart (HLRS).

Although we were planning originally to do simulations with the NCAR CCSM version 3.0, both the model performance as well as the platform motivated us to use the ECHAM5/MPI-OM coupled climate model developed at the Max Planck Institute for Meteorology in Hamburg which is running it at DKRZ on a NEC-SX6. Hence, there were not many technical problems; the model was run on one node (8 processors) and 6 to 8 nodes were used simultaneously for the different ensemble members. As the capacity on the NEC-SX8 for accommodating this project was limited it took us eventually about 6 months (July 2005 - January 2006) to run the ESSENCE simulations.

The ECHAM5/MPI-OM version used here is the same that has been used for climate scenario runs in preparation of AR4. ECHAM5 [11] is run at a horizontal resolution of T63 and 31 vertical hybrid levels with the top level at 10 hPa. The ocean model MPI-OM [12] is a primitive equation z-coordinate model. It employs a bipolar orthogonal spherical coordinate system in which the north and south poles are placed over Greenland and West Antarctica, respectively, to avoid the singularities at the poles. The resolution is highest, O(20-40 km), in the deep water formation regions of the Labrador Sea, Greenland Sea, and Weddell Sea. Along the equator the meridional resolution is about 0.5°. There are 40 vertical layers with a thickness ranging from 10 m near the surface to 600 m near the bottom.

The baseline experimental period is 1950-2100. For the historical part of this period (1950-2000) the concentrations of greenhouse gases and tropospheric sulfate aerosols are specified from observations, while for the future part (2001-2100) they follow again the SRES A1b scenario [13]. Stratospheric aerosols from volcanic eruptions are not taken into account, and the solar constant is fixed. The runs are initialized from a long run in which historical greenhouse gas concentrations have been used until 1950. Different ensemble members are generated by disturbing the initial state of the atmosphere. Gaussian noise with an amplitude of 0.1 K is added to the initial temperature field. The initial ocean state is not perturbed.

The basic ensemble consists of 17 runs driven by a time-varying forcing as described above. Additionally, three experimental ensembles have been performed to study the impact of some key parameterizations, again making use of the ensemble strategy to increase the signal-to-noise ratio. Nearly 50 TB of data have been saved from the runs. While most 3-dimensional fields are stored as monthly means, some atmospheric fields are also available as daily means. Some surface fields like temperature and wind speed are available at a time resolution of 3 hours. This makes a thorough analysis of weather extremes and their possible variation in a changing climate possible. The data are stored at the full model resolution and saved at SARA for further analysis.

The ensemble-spread of the global-mean surface temperature is fairly small (≈ 0.4 K), it encompasses the observations very well. Between 1950 and 2005 the observed global-mean temperature increased by 0.65 K [14], while the ensemble-mean gives an increase of 0.76 K. The discrepancy mainly arises from the period 1950-1965. After that period observed and modelled temperature trends are nearly identical. This gives confidence in the model's sensitivity to changes in greenhouse gas concentrations. The global-mean temperature increases by 4 K between 2000 and 2100, and the statistical uncertainty of the warming is extremely low ($0.4\ \text{K}/\sqrt{17} \approx 0.1$ K). Thus within this particular climate model, i.e., neglecting model uncertainties, the expected global warming of 4 K in 2100 is very robust.

Again the advantage of the relatively large size of the ensemble is the large signal-to-noise ratio. We were able to determine the year in which the forced signal (i.e., the trend) in several variables emerges from the noise. The enhanced signal-to-noise ratio that is achieved by averaging over all ensemble members is reflected in a large number of effective degrees-of-freedom, even for short time periods, that enter the significance test. This makes the detection of significant trends over short periods (10-20 years) possible.

The earliest detection times (Figure 2) for the surface temperature are found off the equator in the western parts of the tropical oceans and the Middle East, where the signal emerges as early as around 2005 from the noise. In these regions the variability is extremely low, while the trend is only modest. A second region with such an early detection time is the Arctic, where the trend is very large due to the decrease of the sea-ice. Over most of Eurasia and Africa detection is possible before 2020. The longest detection times are found along the equatorial Pacific, where, due to El Niño, the variability is very high, as well as in the Southern Ocean and the North Atlantic, where the trend is very low.

Having learned from the Dutch Computing Challenge project, we decided to have a first paper out [15] with a summary of main results and we have a strict publication policy. The website of the project can be found at http://www.knmi.nl/~sterl/Essence/.

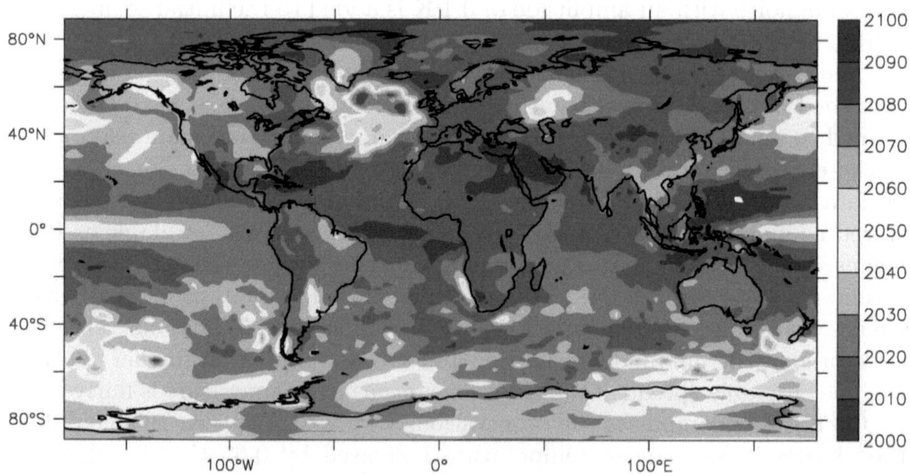

Fig. 2. Year in which the trend (measured from 1980 onwards) of the annual-mean surface temperature emerges from the weather noise at the 95%-significance level, from [15].

4 Summary and Conclusion

This has been a rather more personal account of experiences of a relatively small group of climate researchers with ensemble climate simulations on Teraflop computing resources. I have tried to provide a mix of technical issues, management and results of two large projects carried out by the CKO in 2003 and 2006. I'll summarize these experiences by mentioning several important factors for success of these types of projects.

1. The Teraflop computing resources are essential to carry out this type of work. There is no way that such large projects can be performed efficiently on local clusters because of the computing time and data transport involved.
2. To carry out the specific computations dedicated researchers have to be allocated. In case of the Dutch Computing Challenge project, Michael Kliphuis carried out all simulations. In the ESSENCE project this was done by Andreas Sterl and Camiel Severijns. Without these people the projects could not have been completed successfully.
3. The support from the HPC centers (SARA and HLRS) has been extremely good. Without this support, it would have been difficult to carry out these projects.
4. An adequate management structure is essential to be able to get sufficient scientific output from these type of projects.

Finally, the projects have certainly contributed to the current initiatives in the Netherlands to develop, with other European member states, an independent climate model (EC-Earth). Developments of the development of this model can be seen at http://ecearth.knmi.nl/. We hope to be able to soon carry out ensemble calculations with this model on existing Teraflop (and future Petaflop) systems.

Acknowledgments

The Dutch Computing Challenge project was funded through NCF (Netherlands National Computing Facilities foundation) project SG-122. We thank the DEISA Consortium (co-funded by the EU, FP6 projects 508830/031513), for support within the DEISA Extreme Computing Initiative (www.deisa.org). NCF contributed to ESSENCE through NCF projects NRG-2006.06, CAVE-06-023 and SG-06-267. We thank HLRS and SARA staff, especially Wim Rijks and Thomas Bönisch, for their excellent technical support. The Max-Planck-Institute for Meteorology in Hamburg (http://www.mpimet.mpg.de/) made available their climate model ECHAM5/MPI-OM and provided valuable advice on implementation and use of the model. We are especially indebted to Monika Esch and Helmuth Haak.

References

1. Houghton, J.T., Ding, Y., Griggs, D., Noguer, M., van der Linden, P.J., Xiaosu, D., eds.: Climate Change 2001: The Scientific Basis. Contribution of Working Group I to the Third Assessment Report of the Intergovernmental Panel on Climate Change (IPCC), Cambridge University Press, UK (2001)
2. : Summary for policymakers and technical summary. ipcc fourth assessment report (2001)
3. Selten, F., Kliphuis, M., Dijkstrass, H.A.: Transient coupled ensemble climate simulations to study changes in the probability of extreme events. CLIVAR Exchanges **28** (2003) 11–13
4. Selten, F.M., Branstator, G.W., Dijkstra, H.A., Kliphuis, M.: Tropical origins for recent and future Northern Hemisphere climate change. Geophysical Research Letters **31** (2004) L21205
5. Zelle, H., van Oldenborgh, G.J., Burgers, G., Dijkstra, H.A.: El Niño and Greenhouse warming: Results from Ensemble Simulations with the NCAR CCSM. J. Climate **18** (2005) 4669–4683
6. Haarsma, R.J., Selten, F.M., Weber, S.L., Kliphuis, M.: Sahel rainfall variability and response to greenhouse warming. Geophysical Research Letters **32** (2005) L17702
7. Drijfhout, S.S., Hazeleger, W.: Changes in MOC and gyre-induced Atlantic Ocean heat transport. Geophysical Research Letters **33** (2006) L07707
8. Drijfhout, S.S., Hazeleger, W.: Detecting Atlantic MOC changes in an ensemble of climate change simulations. J. Climate **in press** (2007)
9. Tank, A.M.G.K., Können, G.P., Selten, F.M.: Signals of anthropogenic influence on European warming as seen in the trend patterns of daily temperature variance. International Journal of Climatology **25** (2005) 1–16
10. Stone, D.A., Allen, M., Selten, F., Kliphuis, M., Stott, P.: The detection and attribution of climate change using an ensemble of opportunity. J. Climate **20** (2007) 504–516
11. Roeckner, E., Bäuml, G., Bonaventura, L., Brokopf, R., Esch, M., Giorgetta, M., Hagemann, S., Kirchner, I., Kornblueh, L., Manzini, E., Rhodin, A., Schlese, U., Schulzweida, U., Tompkins, A.: The atmospheric general circulation model echam 5. part i: Model description. Technical Report Report No. 349, Max-Planck-Institut für Meteorologie, Hamburg, Germany (2003)
12. Marsland, S.J., Haak, H., Jungclaus, J., Latif, M., Röskes, F.: The Max-Planck-Institute global ocean/sea ice model with orthogonal curvilinear coordinates. Ocean Modelling **5** (2003) 91–127
13. Nakicenovic, N., Swart, R., eds.: Special Report on Emissions Scenarios: A Special Report of Working Group III of the Intergovernmental Panel on Climate Change, Cambridge University Press, Cambridge, U.K (2000)
14. Brohan, P., Kennedy, J., Haris, I., Tett, S., Jones, P.: Uncertainty estimates in regional and global observed temperature changes: A new data set from 1850. Journal of Geophysical Research **111** (2006) D12106
15. Sterl, A., Severijns, v Oldenborgh, G., Dijkstra, H.A., Hazeleger, W., van den Broeke, M., Burgers, G., van den Hurk, B., van Leeuwen, P., van Velthovens, P.: The essence project - signal to noise ratio in climate projections. Geophys. Res. Lett. (2007) submitted

Applications II

Computational Fluid Dynamics

Current Capability of Unstructured-Grid CFD and a Consideration for the Next Step

Kazuhiro Nakahashi

Department of Aerospace Engineering, Tohoku University
Sendai 980-8579, JAPAN
naka@ad.mech.tohoku.ac.jp

1 Introduction

Impressive progress in computational fluid dynamics (CFD) has been made during the last three decades. Currently CFD has become an indispensable tool for analyzing and designing aircrafts. Wind tunnel testing, however, is still the central player for aircraft developments and CFD plays a subordinate part.

In this article, current capability of CFD is discussed and demands for next-generation CFD are described with an expectation of near future PetaFlops computers. Then, Cartesian grid approach, as a promising candidate for next-generation CFD, is discussed by comparing it with the current unstructured-grid CFD. It is concluded that the simplicity of the algorithms from grid generation to post processing of Cartesian mesh CFD will be a big advantage in the days of PetaFlops computers.

2 Will CFD take over wind tunnels?

More than 20 years ago, I heard an elderly physicist in fluid dynamics say that it was as if CFD were just surging in. Other scientists of the day said that with the development of CFD, wind tunnels would eventually become redundant.

Impressive progress in CFD has been made during the last three decades. In the early stage, one of the main targets of CFD for aeronautical fields was to compute flow around airfoils and wings accurately and quickly. Body-fitted-coordinate grids, commonly known as structured grids, were used in those days.

From the late eighty's, the target was moved to analyzing full aircraft configurations [1]. This spawned a surge of activities in the area of unstructured grids, including tetrahedral grids, prismatic grids, and tetrahedral-prismatic hybrid grids. Unstructured grids provide considerable flexibility in tackling complex geometries as shown in Fig. 1 [2]. CFD has become an indispensable tool for analyzing and designing aircrafts.

The author has been studying various aircraft configurations with his students using the advantages of unstructured-grid CFD as shown in Fig. 2

So, is CFD taking over the wind tunnels as predicted twenty years ago? Today, Reynolds-averaged Navier-Stokes (RANS) computations can accurately predict lift and drag coefficients of a full aircraft configuration. It is, however, still quantitatively not reliable for high-alpha conditions where flow separates. Boundary layer transition is another cause of inaccuracy. These are mainly due to the incompleteness of physical models used in RANS simulations.

Large Eddy Simulation (LES) and Direct Numerical Simulation (DNS) are expected to reduce the physical model dependencies. But we have to wait for the further progress of computers for the use of those large-scale computations in engineering purposes.

For the time being, the wind tunnel is the central player and CFD plays a subordinate part in aircraft developments.

3 Rapid progress of computers

The past CFD progress has been highly supported by the improvements of computer performance. Moorfs Law tells us that the degree of integration of a computer chip has been doubled in 18 months. This basically corresponds to a factor of 100 every 10 years. The latest Top500 Supercomputers Sites [6], on the other hand, tell us that the performance improvement of computers has reached a factor of 1000 in the last 10 years as shown in Fig. 3. Increase in the number of CPUs in a system in addition to the degree of integration contributes to this rapid progress.

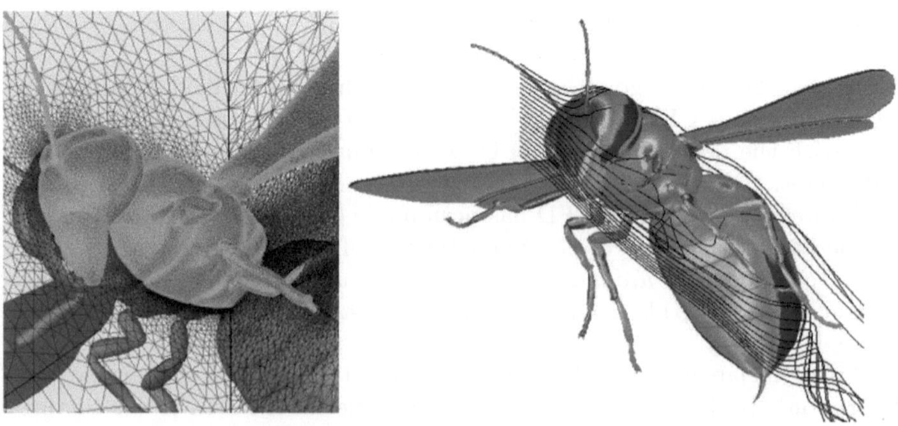

Fig. 1. Flow computation around a hornet by unstructured-grid CFD [2].

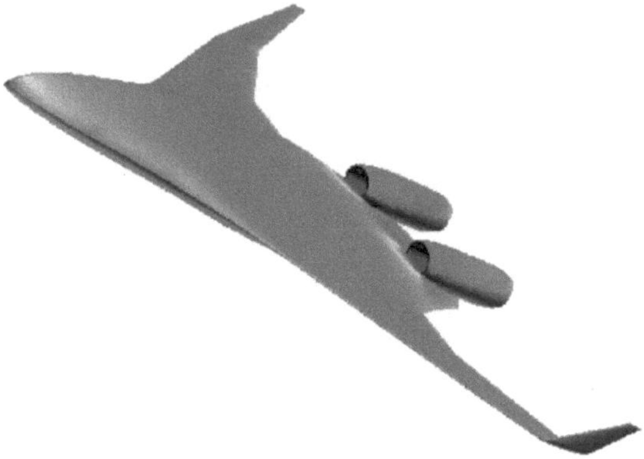

(a) Design of a blended-wing-body airplane [3].

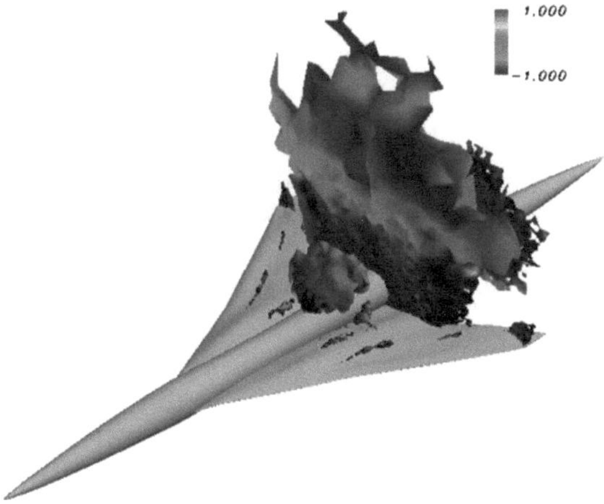

(b) Design and shock wave visualization of a sonic plane [4].

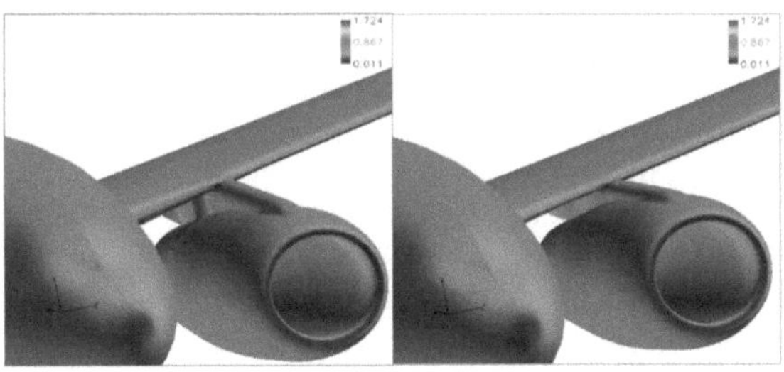

(c) Optimization of wing-nacelle-pylon configuration (left = original, right = optimized) [5].

Fig. 2. Application of unstructured-grid CFD

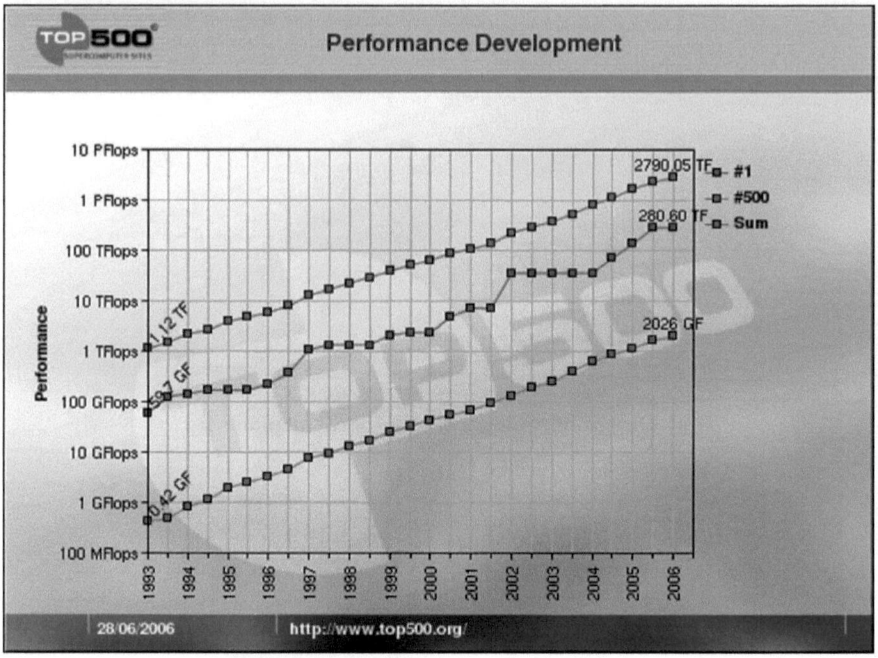

Fig. 3. Performance development in Top500 Super-computers.

With a simple extrapolation of Fig. 3, we can expect to use PetaFlops computers in ten years. This will accelerate the use of 3D RANS computations for the aerodynamic analysis and design of entire airplanes. DNS which does not use any physical models may also be used for engineering analysis of wings. In the not very far future, CFD could take over wind tunnels.

4 Demands for next-generation CFD

So, is it enough for us as CFD researchers to just wait for the progress of computers? Probably it is not. Let us consider demands for next-generation CFD on PetaFlops computers:

1. Easy and quick grid generation around complex geometries.
2. Easy adaptation of local resolution to local flow characteristic length.
3. Easy implementation of spatially higher-order schemes.
4. Easy massively-parallel computations.
5. Easy post processing for huge data output.
6. Algorithm simplicity for software maintenance and update.

Unstructured-grid CFD is a qualified candidate for the demands 1 and 2 as compared to structured grid CFD. However, an implementation of higher-order schemes on unstructured grids is not easy. Post processing of huge data output may also become another bottleneck due to irregularity of the data structure.

Recently, studies of Cartesian grid method were renewed in the CFD community, because of the several advantages such as rapid grid generation, easy higher-order extension, and simple data structure for easy post processing. This is another candidate for the next-generation CFD.

Letfs compare the computational cost of uniform Cartesian grid methods with that of tetrahedral unstructured grids. The most time-consuming part in compressible flow simulations is the numerical flux computations. The number of flux computations on a cell-vertex, finite volume method is proportional to the number of edges in the grid. In a tetrahedral grid, the number of edges is at least twice of that of the edges in a Cartesian grid of the same number of node points. Therefore, the computational costs on unstructured grids are at least twice as large as the costs of Cartesian grids. Moreover, computations of linear reconstructions, limiter functions, and implicit time integrations on tetrahedral grids easily doubles the total computational costs.

For higher-order spatial accuracy, the difference of computational costs between two approaches expands rapidly. In Cartesian grids, the spatial accuracy can be easily increased up to the fourth order without extra computational costs. In contrast, to increase the spatial accuracy from second to third-order on unstructured grids can easily increase tenfold the computational cost.

Namely, for the same computational cost and the same spatial accuracy of third-order or higher, we can use 100 to 1000 times more grid points in the Cartesian grid than in unstructured grid. The increase of grid points improves the accuracy of geometrical representation in computations as well as the spatial solution accuracy.

Although the above estimate is very rough, it is apparent that the Cartesian grid CFD is a big advantage for high resolution computations required for DNS.

5 Building-Cube Method

A drawback of uniform Cartesian grid is the difficulty of changing the mesh size locally. This is critical, especially for airfoil/wing computations, where an extremely large difference in characteristic flow lengths exists between boundary layer regions and far fields. Accurate representation of curved boundaries by Cartesian meshes is another issue.

A variant of the Cartesian grid method is to use the adaptive mesh refinement [7] in space and cut cells or the immersed boundary method [8] on the wall boundaries. However, introduction of irregular subdivisions and cells into Cartesian grids complicate the algorithm for higher-order schemes. The advantages of the Cartesian mesh over the unstructured grid, such as simplicity and less memory requirement, disappear.

The present author proposes a Cartesian grid based approach, named Building-Cube method [9]. Basic strategies employed here are; (a) zoning of a flow field by cubes (squares in 2D as shown in Fig. 4) of various sizes to adapt the mesh size to local flow characteristic length, (b) uniform Cartesian mesh in each cube for easy implementation of higher-order schemes, (c) same grid size in all cubes for easy parallel computations, (d) staircase representation of wall boundaries for algorithm simplicity.

It is similar to a block-structured uniform Cartesian mesh approach [10], but unifying the block shape to a cube simplifies the domain decomposition of a computational field around complex geometry. Equality of computational cost among all cubes significantly simplifies the massively parallel computations. It also enables us to introduce data compression techniques for pre and post processing of huge data [11].

A staircase representation of curved wall boundaries requires a very small grid spacing to keep the geometrical accuracy. But the flexibility of geometrical treatments obtained by it will be a strong advantage for complex geometries and their shape optimizations. An example is shown in Fig. 5 where a tiny boundary layer transition trip attached to an airfoil is included in the computational model. Figure 6 are the computed pressure distributions which show the detailed flow features including the effect of trip wire, interactions between small vortices and the shock wave, and so on.

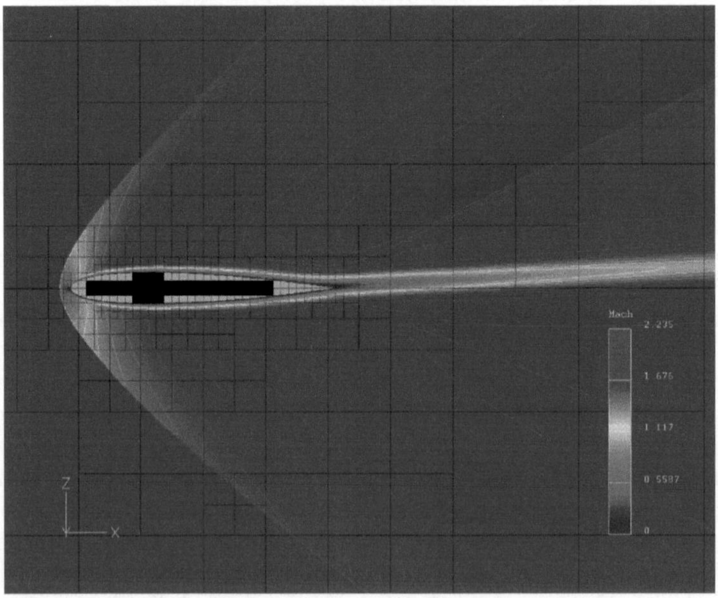

Fig. 4. Computed Mach distribution around NACA0012 airfoil at Re=5000, $M_\infty = 2$, and $\alpha = 3°$

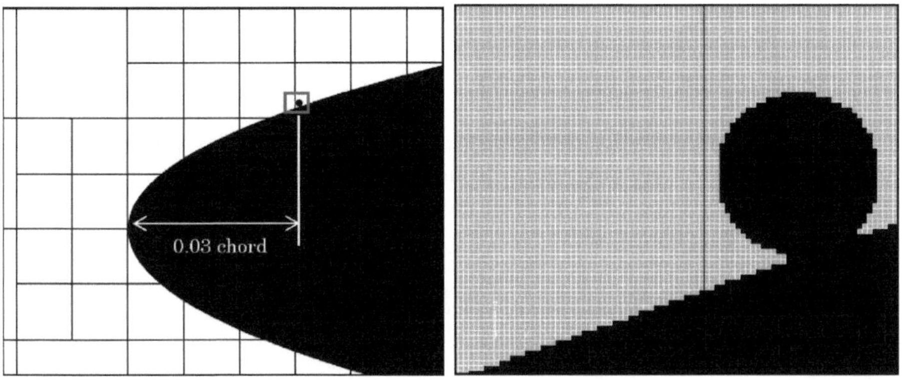

Fig. 5. Cube frames around RAE2822 airfoil (left) and an enlarged view of Cartesian grid near tripping wire (right).

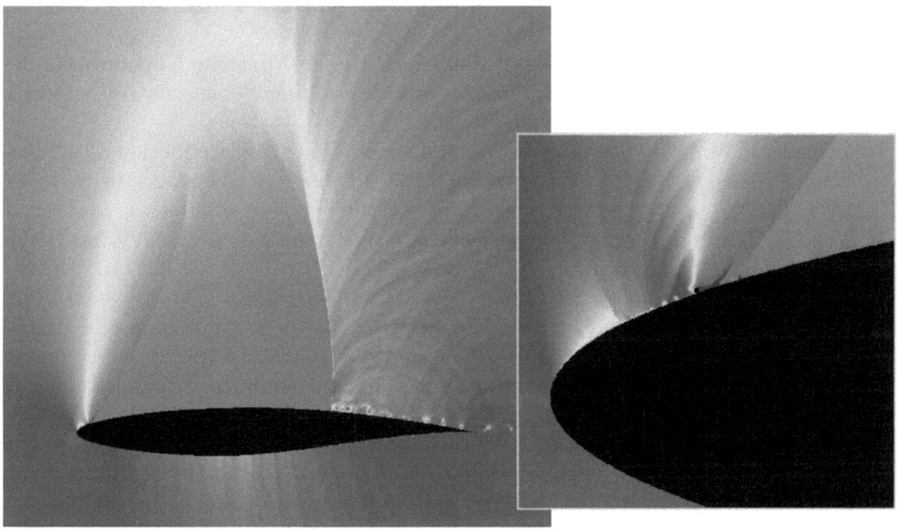

Fig. 6. Computed pressure distributions around RAE2822 airfoil at Re=6.5 $\times$ 10^6, $M_\infty = 0.73$, $\alpha = 2.68°$.

The result was obtained by solving the two dimensional Navier-Stokes equations. We did not use any turbulence models, but just used a high-density Cartesian mesh and a fourth-order scheme. This 2D computation may not describe the correct flow physics, since the three-dimensional flow structures are essential in the turbulent boundary layers for high-Reynolds number flows. However, the result indicates that a high-resolution computation using a high-density Cartesian mesh is very promising with a progress of computers.

6 Conclusion

CFD, using a high-density Cartesian mesh, is still limited in its application due to the computational cost. The predictions about Cartesian mesh CFD and computer progress in this article may be too optimistic. However, it is probably correct to say that the simplicity of the algorithm from grid generation to post processing of Cartesian mesh CFD will be a big advantage in the days of PetaFlops computers.

References

1. Jameson, A. and Baker, T.J. : Improvements to the Aircraft Euler Method. AIAA Paper 1987-452 (1987).
2. Nakahashi, K., Ito, Y., and Togashi, F.: Some challenges of realistic flow simulations by unstructured grid CFD. Int. J. for Numerical Methods in Fluids, **43**, 769–783 (2003).
3. Pambagjo,T.E., Nakahashi, K., Matsushima, K.: An Alternate Configuration for Regional Transport Airplane. Transactions of the Japan Society for Aeronautical and Space Sciences, **45**, 148, 94–101 (2002).
4. Yamazaki, W., Matsushima, K. and Nakahashi, K.: Drag Reduction of a Near-Sonic Airplane by using Computational Fluid Dynamics. AIAA J., **43**, 9, 1870–1877 (2005).
5. Koc, S., Kim, H.J. and Nakahashi, K.: Aerodynamic Design of Wing-Body-Nacelle-Pylon Configuration. AIAA-2005-4856, 17th AIAA CFD Conf. (2005).
6. Top500 Supercomputers Sites, http://www.top500.org/.
7. Berger, M. and Oliger, M.: Adaptive Mesh Refinement for Hyperbolic Partial Differential Equations. J. Comp. Physics, **53**, 561–568 (1984).
8. Mittal, R. and Iaccarino, G.: Immersed Boundary Methods. Annual Review of Fluid Mechanics, **37**, 239–261 (2005).
9. Nakahashi, K.: Building-Cube Method for Flow Problems with Broadband Characteristic Length. *Computational Fluid Dynamics 2002*, edited by S. Armfield, et. al., Springer, 77–81 (2002).
10. Meakin, R.L. and Wissink, A.M.: Unsteady Aerodynamic Simulation of Static and Moving Bodies Using Scalable Computers. AIAA-99-3302, Proc. AIAA 14th CFD Conf. (1999).
11. Nakahashi, K.: High-Density Mesh Flow Computations with Pre-/Post-Data Compressions. AIAA 2005-4876, Proc. AIAA 17th CFD Conf. (2005).

Smart Suction – an Advanced Concept for Laminar Flow Control of Three-Dimensional Boundary Layers

Ralf Messing and Markus Kloker

Institut für Aerodynamik und Gasdynamik, Universität Stuttgart, Pfaffenwaldring 21, 70550 Stuttgart, Germany, e-mail: [last name]@iag.uni-stuttgart.de

A new method combining classical boundary-layer suction with the recently developed technique of Upstream Flow Deformation is proposed to delay laminar-turbulent transition in three-dimensional boundary-layer flows. By means of direct numerical simulations details of the flow physics are investigated to maintain laminar flow even under strongly changing chordwise flow conditions. Simulations reveal that steady crossflow modes are less amplified than in the case of ideal homogeneous suction at equal suction rate.

1 Introduction

The list of reasons for a sustained reduction of commercial-aircraft fuel consumption is getting longer every day: significant environmental impacts of the strongly-growing world-wide air traffic, planned taxes on kerosene and emission of greenhouse gases, and the lasting rise in crude-oil prices. As fuel consumption during cruise is mainly determined by viscous drag its reduction offers the greatest potential for fuel savings. One promising candidate to reduce viscous drag of a commercial aircraft is laminar flow control (LFC) by boundary-layer suction on the wings, tailplanes, and nacelles with a fuel saving potential of 16%. (The other candidate is management of turbulent flow, e.g., on the fuselage of the aircraft, by a kind of shark-skin surface structure that however has a much lower saving potential.)

Suction has been known for decades to delay the onset of the drag-increasing turbulent state of the boundary-layer by significantly enhancing its laminar stability and thus pushing laminar-turbulent transition downstream. However, in case of swept aerodynamic surfaces, boundary-layer suction is not as straightforward and efficient as desired due to a crosswise flow component inherent in the three-dimensional boundary layer. Suction aims here primarily at reducing this crossflow, and not, as on unswept wings, at

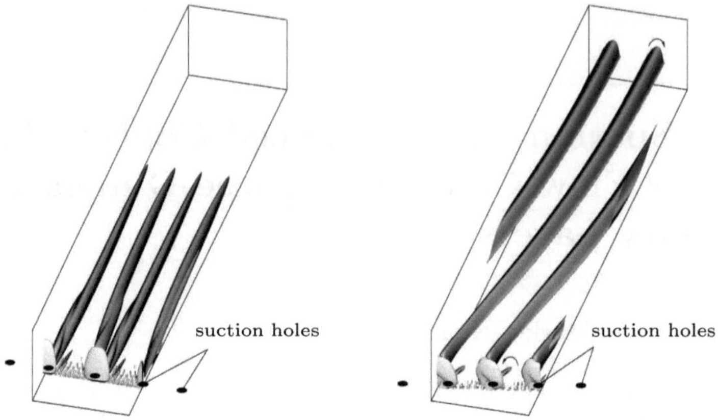

Fig. 1. Visualisation of vortices emanating from a single suction hole row on an unswept (left) and swept wedge (right). Arrows indicate the sense of vortex rotation. On the unswept wedge vortices are damped downstream, on the swept wedge co-rotating vortices are amplified due to crossflow instability.

influencing the wall-normal distribution of the streamwise flow component. The crossflow causes a primary instability of the boundary-layer with respect to three-dimensional disturbances. They can grow exponentially in downstream direction, depending on their spanwise wave number, and lead to co-rotating longitudinal vortices, called crossflow vortices (see Fig. 1), in the front part of, e.g., the wings. Now, on swept wings with their metallic or carbon-fibre-ceramic skins, the discrete suction through groups of micro-holes or -slots with diameters of typically 50 micrometers can excite unwanted, nocent crossflow vortices. The grown, typically steady vortices deform the laminar boundary layer and can cause its breakdown to the turbulent state by triggering a high-frequency secondary instability, occuring now already in the front part of the wing. The onset of such an instability highly depends on the state of the crossflow vortices. A determinant parameter is the spanwise spacing of the vortices, influencing also their strength. The spacing of naturally grown vortices corresponds to the spanwise wavelength of the most amplified eigenmode disturbance of the base flow. Even in most cases of discrete suction through groups of holes or slots with relatively small spanwise and streamwise spacings such a vortex spacing appears on a suction panel as the strong growth of the most amplified disturbance always prevails. This explains why transition can also set in when "stabilizing" boundary-layer suction is applied. If the suction were perfectly homogeneous over the wall, the suction itself would not excite nocent modes, but any surface imperfections like dirt, insects and so on would.

Recently, a new strategy for laminar flow control has been proposed and experimentally [Sar98] and numerically demonstrated [Was02].At a single chordwise location artifical roughness elements are laterally placed and ordered such that they excite relatively closely-spaced, benign crossflow vortices that suppress the nocent ones by a nonlinear mechanism and do not trigger fast secondary instability. If the streamwise variation of flow conditions and stability characteristics is weak this approach has proven to impressively delay transition. A better understanding of the phyiscal background of the effectiveness of this approach has been provided by direct numerical simulations [Was02], who coined the term Upstream Flow Deformation (UFD).

A major shortcoming of UFD with its single excitation of benign vortices is that it works persistently only for flows with non- or weakly varying stability properties. This typically is not the case on swept wings or tail planes where the boundary-layer flow undergoes a varying acceleration. Hence, a new method is proposed to overcome this deficiency. Before introducing this new approach the base flow and its stability characteristics are addressed to highlight the basic challenge.

2 Laminar base flow

The considered baseflow is the flow past the A320 vertical fin as available from the EUROTRANS project. Within this project the flow on the fin of the medium-range aircraft Airbus A320 has been measured and documented with the purpose to provide a database open to the scientific community. In order to demonstrate the feasibility of the new method the EUROTRANS baseflow has been chosen although it causes tremendous additional computational effort compared to generic base flows with constructed freestream-velocity distributions and stability characterictics. The reason is that such a procedure excludes any uncertainties arising from artificial base flows. This constitutes an important step towards realistic numerical simulations and could only be performed within a reasonable time frame on the NEC-SX 8 at the Höchstleistungsrechenzentrum Stuttgart (HLRS).

The freestream velocity perpendicular to the leading edge is $\overline{U}_\infty = 183.85m/s$, resulting from a flight velocity of $240m/s$ and a sweep angle of 40 degrees, the kinematic viscosity is $\overline{\nu} = 3.47 \cdot 10^{-5}m^2/s$ and the reference length is $\overline{L} = 4 \cdot 10^{-4}m$. The integration domain begins at $x_0 = 70$ and ends at $x_N = 500$, covering a chordwise length of $17.2cm$ on the fin. The shape factor is $H_{12} = 2.32$ at the inflow boundary and weakly rises up to $H_{12} = 2.41$ at $x_0 \approx 160$ where it keeps constant for the rest of the integration domain. The local Hartree parameter of a *virtual* Falker-Skan-Cooke flow would be roughly $\beta_H \approx 0.4$ at the inflow boundary

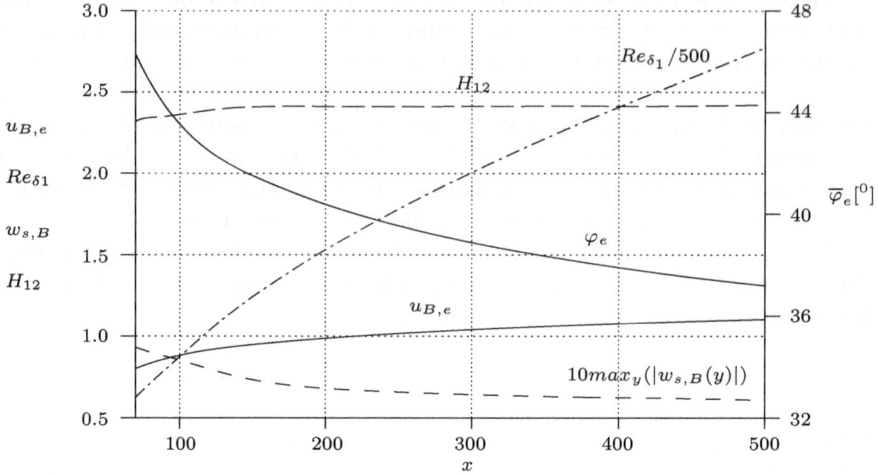

Fig. 2. Boundary-layer parameters of the EUROTRANS baseflow in the plate-fixed coordinate system.

and $\beta_H \approx 0.2$ downstream of $x_0 \approx 160$. The angle of the potential streamline decreases from $\overline{\varphi}_e(x_0) = 46^0$ to $\overline{\varphi}_e(x_n) = 37^0$. The maximum crossflow component reaches its maximum of about 9.3% at the inflow and continously declines to 6.1% at the outflow. The most important parameters of the Eurotrans baseflow are summarised in Fig. 2.

The resulting stability diagram for steady modes $\beta = 0$ gained by primary Linear Stability Theory based on the Orr-Sommerfeld equation is shown in Fig. 3 and reflects the strong stability variation which the boundary-layer flow undergoes in streamwise direction. Disturbances are amplified inside the curves with $\alpha_i = 0$ while the spanwise wavelength range of amplified steady crossflow modes extends from $800\mu m \leq \overline{\lambda}_z \leq 4740\mu m$ at $x = 70$ it shifts to $1850\mu m \leq \overline{\lambda}_z \leq 19330\mu m$ at $x = 500$. The value of the locally most amplified mode increases almost by a factor of 2.5 from $1480\mu m$ to $\overline{\lambda}_z(x = 500) = 3590\mu m$. For a control strategy based on the principle of the Upstream Flow Deformation technique this significant shift in the wave-number range implies that a single excitation of UFD-vortices only can successfully be applied on short downstream distances. A UFD vortex excited in the upstream flow region can only act as a stabilizer on a limited streamwise distance as the range of amplified disturbances shifts to smaller wave numbers and the UFD-vortex is damped loosing his stabilising influence. Without taking further measures laminar flow control by single excitation of UFD vortices provides no additional benefit downstream of this streamwise location. At this point an advanced control strategy has to set in. Details of how the proposed

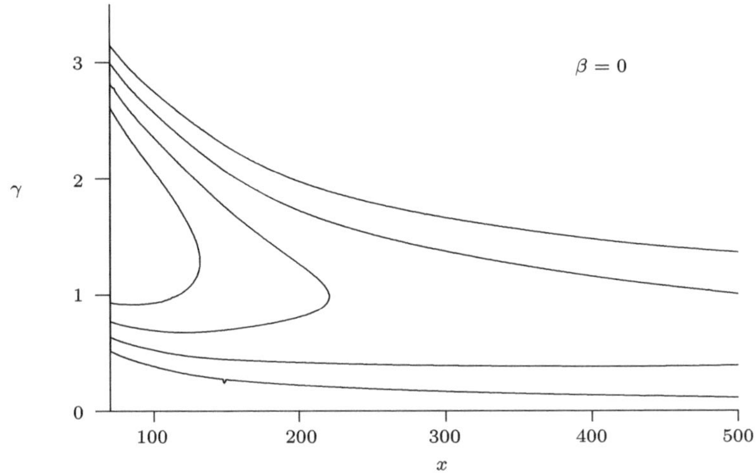

Fig. 3. Spatial amplification rates α_i of steady disturbances $\beta = 0$ according to linear stability theory for EUROTRANS baseflow; $\alpha_i = -d/dx\, ln(A/A_0)$, A-disturbance amplitude; $-\Delta\alpha_i = -0.01$ starting with the neutral curve.

method handles this issue and occuring difficulties are presented in the next section.

3 Smart Suction

The main idea of our approach is to combine two laminar-flow techniques that have already proven to delay transition in experiments and free-flight tests, namely boundary-layer suction and UFD. The suction orifices serve as excitation source and are ordered such that benign, closely-spaced UFD vortices are generated and maintained on a beneficial amplitude level. A streamwise variation of flow conditions and stability characteristics can be taken into account by adapting the spacing of the suction orifices continuously or in discrete steps. In this way we overcome the shortcomings of the single excitation of UFD vortices. However we note that this is not at all a trivial task because it is not clear a priori which direction the vortices follow – the flow direction depends on the wall-normal distance -, and improper excitation can lead to destructive nonlinear interaction with benign vortices from upstream, or nocent vortices. For illustration of a case where the adaptation to the chordwise varying flow properties has been done in an improper way see Fig. 4. On the left side of the figure a visualisation of the vortices in the flow field, and on the right side the location of the suction orifices at the wall, in this case spanwise slots, are shown. At about one-quarter of the domain

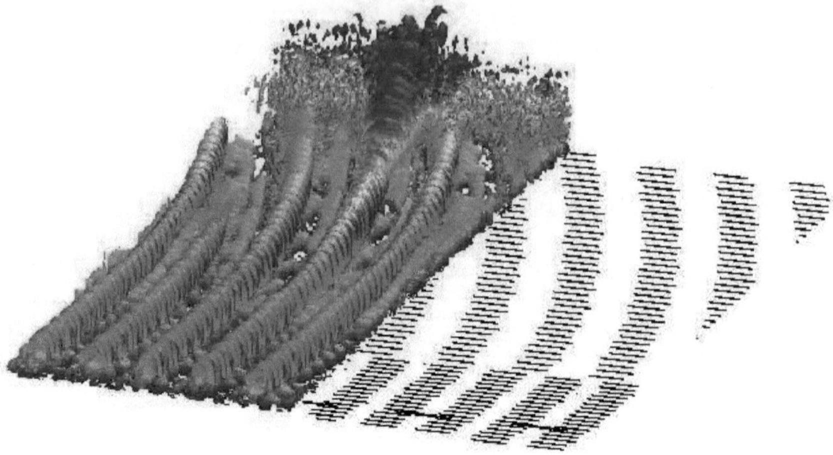

Fig. 4. Visualisation of vortices (left) and suction orifices/slots at the wall (right) in perspective view of the wing surface for smart suction with improper adaptation of suction orifices. Flow from bottom to top, breakdown to turbulence at the end of domain. In crosswise (spanwise) direction about $1.25\lambda_z$ is shown.

the spanwise spacing of the slots is increased in a discrete step from four slots per fundamental spanwise wavelength λ_z corresponding to a spanwise wave number $\gamma = 1.6$ (refer to Fig. 3) to three slots per λ_z corresponding to a spanwise wave number $\gamma = 1.2$ to adapt to the changing region of unstable wave numbers. In this case adaptation fails as transition to turbulent flow takes place at the end of the considered domain. A more detailed analysis reveals that nonlinear interactions between the both excited UFD vortices lead to conditions triggering secondary instability in the vicinity of the slot-spacing adaptation.

A successful application of the new method is shown in Fig. 5. The slots are ordered to excite UFD vortices with a spanwise wave number $\gamma = 1.6$. A puls-like disturbance has been excited to check whether an instability leads to the breakdown of laminar flow but indeed no transition is observed and further analysis reveals that unstable steady disturbances are even more effectively stabilised compared to ideal homogeneous suction at equal suction rate. If properly designed (see Fig. 5) the proposed method unifies the stabilising effects of bounday-layer suction and UFD. Consequently, the new method strives for (i) securing the working of suction on swept surfaces, and (ii) an additional stabilisation of the boundary-layer flow compared to classical suction alone, or, alternatively, it allows to reduce the suction rate for the same degree of stabilisation. By the excitation of selected crossflow modes being exponentially amplified and finally forming crossflow vortices not triggering turbulence, the stability of the flow is enhanced as would the suction rate of a

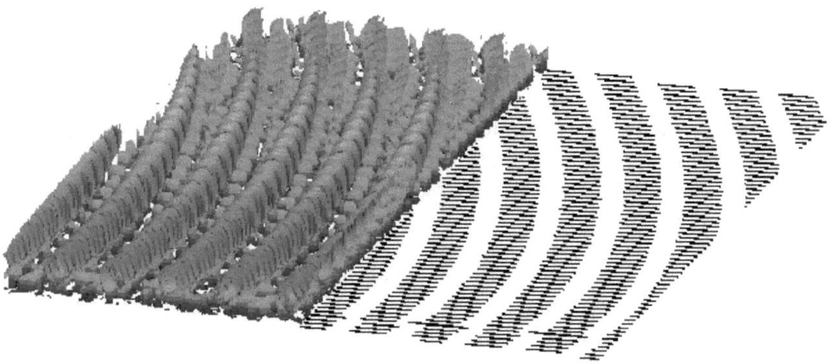

Fig. 5. As for Fig. 4 but with proper adaptation of suction orifices for a simple case.

conventional suction system have been risen. The reason is that the vortices generate by nonlinear mechanisms a mean-flow distortion not unlike suction, cf. [Was02], influencing the stability in an equally favourable manner as suction itself. The new method is termed *smart suction* as the instability of the laminar flow is exploited to enhance stability rather than increasing the suction rate.

4 Conclusions and Outlook

By means of direct numerical simulations it could be shown that a new technique combining boundary-layer suction and the method of Upstream Flow Deformation has the potential to significantly enhance the stability of three-dimensional boundary layers and therefore to delay laminar-turbulent transition on wings and tailplanes. The major challenge is the appropriate adaption of the excitation of benign crossflow vortices to the changing flow conditions. If this issue is properly mastered it can be expected that extended areas of the wing or tailplane can be kept laminar. The concept of smart suction offers a promising approach and an European and US-patent have been filed. Naturally, when dealing with such complex problems several points need further clarification and the corresponding investigations are ongoing. On the other hand the benefits are overwhelming. Fuel and involved exhaust gas savings of about 16% can be expected using properly working suction on the wings, the tailplanes and the nacelles (see also [Schra05]). Furthermore, suction orifices are one option as excitation source. Others would be artifical roughness, bumps, dips, or localized suction/blowing actuators and corresponding investigations are planned. Moreover, the applicability of the technique to LFC on wind turbine rotors is scrutinized.

References

[Sar98] SARIC, W.S., CARRILLO, JR. & REIBERT, M.S. 1998a Leading-Edge Roughness as a Transition Control Mechanism. AIAA-Paper 98-0781.

[Mes05] MESSING, R. & KLOKER, M. 2005 DNS study of spatial discrete suction for Laminar Flow Control. In: High Performance Computing in Science and Engineering 2004 (ed. E. Krause, W. Jäger, M. Resch), 163–175, Springer.

[Mar06] MARXEN, O. & RIST, U. 2006 DNS of non-linear transitional stages in an experimentally investigated laminar separation bubble. In: High Performance Computing in Science and Engineering 2005 (ed. W.E. Nagel, W. Jäger, M. Resch), 103–117, Springer.

[Schra05] SCHRAUF, G. 2005 Status and perspectives of laminar flow. *The Aeronautical Journal (RAeS)*, **109**, no. 1102, 639–644.

[Was02] WASSERMANN, P. & KLOKER, M. J. 2002 Mechanisms and passive control of crossflow-vortex induced transition in a three-dimansional boundary layer. J. Fluid Mech., **456**, 49–84.

Supercomputing of Flows with Complex Physics and the Future Progress

Satoru Yamamoto

Dept. of Computer and Mathematical Sciences, Tohoku University,
Sendai 980-8579, Japan
yamamoto@caero.mech.tohoku.ac.jp

1 Introduction

Current progress of Computational Fluid Dynamics(CFD) researches proceeded in our laboratory (Mathematical Modeling and Computation), is presented in this article. In our labo., mainly three projects are running. First one is the project: Numerical Turbine(NT Project). A parallel computational code which can simulate two- and three-dimensional multistage stator-rotor cascade flows in gas and steam turbines is being developed in this project with the development of pre- and post-processing softwares. This code can calculate not only those air flows, but also those flows of moist air and of wet steam. Second one is the project: Supercritical-Fluids Simulator(SFS Project). A computational code for simulating flows of arbitrary substance in arbitrary conditions, such in gas, liquid, and supercritical states is being developed. Third one is the project for making a custom computing machine optimized for CFD. A systric computational-memory architecture for high-performance CFD solvers is designed on a FPGA board. In this article, the NT and SFS projects are briefly introduced and some typical calculated results are shown as visualized figures.

2 The Project: Numerical Turbine(NT)

In recent ten years, a number of numerical methods and the computational codes for simulating flows with complex physics have been developed in our laboratory. For examples, a numerical method for hypersonic thermo-chemical nonequilibrium flows [1], a method for supersonic flows in magnet-plasma dynamic(MPD) thrusters [2], and a method for transonic flows of moist air with nonequilibrium condensation [3], have been proposed for simulating compressible flows with such complex physics. These three methods are based on compressible flow solvers. The approximate Riemann solvers, the total-variation diminishing(TVD) limiter and robust implicit schemes are employed for accurate and robust calculations.

The computational code for the NT project is developed from the code for condensate flows. Condensation occurs in fan rotors or compressors if moist air streams in them. Also wet-steam flows occasionally condense in steam turbines. The phase change from water vapor to water liquid is governed by homogeneous nucleation and the nonequilibrium process of condensation. The latent heat in water vapor is released to surrounding non-condensed gas when the phase change occurs, increasing temperature and pressure. This non-adiabatic effect induces a nonlinear phenomenon, the so-called "condensation shock". Finally, condensed water vapor affects the performance.

We developed two- and three-dimensional codes for transonic condensate flows assuming homogeneous and heterogeneous nucleations. For examples, 3-D flows of moist air around the ONERA M6[3] airfoil and condensate flows over a 3-D delta-wing in atmospheric flight conditions [4] have been already studied numerically by our group. Figure 1 shows a typical calculated condensate mass fraction contours over the delta-wing at uniform Mach number 0.6. This figure indicates that condensation occurs in a streamwise vortex, that is, the so-called "vapor trail". Those codes were applied to transonic flows of wet steam through steam-turbine cascade channels [5]. Figure 2 shows the calcu-

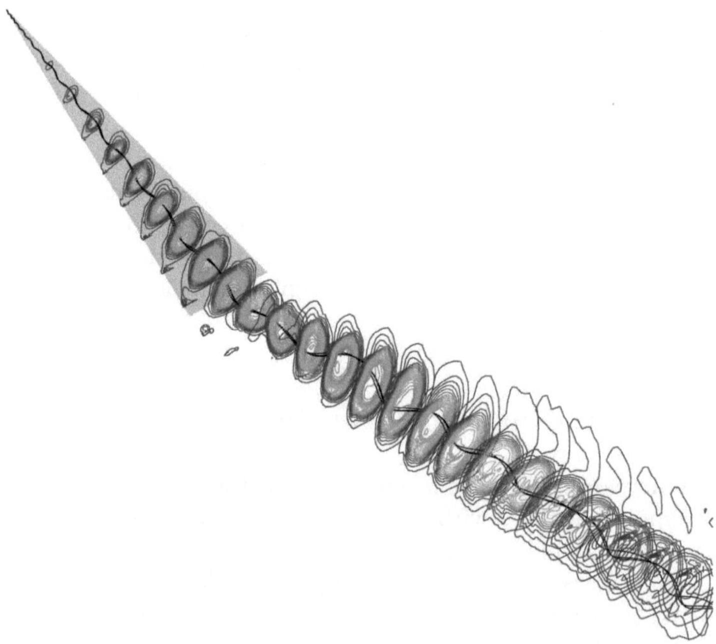

Fig. 1. Vapor trail over a delta-wing

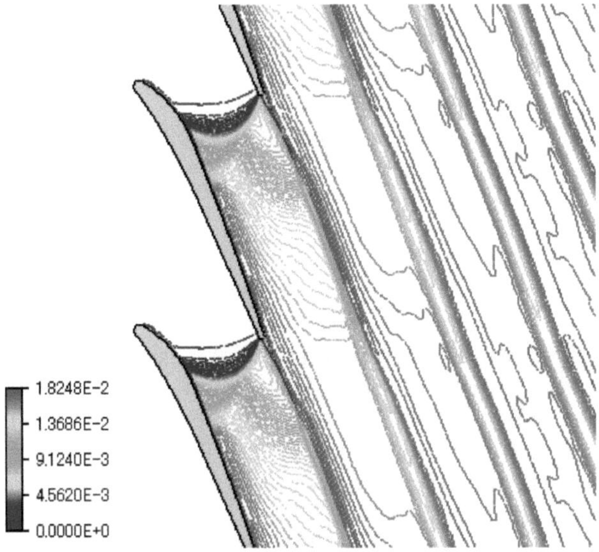

Fig. 2. Condensation in steam-turbine cascade channel

lated condensate mass fraction contours in a steam-turbine cascade channel. Nonquilibrium condensation starts at the nozzle throat of the channel.

In the NT code, the following governing equations are solved.

$$\partial Q/\partial t + \mathcal{F}(\mathcal{Q}) = \frac{\partial Q}{\partial t} + \frac{\partial F_i}{\partial \xi_i} + \frac{1}{Re}S + H = 0 \tag{1}$$

where

$$Q = J \begin{bmatrix} \rho \\ \rho u_1 \\ \rho u_2 \\ \rho u_3 \\ e \\ \rho_v \\ \rho\beta \\ \rho n \\ \rho k \\ \rho\omega \end{bmatrix}, \; F_i = J \begin{bmatrix} \rho U_i \\ \rho u_1 U_i + \partial\xi_i/\partial x_1 p \\ \rho u_2 U_i + \partial\xi_i/\partial x_2 p \\ \rho u_3 U_i + \partial\xi_i/\partial x_3 p \\ (e+p)U_i \\ \rho_v U_i \\ \rho\beta U_i \\ \rho n U_i \\ \rho k U_i \\ \rho\omega U_i \end{bmatrix}, \quad (i=1,2,3) \tag{2}$$

$$S = -J\frac{\partial \xi_i}{\partial x_j}\frac{\partial}{\partial \xi_i}\begin{bmatrix} 0 \\ \tau_{1j} \\ \tau_{2j} \\ \tau_{3j} \\ \tau_{kj}u_k + K\partial T/\partial x_j \\ 0 \\ 0 \\ 0 \\ \sigma_{kj} \\ \sigma_{\omega j} \end{bmatrix}, \quad H = -J\begin{bmatrix} 0 \\ 0 \\ \rho(\omega^2 x_2 + 2\omega u_3) \\ \rho(\omega^2 x_3 - 2\omega u_2) \\ 0 \\ -\Gamma \\ \Gamma \\ \rho I \\ f_k \\ f_\omega \end{bmatrix}$$

This system of equations is composed of compressible Navier-Stokes equations with carioles and centrifugal forces, mass equations for vapor, liquid, and the number density of nucleus, and the equations for the SST turbulence model [6]. The detail of equations is explained in Ref. [5]. The system of equations are solved by using the finite-difference method based on Roe's approximate Riemann solver [7], the fourth-order compact MUSCL TVD scheme [8], and the lower-upper symmetric Gauss Seidel(LU-SGS) implicit scheme [9].

In this article, we focus on the parallel-implicit computation using the LU-SGS scheme. The Gauss-Seidel method is one of relaxation methods for calculating the inverse of a matrix. The matrix is divided to lower and upper parts in the LU-SGS scheme. The inverse calculation is algebraically approximated and the procedure of the LU-SGS scheme is finally written as

$$D\Delta Q^* = RHS^n + \theta_L \Delta t G^+(\Delta Q^*) \tag{3}$$
$$\Delta Q^n = \Delta Q^* - D^{-1}\theta_L \Delta t G^-(\Delta Q^n) \tag{4}$$
$$G^+(\Delta Q^*) = (A_1^+\Delta Q^*)_{i-1,j,k} + (A_2^+\Delta Q^*)_{i,j-1,k} + (A_3^+\Delta Q^*)_{i,j,k-1} \tag{5}$$
$$G^-(\Delta Q^n) = (A_1^-\Delta Q^n)_{i+1,j,k} + (A_2^-\Delta Q^n)_{i,j+1,k} + (A_3^-\Delta Q^n)_{i,j,k+1} \tag{6}$$

where

$$D = I + \theta_L \Delta t[r(A_1) + r(A_2) + r(A_3)]$$
$$r(A) = \alpha \max[\lambda(A)]$$

The computational cost has been relatively saved by employing the algebraic approximation. However, the calculation should be sequentially proceeded, because the calculation at a grid point depends on those at the neighboring points in the same time-step. Therefore, the computational algorithm for the LU-SGS scheme is not suitable for the parallel computation. 39.4% of the total CPU time per one time-iteration was occupied for the LU-SGS routine when a wet-steam flow through a 3-D single channel was calculated using single CPU. In the NT code, flows through multistage stator-rotor cascade channels as shown in Fig. 3 are simultaneously calculated. Since each channel can be calculated separately in each time-iteration, a parallel computation using MPI is preferable. Also the 3-D LU-SGS algorithm may be parallelized on

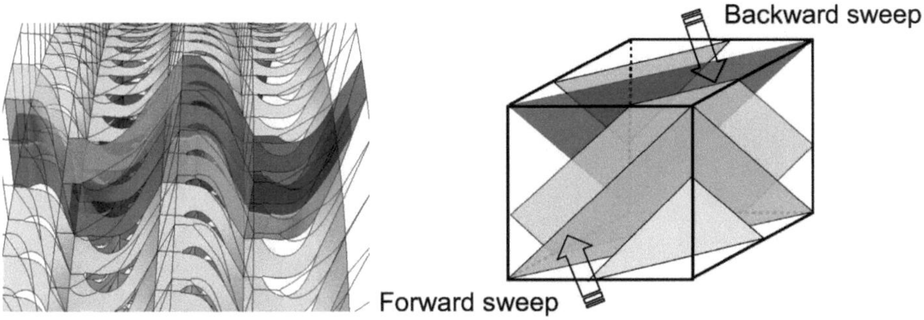

Fig. 3. Left: System of Grids. Right: Hyper-planes of LU-SGS algorithm

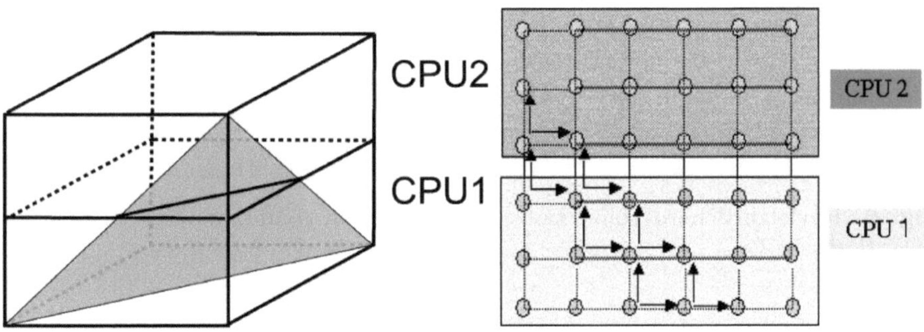

Fig. 4. Left: A Hyper-plane divided to two threads. Right: Pipelined LU-SGS algorithm

the so-called "hyper-plane" in each channel (Fig. 4). We applied the pipeline algorithm [10] to the hyper-plane assisted by the OpenMP directives. Then, the hyper-plane can be divided to a multi-thread.

Here, the pipeline algorithm applied to two threads are taken into account (Fig. 5). The algorithm is explained simply using the 2-D case (Fig. 6). Then, the calculation of the data on the hyper-plane depends on the left and the low grid point. In this example, the data are divided to two blocks. The lower block is calculated by CPU1 and the upper block is calculated by CPU2. CPU1 starts from the low-left corner and the data on the first grid-column is calculated. Then, CPU2 starts the calculation of the upper block from its low-left corner, using the boundary data of the first column in the lower block. Hereafter, CPU1 and CPU2 synchronously perform their calculation toward the right column. The number of CPUs can be increased easily. As increasing the number of threads to 2, 4, 8, and 16, the speedup ratio increases. But, the ratio is not always improved, even though the number of threads is increased up to 32 and 64. Consequently, 4CPUs may be the most effective and economical number for the pipelined LU-SGS calculation [11]. In the NT code, the calculation of each passage through turbine blades are assisted by the OpenMP and the set of passages is parallelized using MPI.

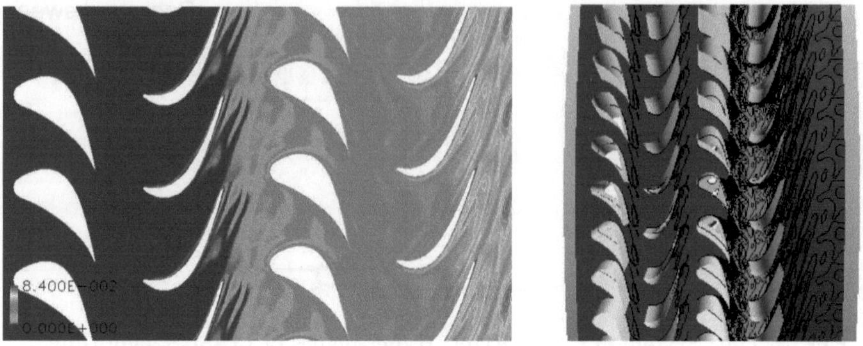

Fig. 5. Calculated instantaneous condensate mass fractions. Left: 2D code. Right: 3D code

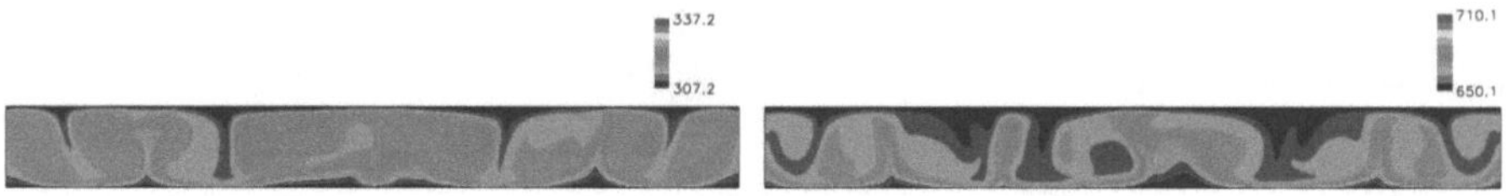

Fig. 6. Rayleigh-Bènard convections in supercritical conditions (Ra=3E5). Left: CO_2. Right: H_2O

Figure 5 shows the typical calculated results for the 2D and 3D codes. Contours of condensate mass fraction are visualized in both figures.

3 The Project: Supercritical-Fluids Simulator(SFS)

The above mentioned codes including those developed in the NT project are fundamentally compressible flow solvers. On the other hand, incompressible flow solvers are usually constructed by their own computational algorithms. The MAC method is one of the standard methods. In such the so-called "pressure-based methods", the Poisson equation for pressure should be solved with the calculation of incompressible Navier-Stokes equations. However, when an nonlinear property associated with the rapid change of thermal properties dominates the solution, incompressible flow solvers may be broken due to the instabilities. Therefore, a robust scheme overcoming the instabilities is absolutely necessary. Compressible flow solvers based on the Riemann solver [7] and the TVD scheme [8] are quite robust. Most of discontinuities such as shocks and contact surfaces can be calculated without any instability. But, the Riemann solvers have a weak point when very slow flows are calculated by them. Since the speed of sound is two or three orders of magnitude faster than that of convection, stiffness of solution may be emerged and the accurate solution may not be obtained.

In our approach, the preconditioning method developed by Choi and Merkle [12], and Weiss and Smith [13] are applied to the compressible flow solvers for condensate flows. The preconditioning method can enable the solvers to calculate both high-speed flows and very slow flows using the preconditioning matrix, which switches Navier-Stokes equations from compressible equations to incompressible equations automatically when a very slow flow is calculated. A preconditioned flux-vector splitting scheme [14] which can apply further to a static field(zero-velocity flow) has been also proposed. The SFS code employs the preconditioning code to simulate very slow flows at the Mach number far less than 0.1.

Supercritical fluids appear if both the bulk pressure and the temperature increase beyond the critical values. It is known that some anomalous properties are observed especially near the critical points. For examples, the density, the thermal conductivity, and the viscosity are rapidly changed near the critical points. These values and the variations are different among substances. Therefore, we should define all the thermal properties in each substance if we calculate supercritical fluids accurately. In the present SFS code, the database for thermal properties: PROPATH [15], developed by Kyushu university has been employed and been applied to the preconditioning method. Then, flows of arbitrary substance not only in supercritical conditions but also atmospheric and cryogenic conditions can be accurately calculated.

As a typical comparison for different substances in supercritical conditions, the calculated results of two-dimensional Rayleigh-Bènard(R-B) convections in supercritical CO_2 and H_2O are only shown here [16]. The aspect ratio of the flow field is fixed at 9 and 217×25 grid points are generated for the computational grid. The Rayleigh number is fixed to $Ra = 3 \times 10^5$ in both cases. It is known that the flow properties are fundamentally same if the R-B convections at a same Rayleigh number are calculated by assuming ideal gas. However, even though the flows of CO_2 and H_2O in near-critical conditions are calculated assuming the same Rayleigh number, the calculated instantaneous temperature contours are compared as quite different flow patterns. As obtained in H_2O case, the flow field is dominated by that with relatively lower temperature than that in CO_2 case.

4 Other Projects

The SFS code is now extending to a three-dimensional code. The SFS code is based on the finite-difference method and the governing equations in general curvilinear coordinates are solved. One of big problems using this approach may be how flows around a complex geometry should be solved. Recently we developed an immersed boundary(IB) method [17] which calculates flows using a rectangular grid for the flow field and surface meshes for complex geometries. Even though a number of the IB methods have been proposed in recent 10 years, the present IB method developed by us may be the simplest method.

A typical problem already calculated is only introduced here. A low Reynolds number flow over a sphere with a cylindrical projection was calculated. Figure 7 (left) shows the computational rectangular grid for the flow field and the surface mesh composed of a set of triangle polygons for the sphere. Figure 7 (right) shows the calculated stream lines over the sphere. A flow recirculation is formed behind the sphere and the cylindrical projection on the sphere forces the flow to an asymmetrical one. The present IB method will be applied to the 3-D SFS code in future.

5 Concluding Remarks

Two projects: Numerical Turbine(NT), and Supercritical-Fluids Simulator(SFS), proceeded in our laboratory and the future perspectives were briefly introduced. Both projects are strongly assisted by the supercomputer SX-7 of Information Synergy Center(ISC) in Tohoku University. The NT project has been also collaborated with private companies, Mitsubishi Heavy Industry at Takasago, and Toshiba Cooperation, and with Institute of Fluid Sciences of Tohoku University and the ISC of Tohoku University. The SFS project has been collaborated with private companies, Idemitsu Kousan and JFE Engineering, and with Institute of Multidisciplinary Research for Advanced Materials of Tohoku University. Also the SFS project will be supported by the Grand-in-Aid for Scientific Research(B) of JSPS, for next three years. Although the third project for a custom computing machine for CFD was not explained here. The research activity will be presented at the IEEE Symposium on Field-Programmable Custom Computing Machines(FCCM 2007)-Napa, 2007 [18].

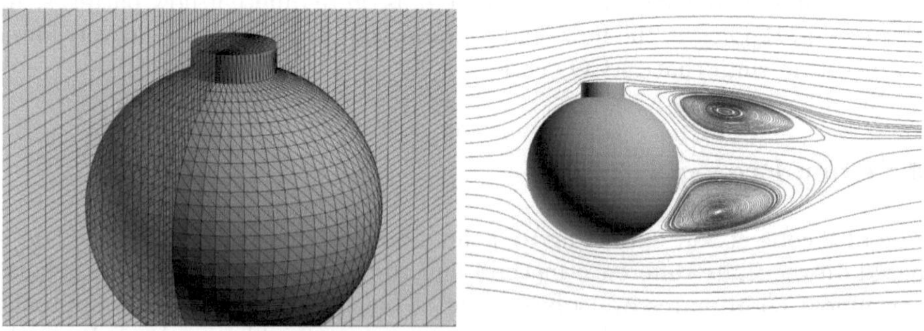

Fig. 7. Immersed Boundary Method. Left: Computational grids. Right: Calculated streamlines

References

1. S. Yamamoto, N. Takasu, and H. Nagatomo, Numerical Investigation of Shock/Vortex Interaction in Hypersonic Thermochemical Nonequilibrium Flow," *J. Spacecraft and Rockets*, **36**-2(1999), 240-246.
2. H. Takeda and S. Yamamoto, Implicit Time-marching Solution of Partially Ionized Flows in Self-Field MPD Thruster," *Trans. the Japan Society for Aeronautical and Space Sciences*, **44**-146 (2002), 223-228.
3. S. Yamamoto, H. Hagari and M. Murayama, Numerical Simulation of Condensation around the 3-D Wing," *Trans. the Japan Society for Aeronautical and Space Sciences*, **42**-138(2000), 182-189.
4. S. Yamamoto, Onset of Condensation in Vortical Flow over Sharp-edged Delta Wing," *AIAA Journal*, **41**-9 (2003), 1832-1835.
5. Y. Sasao and S. Yamamoto, Numerical Prediction of Unsteady Flows through Turbine Stator-rotor Channels with Condensation," *Proc. ASME Fluids Engineering Summer Conference*, FEDSM2005-77205(2005).
6. F.R. Menter, Two-equation Eddy-viscosity Turbulence Models for Engineering Applications," *AIAA Journal*, **32**-8(1994), 1598-1605.
7. P.L. Roe, Approximate Riemann Solvers, Parameter Vectors, and Difference Schemes," *J. Comp. Phys.*, **43**(1981), 357-372.
8. S.Yamamoto and H.Daiguji, Higher-Order-Accurate Upwind Schemes for Solving the Compressible Euler and Navier-Stokes Equations," *Computers and Fluids*, **22**-2/3(1993), 259-270.
9. S.Yoon and A.Jameson, Lower-upper symmetric-Gauss-Seidel method for the Euler and Navier-Stokes equations," *AIAA Journal* , **26**(1988), 1025-1026.
10. M.Yarrow and R. Van der Wijngaart, Communication Improvement for the NAS Parallel Benchmark: A Model for Efficient Parallel Relaxation Schemes," Tech. Report NAS RNR-97-032, NASA ARC, (1997).
11. S.Yamamoto, Y.Sasao, S.Sato and K.Sano, Parallel-Implicit Computation of Three-dimensional Multistage Stator-Rotor Cascade Flows with Condensation," *Proc.18th AIAA CFD Conf.-Miami*, (2007).
12. Y.-H.Choi and C.L. Merkle, The Application of Preconditioning in Viscous Flows," *J. Comp. Phys.*, **105**(1993), 207-223.
13. J.M. Weiss and W.A. Smith, Preconditioning Applied to Variable and Constant Density Flows," *AIAA Journal*, **33**(1995), 2050-2056.
14. S. Yamamoto, Preconditioning Method for Condensate Fluid and Solid Coupling Problems in General Curvilinear Coordinates," *J. Comp. Phys.*, **207**-1(2005), 240-260.
15. A Program Package for Thermophysical Properties of Fluids(PROPATH), Ver.12.1, PROPATH GROUP.
16. S. Yamamoto and A. Ito, Numerical Method for Near-critical Fluids of Arbitrary Material," *Proc. 4th Int. Conf. on Computational Fluid Dynamics-Ghent*, (2006).
17. S.Yamamoto and K.Mizutani, A Very Simple Immersed Boundary Method Applied to Three-dimensional Incompressible Navier-Stokes Solvers using Staggered Grid," *Proc. 5th Joint ASME_JSME Fluids Engineering Conference-San Diego*, (2007).
18. K.Sano, T.Iizuka and S.Yamamoto, Systolic Architecture for Computational Fluid Dynamics on FPGAs," *Proc. IEEE Symp. on Field-Programmable Custom Computing Machines-Napa*, (2007).

Large-Scale Computations of Flow Around a Circular Cylinder

Jan G. Wissink and Wolfgang Rodi

Institute for Hydromechanics,
University of Karlsruhe,
Kaisertstr. 12,
D-76128 Karlsruhe, Germany,
wissink@ifh.uka.de, rodi@ifh.uka.de

Summary. A series of Direct Numerical Simulations of three-dimensional, incompressible flow around a circular cylinder in the lower subcritical range has been performed on the NEC SX-8 of HLRS in Stuttgart. The Navier-Stokes solver, that is employed in the simulations, is vectorizable and parallellized using the standard Message Passing Interface (MPI) protocol. Compared to the total number of operations, the percentage of vectorizable operations exceeds 99.5% in all simulations. In the spanwise direction Fourier transforms are used to reduce the originally three-dimensional pressure solver into two-dimensional pressure solvers for the parallel (x, y) planes. Because of this reduction in size of the problem, also the vectors lengths are reduced which was found to lead to a reduction in performance on the NEC. Apart from the performance of the code as a function of the average vector-lenght, also the performance of the code as a function of the number of processors is assessed. An increase in the number of processors by a factor of 5 is found to lead to a decrease in performance by approximately 22%.

Snapshots of the instantaneous flow field immediately behind the cylinder show that free shear-layers from as the boundary layers along the top an bottom surface of the cylinder separate. Alternatingly, the most upstream part of the shear layers rolls up, becomes three-dimensional and undergoes rapid transition to turbulence. The rolls of rotating turbulent flow are subsequently convected downstream to form a von Karman vortex street.

1 Introduction

Two-dimensional flows around circular and square cylinders have always been popular test cases for novel numerical methods, see for instance [Wis97]. With increasing Reynolds number, Re, based on the inflow velocity and the diameter of the cylinder, the topology of the flow around a circular cylinder changes.

For low Reynolds numbers, the flow is two-dimensional and consists of a steady separation bubble behind the cylinder. As the Reynolds number increases, the separation bubble becomes unstable and vortex-shedding commences (see Williamson [Wil96a]). At a critical Reynolds number of $Re_{crit} \approx 194$, the wake flow changes from two-dimensional to three-dimensional and a so-called "mode A" instability (with large spanwise modes) emerges. When the Reynolds number becomes larger, the type of the instability changes "mode A" to "mode-B", which is characterized by relatively small spanwise modes which scale on the smaller structures in the wake, see Thompson et al. [Thom96] and Williamson [Wil96b].

The subcritical regime, where $1000 \leq Re \leq 200\,000$, is characterized by the fact that the entire boundary layer along the surface of the cylinder is laminar. Immediately behind the cylinder, two-dimensional laminar shear layers can be found, which correspond to the separating boundary layers along the top and the bottom surface of the cylinder. Somewhere downstream of the location of separation, the free shear-layers alternatingly roll up, become three-dimensional and undergo transition to turbulence. As a result, a turbulent vortex street is formed downstream of the location of transition. For even higher Reynolds number, the boundary layer along the surface of the cylinder becomes partially turbulent. Because of that, separation is delayed and both the size of the wake and the drag forces on the cylinder are reduced.

Many experiments exist for flow around a circular cylinder in the lower subcritical range ($Re = 3900$), see for instance [Lour93, Nor94, Ong96, Dong06]. A very detailed review of further experiments is provided in Norberg [Nor03]. Because of the availability of experimental data, the flow at $Re = 3900$ has become a very popular test case for numerical methods. Though mostly Large-Eddy Simulations (LES) were performed, also some Direct Numerical Simulations (DNS) were reported, see for instance Beaudan and Moin [Beau94], Mittal and Moin [Mit97], Breuer [Breu98], Fröhlich et al. [Froe98], Kravchenko and Moin [Krav00], Ma et al. [Ma00] and Dong et al. [Dong06].

In most of the numerical simulations, a spanwise size of πD – where D is the diameter of the cylinder – or less was used. Only in one of the simulations performed by Ma et al. [Ma00], a spanwise size of $2\pi D$ was used. In this simulation the time-averaged streamwise velocity profile in the near wake ("V-shaped" profile) was found to differ from the velocity profile ("U-shaped" profile) obtained in the other, well-resolved simulations with a spanwise size of πD. Kravchenko and Moin [Krav00] argued that the "V-shaped" profile was probably an artefact that only appeared when the grid resolution was unsatisfactory or in the presence of background noise.

So far, the importance of the spanwise size has been an open question and, therefore, is one of the reasons for performing the present series of DNS of flow around a circular cylinder in the lower subcritical range at $Re = 3300$.

The second reason for performing these DNS, is to generate realistic wake data that can be used in subsequent simulations of flow around a turbine or compressor blade with incoming wakes. In the simulations performed so far – see for instance [Wis03, Wis06] – only artificial wake data – with turbulence statistics that resemble the far-field statistics of a turbulent cylinder wake –

were employed. This data, however, may not contain all relevant length-scales that are typical for a near wake flow. With the availability of new, high-quality data from the near wake of a circular cylinder, we hope to be able to resolve this issue.

1.1 Numerical Aspects

The DNS were perfomed using a finite-volume code with a collocated variable arrangement which was designed to be used on curvi-linear grids. A second-order central discretization was employed in space and combined with a three-stage Runge-Kutta method for the time-integration. To avoid the decoupling of the velocity field and the pressure field, the momentum interpolation procedure of Rhie and Chow [Rhie83] was employed. The momentum interpolation effectively replaced the discretization of the pressure by another one with a more compact numerical stencil. The code was vectorizable to a high degree and was also parallellized. To obtain a near-optimal load balancing, the computational mesh was subdivided into a series of blocks which all contained an equal number of grid points. Each block was assigned to its own unique processor and communication between blocks was performed by using the standard message passing interface (MPI) protocol.

The Poisson equation for the pressure from the original three-dimensional problem was reduced to a set of equations for the pressure in two-dimensional planes by employing a Fourier transform in the homogeneous, spanwise direction. This procedure was found to significantly speed up the iterative solution of the pressure field on scalar computers. Because of the reduction of the original three-dimensional problem into a series of two-dimensional problems, the average vector-length was reduced, which might lead to a reduction in performance on vector computers. For more information on the numerical method see Breuer and Rodi [Breu96].

Figure 1 shows a spanwise cross-section through the computational geometry. Along the top and bottom surface, a free-slip boundary condition was employed and along the surface of the cylinder a no-slip boundary condition was used. At the inflow plane, a uniform flow-field was prescribed with $u = U_0$ and $v = w = 0$, where u, v, w are the velocities in the x, y, z-directions, respectively. At the outflow plane, a convective outflow boundary condition was employed that allows the wake to leave the computational domain without any reflections. In the spanwise direction, finally, periodic boundary conditions were employed.

The computational mesh in the vicinity of the cylinder is shown in Fig. 2. Because of the dense grid, only every eighth grid line is displayed. As illustrated in the figure, "O"-meshes were employed in all DNS of flow around a circular cylinder. Near the cylinder, the mesh was only stretched in the radial direction, and not in the circumpherential direction. Table 1 provides an overview of the performed simulations and shows some details on the performance of the code on the NEC which will be analyzed in the next section. The

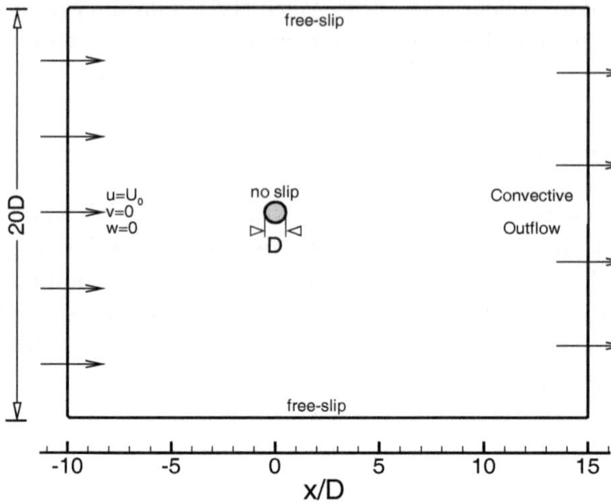

Fig. 1. Spanwise cross section through the computational geometry

maximum size of the grid cells adjacent to the cylinder's surface in Simulations B-F (in wall units) was smaller ore equal to $\Delta\phi^+ = 3.54$ in the circumpherantial direction, $\Delta r^+ = 0.68$ in the radial direction and $\Delta z^+ = 5.3$ in the spanwise direction.

Table 1. Overview of the performed simulations of flow around a circular cylinder at $Re = 3300$. $n\phi, nr, nz$ is the number of grid points in the circumpherential, radial and spanwise direction, respectively, D is the diameter of the cylinder, l_z is the spanwise size of the computational domain, nproc is the number of processors, points/block gives the number of grid points per block. The performance obtained for each of the simulations is calculated by averaging the performances over series of ten runs.

Sim.	$n\phi \times nr \times nz$	l_z	nproc	points/block	performance (Mflops/proc)
A	$406 \times 156 \times 256$	$4D$	24	821632	2247.9
B	$606 \times 201 \times 512$	$4D$	24	2979018	3298.1
C	$606 \times 201 \times 512$	$4D$	48	3398598	3122.6
D	$606 \times 201 \times 512$	$4D$	64	4447548	3105.5
E	$606 \times 201 \times 512$	$8D$	128	4672080	2584.5
F	$606 \times 201 \times 512$	$4D$	80	4447548	2576.1

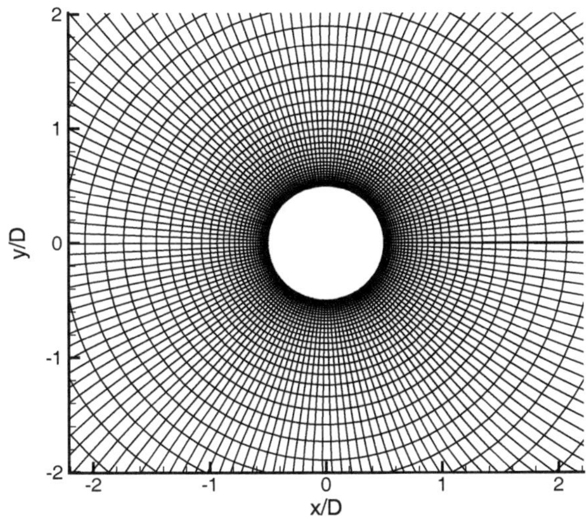

Fig. 2. Spanwise cross-section through the computational mesh near the cylinder as used in Simulations D and E (see Table 1), displaying every 8th grid line

2 Performance of the code on the NEC SX-8

In this section, the data provided in Table 1 will be analyzed. In all simulations, the ratio of the vectorized operations compared to the total number of operations is better than 99.5%. In Fig. 3, the performance of the code - measured in GFlops per processor - is plotted against the average vector-length. The capital letters $A - F$ identify from which simulation the data originate (see also Table 1). A very clear positive correlation was found between the average performance of the code and the average vector-length. Especially in Simulation A, the relatively small number of grid points per block was found to limit the average vector-length and, because of that, reduced the performance of the code by almost 50% as compared to Simulation B, which was performed using the same number of processors but with a significantly larger number of points per block. As already mentioned in the section above, in general the average vector-length in these numerical simulations is rather small because of the usage of a Fourier method to reduce the three-dimensional Poisson problem to a number of two-dimensional problems. When a three-dimensional Poisson solver would have been used instead, we expect that the performance would have significantly improved, though it is unlikely that it also would have resulted in a reduction of the effective computing time. On a scalar computer, for instance, the computing time of the Poisson solver using Fourier transforms in the spanwise direction was found to be a factor of two smaller as compared to the computing time needed when using the standard three-dimensional solver.

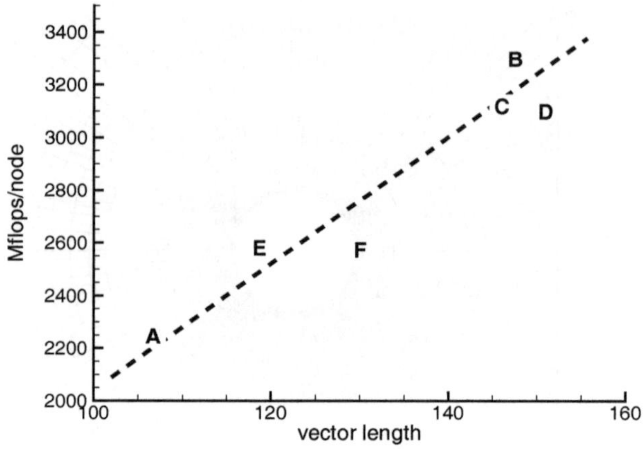

Fig. 3. Performance per processor as a function of the average vector-length, the capital letters identify the simulations (see also Table 1)

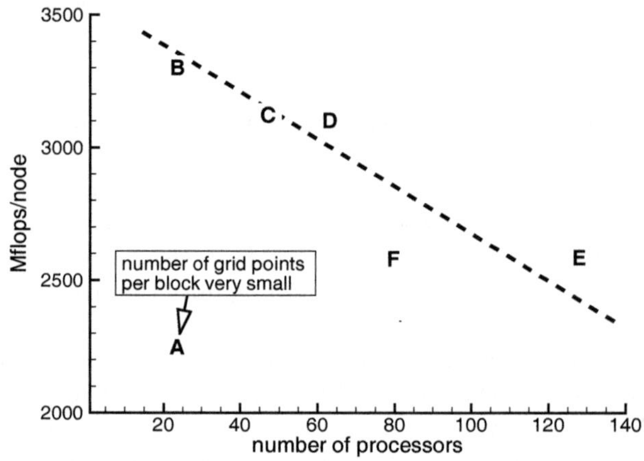

Fig. 4. Performance per processor as a function of the number of processors employed, the capital letters identify the simulations (see also Table 1)

Figure 4 shows the performance per processor as a function of the number of processors employed in the simulation. Again, the capital letters identify the simulation which are listed in Table 1. With the exception of Simulation A, which has an exeptionally small number of points per block, a negative correlation is obtained between the mean performance and the number of processors used. The results shown in this figure give an impression of the scaling behaviour of the code with an increasing number of processors: If the number of processors used increases by a factor of 5, the average performance of the code per processor reduces by about 22%. The obvious reason for this is the increase in communication when the number of blocks (which is the same as the number of processors) increases. Communication is needed to exchange information between the outer parts of the blocks during every timestep. Though the amount of data exchanged is not very large, the frequency with which this exchange takes place is relatively high and, therefore, slows down the calculations.

As an illustration of what happens when the number of points in the spanwise direction is reduced by a factor of 4 (compared to Simulations B,C,D,F), we consider a simulation of flow around a wind turbine blade. In this simulation, the mesh consisted of $1510 \times 1030 \times 128$ grid points, in the streamwise, wall-normal and spanwise direction, respectively. The grid was subdivided into 256 blocks, each containing 777650 computational points. The average vector-length was 205 and the vector operation ratio was 99.4%. The mean performance of code reached values of approximately 4 GFlops per processor, such that the combined performance of the 256 processors reached a value in excess of 1 Tflops for this one simulation.

From the above we can conclude that optimizing the average vector-length of the code resulted in a significant increase in performance and that it is helpful to try to increase the number of points per block. The latter will reduce the amount of communication between blocks and may also increase the average vector-length.

3 Instantaneous flow fields

Figure 5 shows a series of snapshots with iso-sufaces of the spanwise vorticity. The blue iso-surface corresponds to negative spanwise vorticity originating from the boundary layer along the top surface of the cylinder and the orange/red iso-surface corresponds to positive vorticity originating from the bottom surface of the cylinder. For transitional flows that are homogeneous in the spanwise direction, spanwise vorticity isosurfaces are used to identify boundary layers and horizontal, two-dimensional shear layers. The snapshots clearly illustrate the presence of 2D laminar shear layers immediately downstream of the cylinder. These shear layers alternatingly roll-up and become three-dimensional, followed by a rapid transition to turbulence. As the rolls are washed downstream, a von Karman vortex street is formed consisting of rolls of rotating, turbulent fluid with an alternating sense of rotation.

The rotating turbulent fluid that originates from the rolled-up shear layer is visualized using the sequence of iso-surfaces of the fluctuating pressure shown in Fig. 6. The iso-surfaces identify those regions of the flow that rotate. It can be seen that there are many three-dimensional structures superposed on the spanwise rolls of recirculating flow. Because of the re-circulating flow, some of the turbulence generated as the rolls of recirculating flow undergo transition will enter the dead air region immediately behind the cylinder. The turbulence level in this region, however, is found to be very low.

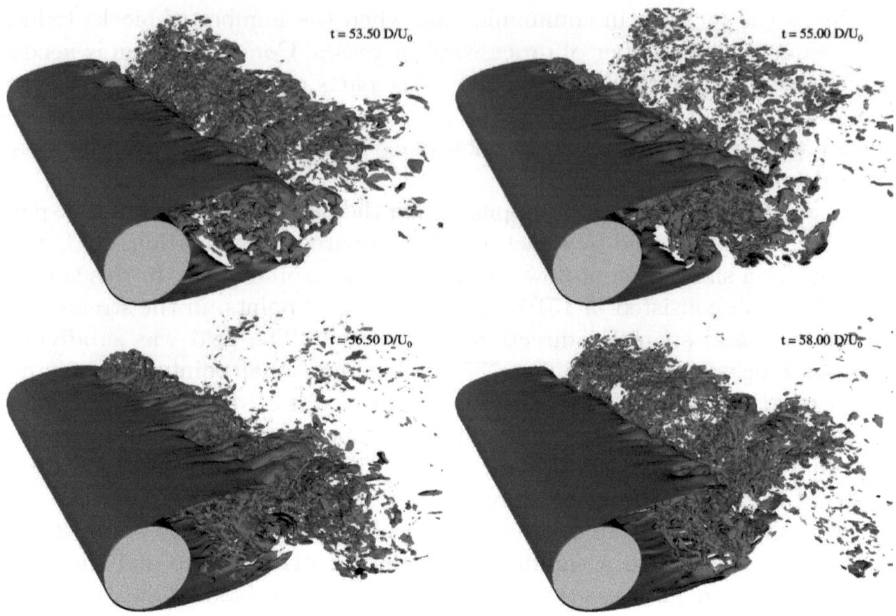

Fig. 5. Snapshots showing iso-surfaces of the spanwise vorticity at $t = 53.5D/U_0$, $t = 55.00D/U_0$,$t = 56.50D/U_0$ and $t = 58.00D/U_0$.

4 Discussion and Conclusions

Direct numerical simulations of three-dimensional flow around a circular cylinder in the lower subcritical range at $Re = 3300$ were performed on the NEC SX-8. The numerical code used is both vectorizable and parallellized. The ratio of vector operation compared to the total number of operations was found to be more than 99.5% for all simulations.

A series of snapshots, showing instantaneous flow fields, illustrates the presence of two laminar free shear-layers immediately downstream of the cylinder which correspond to the separating boundary layers from the top and bottom surface of the cylinder. In order to correctly predict the flow-physics it is very important that these layers are well resolved. The downstream part of the shear layers is observed to roll-up, become three-dimensional and undergo rapid transition to turbulence, forming rolls of re-circulating turbulent flow. Rolls with alternating sense of rotation are shed from the upper and lower shear layer, respectively, and form a von Karman vortex street.

In order to assess the quality of the grids, a convergence study was performed employing by gradually increasing the number of grid points. From this series of simulations, also performance of the numerical code on the NEC could be assessed. The performance of the code was found to be adversely affected by the usage of a Fourier method in the pressure solver. This method reduces the originally three-dimensional set of equations for the pressure into

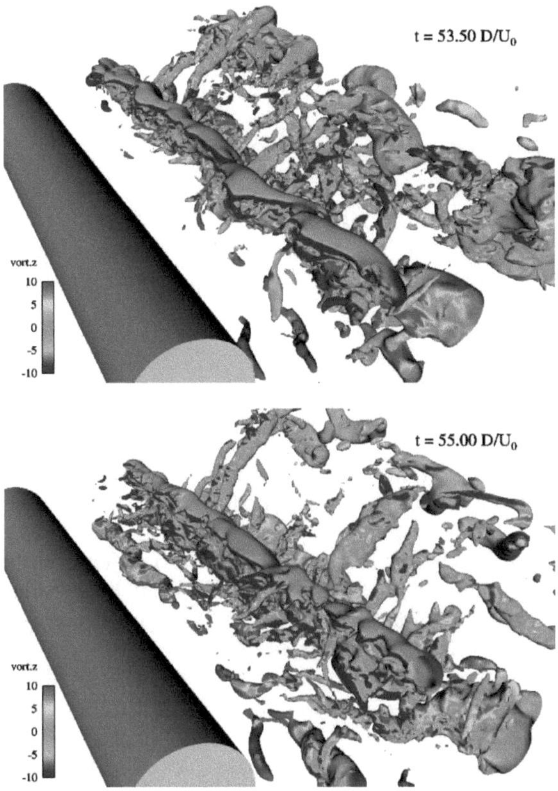

Fig. 6. Snapshots showing iso-surfaces of the fluctuating pressure $p' = p - \bar{p}$ at $t = 53.5D/U_0$ and $t = 55.00D/U_0$. The iso-surfaces are coloured with spanwise vorticity

a series of two-dimensional sets of equations to be solved in parallel grid planes ($z/D = const.$). Because of this reduction in size of the problem, also the average vector length was reduced which resulted in a decrease in the performance achieved on the NEC.

The mesh was subdivided into a number of blocks, each containing the same number of computational points. On average, the performance of the simulations with a large number of computational points per block was found be better than the performance of the computations with a smaller number of points per block. The main reasons for this were 1) the reduced communication when using larger blocks and 2) the fact that larger blocks often lead to larger average vector lengths.

An estimate for the scaling of the code on the NEC was obtained by assessing the performance of the code as function of the number of processors used in the simulation. It was found that the performance reduces by a factor of approximately 22% when the number of processors was increased by a factor of 5.

Acknowledgements

The authors would like to thank the German Research Council (DFG) for sponsoring this research and the Steering Committee of the Supercomputing Centre (HLRS) in Stuttgart for granting computing time on the NEC SX-8.

References

[Beau94] Beaudan, P., Moin, P.: Numerical experiments on the flow past a circular cylinder at subcritical Reynolds number, In Report No. TF-62, Thermosciences Division, Department of Mechanical Engineering, Stanford University, pp. 1–44 (1994)

[Breu96] Breuer, M., Rodi, W.: Large eddy simulation for complex turbulent flow of practical interest, In E.H. Hirschel, editor, Flow simulation with high-preformance computers II, Notes in Numerical Fluid Mechanics, volume 52, Vieweg Verlag, Braunschweig, (1996)

[Breu98] Breuer, M.: Large eddy simulations of the subcritical flow past a circular cylinder: numerical and modelling aspects, Int. J. Numer. Meth. Fluids, **28**, 1281–1302 (1998)

[Dong06] Dong, S., Karniadakis, G.E., Ekmekci, A., Rockwell, D.: A combined direct numerical simulation-partical image velocimetry study of the turbulent near wake, J. Fluids Mech., **569**, 185–207 (2006)

[Froe98] Fröhlich, J., Rodi, W., Kessler, Ph., Parpais, S., Bertoglio, J.P., Laurence, D.: Large eddy simulation of flow around circular cylinders on structured and unstructured grids, In E.H. Hirschel, editor, Flow simulation with high-preformance computers II, Notes in Numerical Fluid Mechanics, volume 66, Vieweg Verlag, Braunschweig, (1998)

[Krav00] Kravchenko, A.G., Moin, P.: Numerical studies of flow around a circular cylinder at $Re_D = 3900$, Phys. Fluids, **12**, 403–417 (2000)

[Lour93] Lourenco, L.M., Shih, C.: Characteristics of the plane turbulent near wake of a circular cylinder, a partical image velocimetry study, Published in [Beau94], data taken from Kravchenko and Moin [Krav00] (1993)

[Ma00] Ma, X., Karamonos, G.-S., Karniadakis, G.E.: Dynamics and low-dimensionality of a turbulent near wake, J. Fluids Mech., **410**, 29–65 (2000)

[Mit97] Mittal, R., Moin, P.: Suitability of upwind-biased finite-difference schemes for large eddy simulations of turbulent flows, AIAA J., **35**(88), 1415–1417 (1997)

[Nor94] Norberg, C.: An experimental investigation of flow around a circular cylinder: influence of aspect ratio, J. Fluid Mech., **258**, 287–316 (1994)

[Nor03] Norberg, C.: Fluctuating lift on a circular cylinder: review and new measurements, J. Fluids and Structures, **17**, 57–96 (2003)

[Ong96] Ong, J., Wallace, L.: The velocity field of the turbulent very near wake of a circular cylinder, Exp. in Fluids, **20**, 441–453 (1996)

[Rhie83] Rhie, C.M., Chow, W.L.: Numerical study of the turbulent flow past an airfoil with trailing edge separation, AIAA J, **21**(11), 1525–1532 (1983)

[Stone68] Stone, H.L.: Iterative solutions of implicit approximations of multidimensional partial differential equations, SIAM J Numerical Analysis, **5**, 87–113 (1968)

[Thom96] Thompson, M., Hourigan, M., Sheridan, J.: Three-dimensional instabilities in the wake of a circular cylinder, Exp. Thermal Fluid Sci., **12**(2), 190–196 (1996)

[Wis97] Wissink, J.G.: DNS of 2D turbulent flow around a square cylinder, Int. J. for Numer. Methods in Fluids, **25**, 51–62 (1997)

[Wis03] Wissink, J.G.: DNS of a separating low Reynolds number flow in a turbine cascade with incoming wakes, Int. J. of Heat and Fluid Flow, **24**, 626–635 (2003).

[Wis06] Wissink, J.G. and Rodi, W.: Direct Numerical Simulation of flow and heat transfer in turbine cascade with incoming wakes, J. Fluid Mech., **569**, 209–247 (2006).

[Wil96a] Williamson, C.H.K.: Vortex dynamics in the cylinder wake, Ann. Rev. Fluid Mech., **28**, 477–539 (1996)

[Wil96b] Williamson, C.H.K.: Three-dimensional wake transition, J. Fluid Mech., **328**, 345–407 (1996)

Performance Assessment and Parallelisation Issues of the CFD Code NSMB

Jörg Ziefle, Dominik Obrist and Leonhard Kleiser

Institute of Fluid Dynamics, ETH Zurich, 8092 Zurich, Switzerland

Summary. We present results of an extensive comparative benchmarking study of the numerical simulation code NSMB for computational fluid dynamics (CFD), which is parallelised on the level of domain decomposition. The code has a semi-industrial background and has been ported to and optimised for a variety of different computer platforms, allowing us to investigate both massively-parallel microprocessor architectures (Cray XT3, IBM SP4) and vector machines (NEC SX-5, NEC SX-8).

The studied test configuration represents a typical example of a three-dimensional time-resolved turbulent flow simulation. This is a commonly used test case in our research area, i. e., the development of methods and models for accurate and reliable time-resolved flow simulations at relatively moderate computational cost. We show how the technique of domain decomposition of a structured grid leads to an inhomogeneous load balancing already at a moderate CPU-to-block-count ratio. This results in severe performance limitations for parallel computations and inhibits the efficient usage of massively-parallel machines, which are becoming increasingly ubiquitous in the high-performance computing (HPC) arena.

We suggest a practical method to alleviate the load-balancing problem and study its effect on the performance and related measures on one scalar (Cray XT3) and one vector computer (NEC SX-8). Finally, we compare the results obtained on the different computation platforms, particularly in view of the improved load balancing, computational efficiency, machine allocation and practicality in everyday operation.

Key words: performance assessment, parallelisation, domain decomposition, multi-block grid, load-balancing, block splitting, computational fluid dynamics

1 Introduction and Background

Our computational fluid dynamics research group performs large-scale simulations of time-dependent three-dimensional turbulent flows. The employed simulation methodologies, mainly direct-numerical simulations (DNS) and large-eddy simulations (LES), require the extensive use of high-performance computing (HPC) infrastructure, as well as parallel computing and optimisation of the simulation codes. As a part of the ETH domain, the Swiss National Supercomputing Centre (CSCS) is the main provider of HPC infrastructure to academic research in Switzerland. Due to our excellent experience with vector machines, we developed and optimised our CFD codes primarily for this architecture. Our main computational platform has been the NEC SX-5 vector computer (and its predecessor machine, a NEC SX-4) at CSCS. However, the NEC SX-5 was decommissioned recently, and two new scalar machines were installed, a Cray XT3 and an IBM SP5.

This shift in the computational infrastructure at CSCS from vector to scalar machines prompted us to assess and compare the code performance on the different HPC systems available to us, as well as to investigate the potential for optimisation. The central question to be answered by this benchmarking study is, if — and how — the new Cray XT3 is capable to achieve a code performance that is superior to the NEC SX-5 (preferably in the range of the NEC SX-8). This report tries to answer this question by presenting results of our benchmarking study for one particular, but important, flow case and simulation code.

It thereby adheres to the following structure. In Sect. 2 we outline the key properties of our simulation code. After the description of the most prominent performance-related characteristics of the machines that were investigated in the benchmarking study in Sect. 3, we will give some details about the test configuration and benchmarking method in Sect. 4. The main part of this paper is comprised by Section 5, where we present the performance measurements. In the first subsection, Sect. 5.1, we show data that was obtained on the two vector systems. After that, we discuss the benchmarking results from two massively-parallel machines in Sect. 5.2. In Sect. 5.3, we discuss a way to alleviate the problem of uneven load-balancing, a problem which is commonly observed in the investigated simulation scenario and severely inhibits performance in parallel simulations. First, we demonstrate its favourable impact on the performance on the massively-parallel computer Cray XT3, and study its effect on the performance and performance-related measures on the NEC SX-8 vector machine. Finally, in Sect. 5.4 we compare the results of all the platforms and return to the initial question, how the code performance on the Cray XT3 compares to the two vector machines NEC SX-5 and SX-8.

2 Simulation Code NSMB

The NSMB (Navier-Stokes Multi-Block) code [1, 2] is a cell-centred finite-volume solver for compressible flows using structured grids. It supports domain decomposition into grid blocks (multi-block approach) and is parallelised using the MPI library on the block level. This means that the individual grid blocks are distributed to the processors of a parallel computation, but the processing of the blocks on each CPU is performed in a serial manner. NSMB incorporates a number of different RANS and LES models, as well as numerical (central and upwind) schemes with an accuracy up to fourth order (fifth order for upwind schemes). The core routines are available in two versions, optimised for vector and scalar architectures, respectively. Furthermore, NSMB offers convergence acceleration by means of multi-grid and residual smoothing, as well as the possibility of moving grids and a large number of different boundary conditions. Technical details about NSMB can be found in [3], and examples of complex flows simulated with NSMB are published in [4, 5, 6, 7, 8].

3 Description of the Tested Systems

The tested NEC SX-8 vector computer is installed at the "High Performance Computing Center Stuttgart" (HLRS in Stuttgart, Germany). It provides 72 nodes with 8 processors each running at a frequency of 2 Gigahertz. Each CPU delivers a peak floating-point performance of 16 Gigaflops. With a theoretical peak performance of 12 Teraflops, the NEC SX-8 is on place 72 of the current (November 2006) Top500 performance ranking [9] and thus the second-fastest vector supercomputer in the world (behind the "Earth Simulator", which is on rank 14). The total installed main memory is 9.2 Terabytes (128 Gigabytes per node). The local memory is shared within the nodes and can be accessed by the local CPUs at a bandwidth of 64 Gigabytes per second. The nodes are connected through an IXS interconnect (a central crossbar switch) with a theoretical throughput of 16 Gigabytes per second in each direction.

The NEC SX-5 was a single-node vector supercomputer installation at the Swiss National Supercomputing Centre (CSCS) in Manno, Switzerland. Each of its 16 vector processors delivered a floating-point peak performance of 8 Gigaflops, yielding 128 Gigaflops for the complete machine. The main memory of 64 Gigabytes was shared among the CPUs through a non-blocking crossbar and could be accessed at 64 Gigabytes per second. As mentioned in the introduction, the machine was decommissioned in March of 2007.

The Cray XT3, also located at CSCS, features 1664 single-core AMD Opteron processors running at 2.6 Gigahertz. The theoretical peak performance of the machine is 8.65 Teraflops, placing it currently on rank 94 of the Top500 list [9]. The main memory of the processors (2 Gigabytes per CPU, 3.3 Terabytes in total) can be accessed by the CPUs with a bandwidth of 64 Gigabytes per second and is connected through a SeaStar interconnect with a maximum throughput of 3.8 Gigabytes per second in each direction (2 Gigabytes per second sustained bandwidth). The network topology is a 3D torus of size $9 \times 12 \times 16$.

At CSCS also, an IBM p5-575 consisting of 48 nodes with 16 CPUs each has recently been installed. The 768 IBM Power5 processors are running at 1.5 Gigahertz, leading to a theoretical peak performance of 4.5 Teraflops. The 32 Gigabytes main memory on each node are shared among the local processors. The nodes are connected by a 4X Infiniband interconnect with a theoretical throughput of 1 Gigabyte per second in each direction. Each Power5 chip can access the memory at a theoretical peak rate of 12.7 Gigabytes per second. During our benchmarking investigation, the machine was still in its acceptance phase, therefore we could not perform any measurements on it.

The IBM SP4 massively-parallel machine at CSCS (decommissioned in May of 2007) consisted of 8 nodes with 32 IBM Power4-CPUs per node. The processors ran at 1.3 Gigahertz and had a peak floating-point performance of 5.2 Gigaflops, yielding 1.3 Teraflops for the complete machine. The main memory was shared within one node, and the total installed memory was 768 Gigabytes (96 Gigabytes per node). The nodes were connected through a Double Colony switching system with a throughput of 250 Megabytes per second.

4 Test Case and Benchmarking Procedure

In order to render the benchmarking case as realistic as possible, we used one of our current mid-sized flow simulations for the performance measurements. In this numerical simulation of film cooling, the ejection of a cold jet into a hotter turbulent boundary layer crossflow is computed, see Fig. 1. The jet originates from a large isobaric plenum and is lead into the boundary layer through an oblique round nozzle.

The computational domain consists of a total of 1.7 million finite-volume cells, which are (in the original configuration) distributed to 34 sub-domains (in the following called blocks) of largely varying dimensions and cell count. The ratio of the number of cells between the largest and the smallest block is approximately 11. The mean number of cells in a block is about 50 000.

In order to enhance the load balancing and thus the parallel performance, we also conducted benchmarking simulations with a more refined domain de-

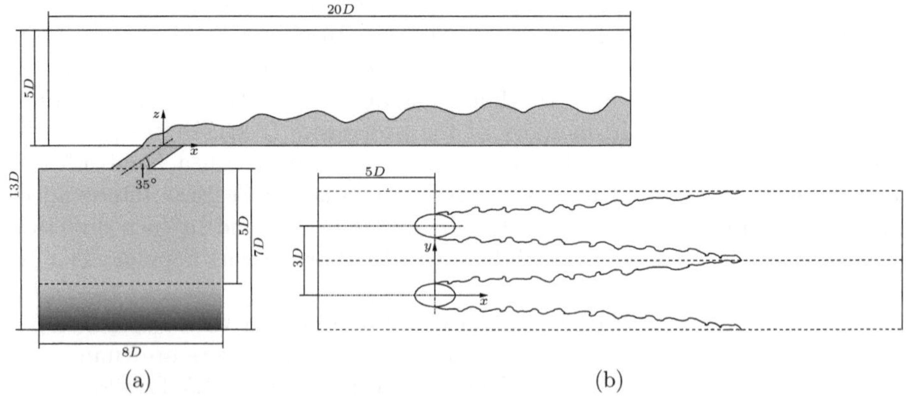

Fig. 1. Schematic of the jet-in-crossflow configuration. (**a**) Side view and (**b**) top view. The gray areas symbolise the jet fluid

composition (see Sect. 5.3). This was done by splitting the original topology of 34 blocks into a larger number of up to 680 blocks. Table 1 lists the properties of the investigated block configurations. Further information about this flow case and the simulation results will be available in [10].

Unlike in other benchmarking studies, we do not primarily focus on common measures such as parallel efficiency or speedup in the present paper. Rather, we also consider quantities that are more geared towards the pragmatic aspects of parallel computing. This study was necessitated by the need to evaluate new simulation platforms in view of our practical needs, such as a quick-enough simulation throughput to allow for satisfactory progress in our research projects.

To this end, we measured the elapsed wall-clock time which is consumed by the simulation to advance the solution by a fixed number of time steps. All computations were started from the exactly same initial conditions, and the fixed amount of time steps was chosen so that the fastest resulting simulation time would still be long enough to keep measuring errors small. For better illustration, the elapsed wall-clock time is normalised with the number of time steps. Therefore, in the following, our performance charts display the performance in terms of elapsed time per time step (abbreviated "performance measure"). Since in our work typically the simulation time interval and thus the number of time steps is known a priori, this performance measure allows for a quick estimate of the necessary computational time (and simulation turnover time) for similar calculations.

As we wanted to get information about the real-world behaviour of our codes, we did not artificially improve the benchmarking environment by obtaining exclusive access to the machine and other privileged priorities, such as performing the computations in times with decreased general system load. To ensure statistically significant results, all benchmarking calculations were repeated several times and only the best result was used. In this sense, our data reflects the maximum performance that can be expected from the given machines in everyday operation. Of course, the simulation turnover time is typically significantly degraded by the time the simulation spends in batch

Table 1. Characteristics of the employed block configurations

number of blocks	max./min. block size	median block size	mean block size	std. dev. of block size
34	11.14	40 112	50 692	40 780
68	2.19	24 576	25 346	5 115
102	2.00	15 200	16 897	3 918
136	1.97	12 288	12 673	2 472
170	1.99	9 728	10 138	2 090
340	2.00	4 864	5 069	1 047
680	2.11	2 560	2 535	531

queues, but we did not include this aspect into our study as we felt that the queueing time is a rather arbitrary and heavily installation-specific quantity which would only dilute the more deterministic results of our study.

5 Benchmarking Results

5.1 Performance Evaluation of the Vector Machines NEC SX-5 and NEC SX-8

In Fig. 2, we compare the performance of the NEC SX-5 at CSCS to the NEC SX-8 at HLRS. Both curves exhibit the same general behaviour, with an almost linear scaling up to four processors, followed by a continuous flattening of the curve towards a higher number of CPUs. As shown in Fig. 2(b), the NEC SX-5 scales somewhat worse for 8 processors than the NEC SX-8. Note that we could not obtain benchmarking information on the NEC SX-5 for more than eight CPUs, since this is the maximum number of processors which can be allocated in a queue.

At first glance, the further slowdown from 8 to 16 processors on the SX-8 could be considered a typical side effect due to the switch from single-node to multi-node computations. As mentioned in Sect. 3, one node of the NEC SX-8 is composed of eight CPUs, with a significantly better memory performance within a node than in-between nodes. When using 16 processors, the job will be distributed to two nodes, and only the intra-node communication will adhere to shared-memory principles. The slow MPI communication between the two nodes partially compensates the performance gain due to the higher degree of parallelism.

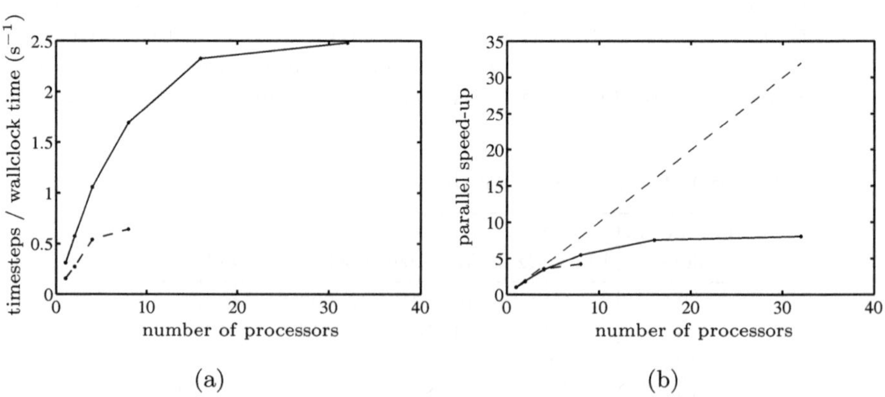

Fig. 2. Dependence of (**a**) performance measure and (**b**) parallel speed-up of vector machines on number of processors (34 blocks, i. e., no block splitting). ----NEC SX-5, ——NEC SX-8, ----ideal scaling in (**b**)

However, as the performance further stagnates when going from 16 to 32 processors, it is obvious that there is another factor that prevents a better exploitation of the parallel computation approach, namely suboptimal load-balancing. This issue plays an even bigger role in the performance optimisation for massively-parallel machines such as the Cray XT3. Its effect on the parallel performance and a method for its alleviation will be investigated in detail for both the Cray XT3 and the NEC SX-8 in the following sections.

The performance ratio of the NEC SX-8 to the NEC SX-5 in Fig. 2(a) corresponds roughly to the floating-point performance ratio of their CPUs, 16 Gigaflops for the NEC SX-8 versus 8 Gigaflops for the NEC SX-5. The reason why the NEC SX-8 performance does not come closer to twice the result of the NEC SX-5 is probably related to the fact that the CPU-to-memory bandwidth of the NEC SX-8 did not improve over the NEC SX-5 and is nominally 64 Gigabytes per second for both machines, see Sect. 3. Although floating-point operations are performed twice as fast on the NEC SX-8, it cannot supply data from memory to the processors with the double rate.

5.2 Performance Evaluation of the Scalar Machines Cray XT3 and IBM SP4

Figure 3 shows the benchmarking results for the scalar machine Cray XT3. The Cray compiler supports an option which controls the size of the memory pages ("small pages" of 4 Kilobytes versus "large pages" comprising 2 Megabytes) that are used at run time. Small memory pages are favourable if the data is heavily distributed in physical memory. In this case the available memory is used more efficiently, resulting in generally fewer page faults than with large pages, where only a small fraction of the needed data can be stored in memory at a given time. Additionally, small memory pages can be loaded faster than large ones. On the other hand, large memory pages are beneficial if the data access pattern can make use of data locality, since fewer pages have to be loaded. We investigated the influence of this option by performing all benchmarking runs twice, once with small and once with large memory pages. In all cases, the performance using small memory pages was almost twice as good with a very similar general behaviour. Therefore, we will only consider benchmarking data obtained with this option in the following.

The performance development of the Cray XT3 with small memory pages for an increasing number of processors also exhibits an almost-linear initial scaling, just as observed for the NEC SX-5 and SX-8 vector machines. (With large memory pages, the scaling properties are less ideal, and notable deviations from linear scaling already occur for more than four processors, see 3(b). Furthermore, the parallel speed-up stagnates earlier and at a lower level than for small memory pages.) For 16 or more processors on the Cray XT3, the performance stagnates and exhibits a similar saturation behaviour originating from bad load-balancing as already observed for the two vector machines. This is not surprising, since load-balancing is not a problem specific to the

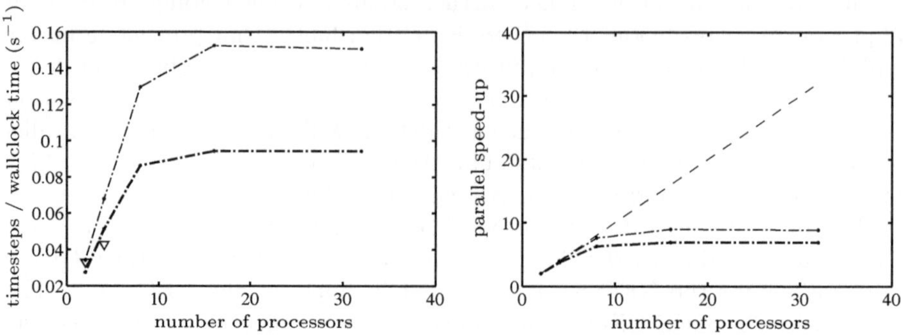

Fig. 3. Dependence of (**a**) performance measure and (**b**) parallel speed-up of scalar machines on number of processors (34 blocks, i. e., no block splitting). —·— Cray XT3 with small memory pages, —·— Cray XT3 with large memory pages, ▽ IBM SP4, - - - - ideal scaling in (**b**)

computer architecture, but rather to the decomposition and distribution of the work to the parallel processing units. Therefore, as expected, both vector and scalar machines are similarly affected by the suboptimal distribution of the computational load. The computational domain of the benchmarking case consists of only 34 blocks of different sizes (see Table 1), and the work load cannot be distributed in a balanced manner to an even moderate number of CPUs. The CPU with the largest amount of work will be continuously busy while the other ones with a smaller total amount of finite-volume cells are idle for most of the time.

Note that the stagnation levels of the parallel speed-up for the Cray XT3, and therefore also its parallel efficiency, are only slightly higher compared to the NEC SX-8, cf. Figs. 3(b) and 2(b). However, the higher processor performance of the NEC SX-8 give it a more than 15-fold advantage over the Cray XT3 in terms of absolute performance. The similar levels of the parallel efficiency for the scalar Cray XT3 and the NEC SX-8 vector machine for 16 and 32 processors is a quite untypical result in view of the two different architectures. Since under normal circumstances the scalar architecture is expected to exhibit a considerably higher parallel efficiency than a vector machine, this observation further indicates the performance limitation on both machines because of inhomogeneous load balancing. When further comparing the two machines, it becomes evident that the range of almost-linear parallel scaling extends even a bit farther (up to eight CPUs) on the Cray XT3, while larger deviations from an ideal scaling already occur for more than four CPUs on the two vector machines. Since the Cray XT3 is a scalar machine and has single-processor nodes, there is no change in memory performance and memory access method (shared memory versus a mix of shared and distributed memory) as on the NEC SX-8, when crossing nodes. Therefore, the general scaling behaviour is expected to be better and more homogeneous on a scalar machine. This is, however, not an explanation for the larger linear scaling range of the Cray, as computations on the NEC are still performed within a single node for eight processors, where they first deviate from linear scaling. We rather conclude that the architecture and memory access performance of the Cray XT3, but most importantly its network performance does not notably inhibit the parallel scaling on the Cray XT3, while on the two vector machines there are architectural effects which decrease performance. Potentially the uneven load-balancing exerts a higher influence on the NECs and thus degrades performance already at a lower number of processors than on the Cray XT3. While the exact reasons for the shorter initial linear parallel scaling range on the vector machines remain yet to be clarified, we will at least show in the following two sections that load-balancing effects show up quite differently in both the Cray XT3 and the NEC SX-5/SX-8.

We also obtained some benchmarking results for the (now rather outdated) IBM SP4 at CSCS in Manno, see the triangles in Fig. 3. Since its successor machine IBM SP5 was already installed at CSCS and was in its acceptance phase during our study, we were interested in performance data, especially the general parallelisation behaviour, of a machine with similar architecture. However, due to the heavy loading of the IBM SP4, which led to long queueing times of our jobs, we could not gather benchmarking data for more than four CPUs during the course of this study. For such a low number of CPUs, the IBM SP4 approximately achieves the same performance as the Cray XT3. This does not come as surprise, since the floating-point peak performance of the processors in the IBM SP4 and the Opteron CPUs in the Cray XT3 are nearly identical with 5.2 Gigaflops. However, when increasing the degree of parallelism on the IBM SP4, we expect its parallel scaling to be significantly worse than on the Cray XT3, due to its slower memory access, and slower network performance when crossing nodes (i. e., for more than 32 processors).

5.3 Overcoming Load-Balancing Issues by Block Splitting

It was already pointed out in the previous two sections that the low granularity of the domain decomposition in the benchmarking problem inhibits an efficient load balancing for an even moderate number of CPUs. Suboptimal load balancing imposes a quite strict performance barrier which cannot be broken with the brute-force approach of further increasing the level of parallelism: the performance will only continue to stagnate.

For our typical applications, this quick saturation of parallel scaling is not a problem on the NEC SX-5 and NEC SX-8 vector machines, since their fast CPUs yield stagnation levels of performance which are high enough for satisfactory turnover times already at a low number of CPUs. On the Cray XT3, where the CPU performance is significantly lower, a high number of processors has to be employed in compensation. However, this increased level of parallelism further aggravates the load-balancing problem. Additionally, technical reasons prescribe an upper bound on the number of CPUs: since the simulation code is parallelised on the block level, it is impossible to distribute the work units in form of 34 blocks to more than 34 processors.

The only way to overcome the load-balancing issues and to perform efficient massively-parallel computations of our typical test case with the given simulation code is to split the 34 blocks into smaller pieces. This has two principal effects. First, the more fine-grained domain decomposition allows for a better load balancing, and the performance is expected to improve considerably from the stagnation level that was observed before. The second effect is even more important for massively-parallel computations, specifically to reach the performance range of the NEC SX-8 on the Cray XT3. With the refined block configuration, the stagnation of performance due to bad load balancing will now occur at a largely increased number of processors, and the almost-linear parallel scaling range will be extended as well. This allows for an efficient use of a high number of CPUs on one hand (e. g., on the Cray XT3), but on the other hand this development also improves the parallel efficiency at a lower number of CPUs.

Fortunately, the utility programme MB-Split [11], which is designed for the purpose of block-splitting, was already available. However, in general such a tool is non-trivial to implement, which can make it hard for users with similar simulation setups to achieve an adequate performance especially on the Cray XT3. MB-Split allows to split the blocks by different criteria and algorithms. We used the easiest method, where the largest index directions of the blocks are successively split until the desired number of blocks is obtained.

In order to evaluate the effect of block splitting on the parallel performance, we refined the block topology to even multiples of the original configuration of 34 blocks up to 680 blocks. Some statistical quantities describing the properties of the resulting block distributions are listed in Table 1. More detailed information related to the specific benchmarking computations is available in Table 2. In the following two sections, we will investigate the effect of the block-splitting strategy on the parallel performance for both the Cray XT3 and NEC SX-8 computers.

Effects of Block Splitting on the Cray XT3

On the Cray XT3, we first performed benchmarking computations with the different refined block topologies at a fixed number of 32 processors. This was the highest number of CPUs used with the initial block configuration, see Fig. 3. Moreover, it did not yield a performance improvement over 16 CPUs due to bad load balancing. The result with block splitting is displayed in Fig. 4, together with the initial data obtained without block splitting, i. e., using only 34 blocks. As clearly visible in Fig. 4(a), the more homogeneous

Table 2. Block-related characteristics of the benchmarking runs on the NEC SX-8 (abbreviated "SX-8") and Cray XT3 ("XT3"). Also see Figs. 13 and 14 for a visual representation of the block distribution

Run	Machine	number of blocks	number of CPUs	max./min. block size	median block size	mean block size	std. dev. of block size
1	SX-8	34	2	1.01	861 757	861 757	7 993
2	SX-8	34	4	1.04	427 893	430 879	8 289
3	SX-8	34	8	1.10	211 257	215 439	7 645
4	SX-8	34	16	2.36	101 417	107 720	33 318
5	SX-8	34	32	7.66	44 204	53 860	40 677
6	SX-8	68	16	1.21	105 404	107 720	7 381
7	SX-8	102	16	1.17	104 223	107 720	6 145
8	SX-8	136	16	1.10	107 227	107 720	4 379
9	SX-8	170	16	1.11	110 383	107 720	3 941
10	SX-8	340	16	1.04	107 402	107 720	1 484
11	SX-8	68	32	1.40	51 461	53 860	5 880
12	SX-8	102	32	1.27	51 793	53 860	4 796
13	SX-8	136	32	1.23	52 205	53 860	3 754
14	SX-8	170	32	1.17	52 183	53 860	3 147
15	SX-8	340	32	1.12	54 480	53 860	1 917
16	SX-8	68	64	2.00	25 601	26 930	5 522
17	SX-8	102	64	1.64	28 272	26 930	4 226
18	SX-8	136	64	1.45	25 664	26 930	3 098
19	SX-8	170	64	1.39	29 016	26 930	3 302
20	SX-8	340	64	1.24	26 205	26 930	1 762
21	XT3	34	2	1.01	861 757	861 757	7 993
22	XT3	34	4	1.04	427 689	430 879	8 140
23	XT3	34	8	1.10	211 257	215 439	7 645
24	XT3	34	16	2.36	100 289	107 720	33 261
25	XT3	34	32	7.66	44 204	53 860	40 677
26	XT3	68	32	1.38	51 281	53 860	5 986
27	XT3	102	32	1.24	51 566	53 860	4 750
28	XT3	136	32	1.19	51 715	53 860	3 859
29	XT3	170	32	1.14	51 750	53 860	3 284
30	XT3	340	32	1.07	55 193	53 860	1 726
31	XT3	680	32	1.03	53 467	53 860	704
32	XT3	680	64	1.07	27 533	26 930	808
33	XT3	680	128	1.13	12 993	13 465	742
34	XT3	680	256	1.33	7 262	6 732	799
35	XT3	680	512	1.69	3 192	3 366	593

load balancing due to the refined block configuration leads to an approximately threefold increase in performance, whereas the performance variations between the different block topologies using block splitting are rather small in comparison. The large improvement of the parallel speed-up in Fig. 4(b), which comes quite close to the ideal (linear) scaling, provides further evidence for the beneficial effect of the block splitting.

The optimum number of blocks for the given configuration can be deduced from Fig. 5, where the performance and parallel efficiency are plotted over the number of blocks for the benchmarking computations with 32 CPUs. When splitting the original block configuration of 34 blocks to twice the number of blocks, the performance jumps by a factor of about 2.5, and the parallel efficiency rises from about 25% to about 75%. For a higher number of blocks, both the performance and the parallel efficiency increase slowly. At 340 blocks, which corresponds to a block-to-CPU-count ratio of about 10, the maximum performance and parallel efficiency (about 85%) are reached. A further doubling of the number of blocks to 680 blocks results in a performance that was obtained already with 102 blocks.

The reason why a more refined block topology does not lead to continued performance improvements lies in the increased communication overhead. Data is exchanged between the blocks using additional layers of cells at their faces, so-called ghost cells. The higher the number of blocks, the larger the amount of data that has to be exchanged. On the Cray XT3, inter-block communication is performed over the network for blocks that are located on different processors. Although this network is very fast (see Sect. 3), this data transfer still represents a bottleneck and is considerably slower than the bandwidth which is available between CPU and memory. Additionally, a part of

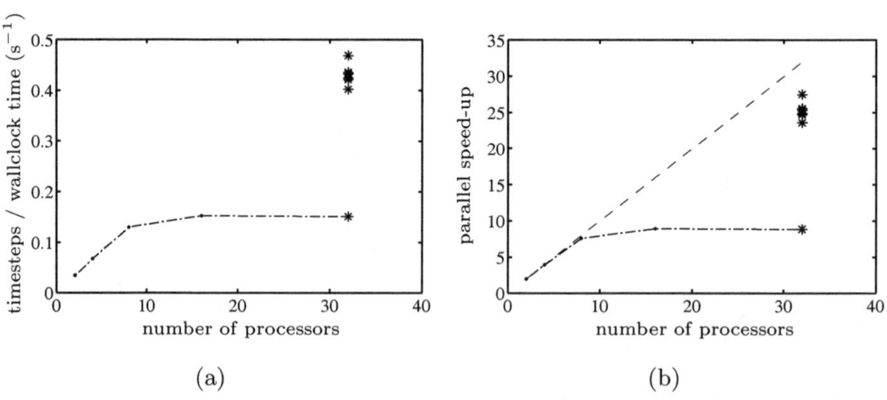

Fig. 4. Effect of block splitting on Cray XT3. Dependence of (**a**) performance measure and (**b**) parallel speed-up on number of processors. —·— Cray XT3 with 34 blocks, $*$ Cray XT3 with varying number of blocks ranging from 34 to 680 and 32 CPUs, - - - - ideal scaling in (**b**)

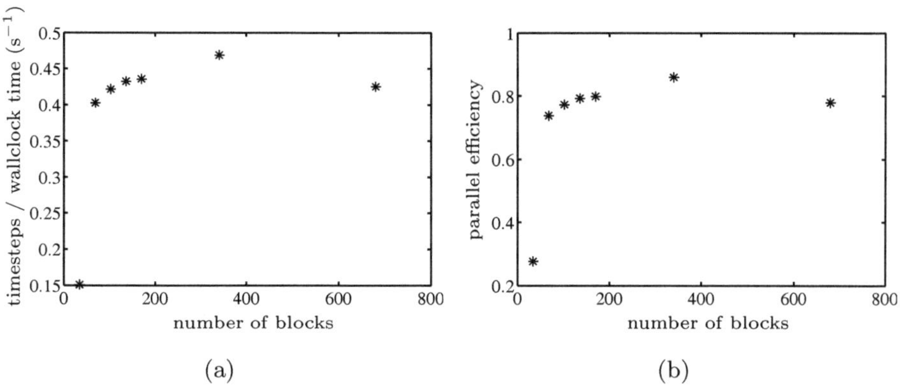

Fig. 5. Effect of block splitting on Cray XT3. Dependence of (**a**) performance measure and (**b**) parallel efficiency on number of blocks for 32 CPUs on Cray XT3

this data transfer would not be necessary when utilising fewer blocks, so it can be considered additional work to obtain the same result. Blocks that reside on the same CPU can communicate faster directly in memory, but the corresponding communication routines still have to be processed. In these parts of the code the data is copied from the source array to a work array, and then from the work array to the target field. In both cases, the computation spends extra time copying data and (in the case of blocks distributed to different CPUs) communicating it over the network. This time could otherwise be used for the advancement of the solution. For very large numbers of processors, this communication overhead over-compensates the performance gain from the more balanced work distribution, and the total performance degrades again. There are additional effects that come along with smaller block sizes, which are mostly related to the locality of data in memory and cache-related effects. However, as their impact is relatively difficult to assess and quantify, we do not discuss them here. Additionally, on vector machines such as the NEC SX-5 and SX-8, the average vector length, which is strongly dependent on the block size, exerts a big influence on the floating-point performance. We will further analyse this topic in the next section.

While the performance gain on the Cray XT3 due to block splitting for 32 processors is quite impressive, the results are still far apart from the levels that were obtained with the NEC SX-8 (see Fig. 2).

Furthermore, since the performance of the Cray XT3 does not depend strongly on the number of blocks (as long as there are enough of them), only a variation of the number of processors can yield further improvements. To this end, we performed a new series of benchmarking computations at a fixed number of blocks with an increasing number of CPUs. Since the optimum block configuration with 340 blocks for 32 processors (see Fig. 5) would limit the maximum number of CPUs to 340, we selected the block configuration with 680 blocks for this investigation. We doubled the number of CPUs successively up to a maximum of 512, starting from the computational setup of 32 processors and 680 blocks that was utilised in the above initial block-splitting investigation. The results are shown as rectangular symbols in Fig. 6. Additionally, the previous performance data for the Cray XT3 is displayed, without block splitting, and with 32 CPUs and block splitting.

Especially in the left plots with linear axes, the considerably extended initial parallel scaling region is clearly visible. In the original configuration without block splitting, it only reached up to eight processors, while with block splitting, the first larger deviation occurs only for more than 32 processors. Furthermore, the simulation performance rises continually with an increasing number of CPUs, and flattens out only slowly. While a perfor-

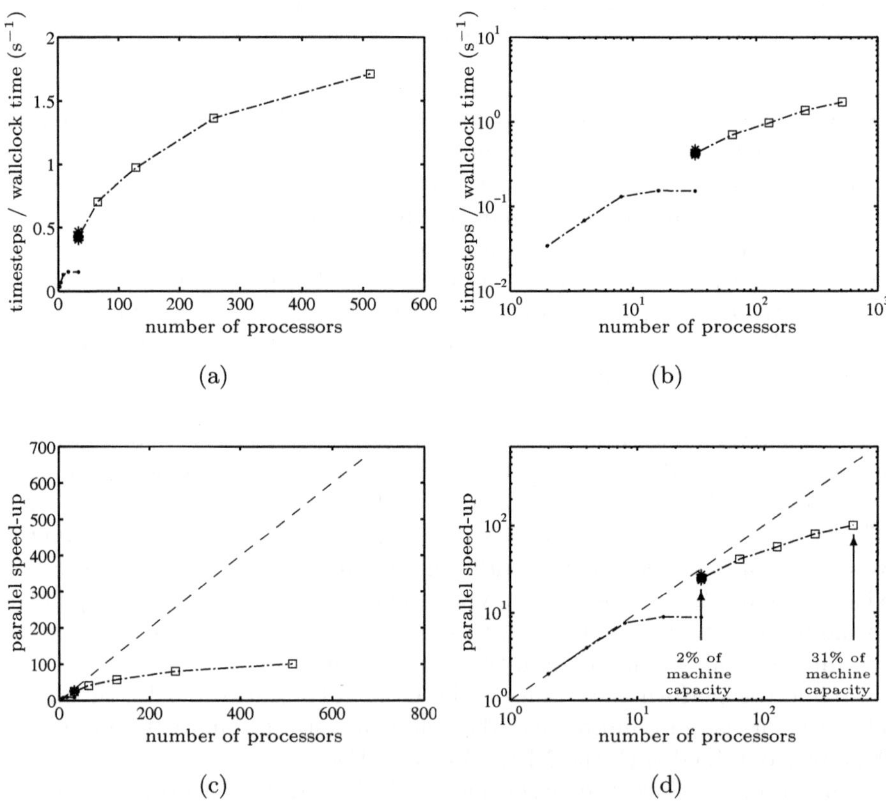

Fig. 6. Effect of block splitting on Cray XT3. Dependence of (**a**)/(**b**) performance measure and (**c**)/(**d**) parallel speed-up on number of processors, plotted in linear axes (left) and logarithmic axes (right). —·—Cray XT3 with 34 blocks, ∗ Cray XT3 with varying number of blocks ranging from 68 to 680 and 32 CPUs, —⊡—Cray XT3 with 680 blocks, - - - -ideal scaling in (**c**) and (**d**)

mance stagnation due to bad load balancing is technically not avoidable, it is delayed by the large number of 680 blocks to a number of CPUs in the order of the total processor count of the machine. (Of course, computations with several hundred CPUs are still rather inefficient due to the large communication overhead, as evident from Figs. 6(c) and (d).) Since the allocation of a complete supercomputer is unrealistic in everyday operation, we conclude that the block-splitting technique provides a means to make much better use of the Cray XT3 at CSCS for massively-parallel applications whose performance is inhibited by load balancing. The maximally obtained computational performance on the Cray XT3 with 512 CPUs (corresponding to 31% of the machine capacity, see Fig. 6(b)) is four times higher than the performance on 32 CPUs (2% of its capacity) with block splitting. This result is achieved with a parallel speed-up of 100 (see Fig. 6(d)), i. e., a parallel efficiency of about 20%, and lies in the range of the performance obtained on the NEC SX-8 in Fig. 2. Although this can be considered a success at first sight, it should be noted that this performance was reached with only 8 CPUs (i. e., slightly over 1% of the machine capacity) and no block splitting on the NEC SX-8. Moreover, we will show in the next section that block splitting also exerts a very favourable effect on the performance of the NEC SX-8 and fully uncovers its performance potential with the given simulation setup.

Load-balancing Issues and Effects of Block Splitting on the NEC SX-8

In the following, we will take a closer look at load-balancing issues on the NEC SX-8 and investigate the effect of block splitting on its parallel performance. For illustration, the distribution of the blocks onto the processors is displayed Fig. 13 for the simulations performed on the NEC SX-8. This is visualised by a stacked bar chart showing the total number of cells per CPU. The black and white parts of each bar denote the cell count of the individual blocks which are located on the corresponding CPU. In this investigation, the total number of cells per processor is of primary importance, because it represents a good measure of the work load of a CPU. The plots are arranged in matrix form, with the horizontal axis denoting the number of processors n_p, and the vertical axis the number of blocks n_b.

Let us focus on the first five figures ①–⑤ on the top left, which show computations with a fixed number of 34 blocks and varying number of CPUs, increasing from 2 in ① up to 32 in ⑤. While the work distribution is still quite homogeneous for up to 8 CPUs, see figures ①–③, the load balancing drastically degrades when doubling the number of CPUs to 16 (figure ④). When considering the sizes of the individual blocks, it becomes clear that the big size difference between the largest and the smallest block inhibits a better load balancing. As the number of CPUs increases in figures ④ and ⑤, the situation gets even worse. There is less and less opportunity to fill the gaps with smaller blocks, as they are needed elsewhere. For the worst investigated case of 32 blocks on 34 CPUs, only two CPUs can hold two blocks — the other 32 CPUs obtain one block only. It is clear that for such a configuration, the ratio of the largest to the smallest block size plays a pivotal role for load balancing. While the CPU containing the largest block is processing it, all the other CPUs idle for most of the time.

The impact of an uneven load balancing on important performance measure can be observed in Fig. 7(a), where the minimum, mean and maximum values of the average floating-point performance per CPU are plotted over the number of CPUs for the above domain decomposition of 34 blocks. Note that the minimum and maximum values refer to the performance reached on one CPU of the parallel simulation. For an information of the floating-point performance of the whole parallel simulation, the mean value needs to be multiplied with the total number of CPUs. This is also the reason for the steadily decreasing floating-point performance as the number of CPUs increases. In a serial simulation, there is a minimal communication overhead, and the simulation code can most efficiently process work routines, where the majority of the floating-point operations take place. For parallel computations, a steadily increasing fraction of the simulation time has to be devoted to bookkeeping and data-exchange tasks. Therefore, the share of floating-point operations decreases, and with it the floating-point performance.

An interesting observation can be made for the graph of the maximum floating-point performance. It declines continuously from one up to eight CPUs, but experiences a jump to a higher level from 8 to 16 CPUs, and continues to climb from 16 to 32 CPUs. When comparing Fig. 7(a) to the corresponding block distributions ①–⑤ in Fig. 13, it is evident that the dra-

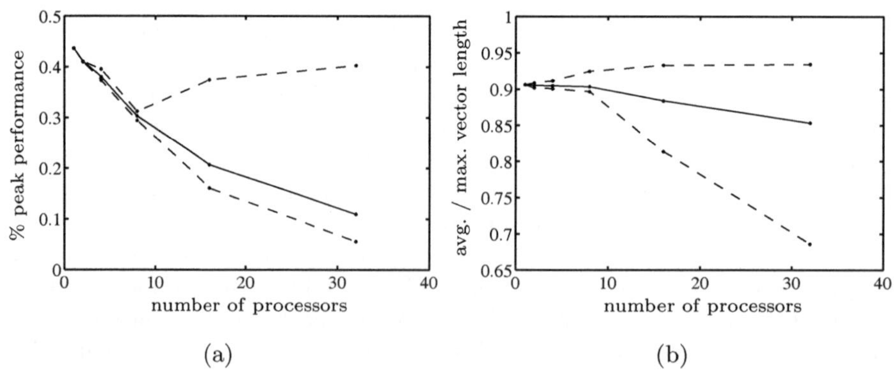

Fig. 7. Dependence of (**a**) average floating-point performance (normalised with the peak performance of 16 Gigaflops) and (**b**) average vector length (normalised with the length of the vector register of 256) on number of processors on NEC SX-8 for a fixed number of 34 blocks. ——— Mean and - - - - maximum/minimum values taken over all processors of a simulation

matic degradation of load-balancing from 8 to 16 CPUs is the reason for this development. When using between two and eight CPUs, the total work is rather evenly distributed, and all processors are working almost all of the time. The situation becomes quite different when utilising more processors. For 16 CPUs, the first CPU has more than twice the work load as the one with the lowest amount of cells. Consequently, it is idling most of the time and waiting for the first CPU to finish processing its data. For CPU number 1 the reverse situation occurs: it is holding up the data exchange and processing. The more pronounced the uneven load balancing, the higher the fraction of floating-point operations in contrast to data communication, and therefore the higher the maximum floating-point performance.

On vector machines such as the NEC SX-5 and SX-8, the average vector length is another important performance-related measure. The advantage of vector processors over scalar CPUs is their ability to simultaneously process large amounts of data in so-called vectors. The CPU of the NEC SX-8 has vector registers consisting of 256 floating-point numbers, and it is obvious that a calculation is more efficient the closer the vector length approaches this maximum in each CPU cycle. Therefore, the average vector length is a good measure of the exploitation of the vector-processing capabilities. In Fig. 7(b), the minimum, mean and maximum average vector length are displayed for 34 blocks and a varying number of processors. Again the minimum and maximum values are CPU-related quantities, and the mean of the average vector length is taken over the average vector lengths of all CPUs in a parallel computation.

As expected, the mean of the average vector length drops slightly with an increasing number of processors. The reason why the maximum average vector length in parallel computations surpasses the value reached in a serial computation can be explained as follows. When all blocks are located on a single CPU, the large blocks generally contribute to high vector lengths, whereas the smaller blocks usually lead to smaller vector lengths. Thus, the resulting average vector length lies between those extreme values. When the blocks are distributed to multiple CPUs, an uneven load-balancing as occurring for 8 or more CPUs favours large vector lengths for some CPUs. In such a case there are isolated processors working on only one or very few of the largest blocks, which results in high vector lengths on them. In contrast to the single-CPU calculation, the vector length on this CPU is not decreased by the processing of smaller blocks. This explanation is supported by plots ①–⑤ in Fig. 13. The fewer the number of small blocks on a processor, the higher its vector length. The optimum case of the largest block on a single processor is reached for 16 CPUs, and the maximum vector length does not increase from 16 to 32 CPUs.

In contrast, the minimum vector length is dependent on the CPU with the smallest blocks. For up to 8 CPUs, all processors contain about the same percentage of large and small blocks, therefore the minimum and average vector lengths are approximately the same as for a single CPU. However, as the load balancing degrades for a higher number of CPUs, there are processors with very small blocks, which result in much smaller vector lengths, and thus in considerably lower minimum vector lengths of this computation.

After this discussion of the CPU-number dependence (using the initial unsplit block configuration with 34 blocks) of the two most prominent performance measures on vector machines, the floating-point performance and the average vector length, we will study the impact of block splitting. Similarly to the investigation on the Cray XT3 in the previous section, we conducted computations with a varying number of blocks (ranging from 34 to 340) at a fixed number of processors. In Fig. 8 we display the variation of the performance measure and the parallel efficiency with the number of blocks for three sets of computations with 16, 32 and 64 processors, respectively.

The graphs can be directly compared to Fig. 5, where the results for the Cray XT3 are plotted. As discussed in Sect. 5.3, its performance curve exhibits a concave shape, i. e., there is an optimum block count yielding a global performance maximum. In contrast, after the initial performance jump from 34 to 68 blocks due to the improved load balancing, all three curves for the NEC SX-8 are continually falling with an increasing number of blocks. The decrease occurs in an approximately linear manner, with an increasing slope for the higher number of CPUs. As expected, the performance for a given block count is higher for the computations with the larger number of processors, but the parallel efficiencies behave in the opposite way due to the increased communication overhead, cf. Figs. 8(a) and (b). Furthermore, the parallel efficiency does not surpass 32%-53% (depending on the number of processors) even after block splitting, which are typical values for a vector machine. In contrast, the parallel efficiency on the Cray XT3 (with 32 processors, see Fig. 5(b)) is improved from about the same value as on the NEC SX-8 (roughly

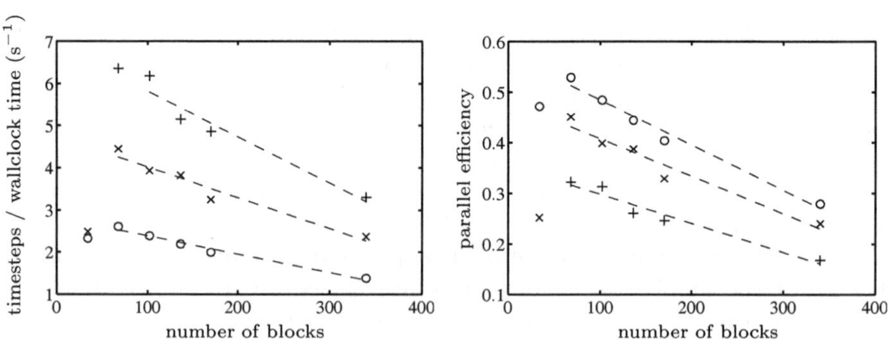

(a) (b)

Fig. 8. Dependence of (**a**) performance measure and (**b**) parallel efficiency on number of blocks on NEC SX-8 for a fixed number of processors. ○ 16 CPUs, × 32 CPUs and + 64 CPUs, ---- linear fit through falling parts of the curves

25% with 34 blocks and 32 CPUs) to about 75% with 68 blocks, whereas only 45% are reached on the NEC SX-8.

The decreasing performance with an increasing block count for the NEC SX-8 can be explained by reconsidering the performance measures that are investigated in the previous section. On the positive side, a higher block count leads to an increasingly homogeneous distribution of the work, cf. plots ④–⑩, ⑤/⑪–⑮ and ⑯–⑳ in Fig. 13. However, at the same time the average block size decreases significantly, as evident from Table 2, and the overhead associated with the inter-block data exchange strongly increases. In the preceding section, it was observed that this results in a degraded mean of the average vector length and floating-point performance. Both quantities are crucial to the performance on vector machines, and thus the overall performance decreases. On the other hand, a smaller dataset size does not yield such unfavourable effects on the Cray XT3 by virtue of its scalar architecture. Only for very high block numbers, the resulting communication overhead over-compensates the improved load balancing, and thus slightly decreases the parallel performance.

The above observations suggest that while block splitting is a suitable method to overcome load-balancing problems also on the NEC vector machines, it is generally advisable to keep the total number of blocks on them as low as possible. This becomes even more evident when looking at the aggregate performance chart in Fig. 11, where the same performance measure is plotted above the number of CPUs for all benchmarking calculations. The solid curve denoting the original configuration with 34 blocks runs into a saturation region due to load-balancing issues for more than 8 CPUs. After refining the domain decomposition by block splitting, the load-balancing problem can be alleviated and delayed to a much higher number of CPUs. As detailed in the discussion of Fig. 10, the envelope curve through all measurements with maximum performance for each CPU count cuts through the points with the lowest number of split blocks, 68 in this case. A further subdivision of the blocks only reduces the performance. For instance, for 32 processors, using the original number of 34 blocks or 680 blocks yields almost the same performance of about 2.5 time-steps/second, while a maximum performance of approximately 4.5 time-steps/second is reached for 68 blocks.

The dependence of the performance-related measures on the total block count can be investigated in Fig. 9. Here the floating-point performance and the average vector length are shown over the total number of blocks for a fixed number of processors. As in Fig. 7, the minimum, mean, and maximum values of a respective computation are CPU-related measures. The three symbols denote three sets of computations, which have been conducted with a fixed number of 16, 32 and 64 processors, respectively.

Both the vector length and the floating-point performance drop with an increasing number of blocks. While this degrading effect of the block size on both quantities is clearly visible, it is relatively weak, especially for the floating-point performance. The number of CPUs exert a considerably higher influence

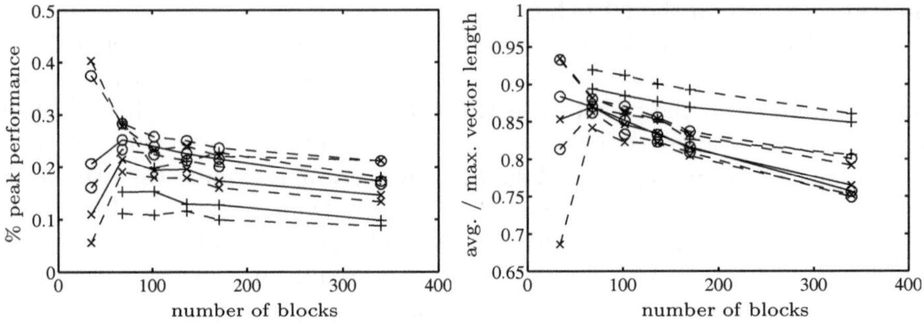

(a) (b)

Fig. 9. Dependence of (**a**) average floating-point performance (normalised with the peak performance of 16 Gigaflops) and (**b**) average vector length (normalised with the length of the vector register of 256) on number of blocks for a fixed number of processors on NEC SX-8. -o- 16 CPUs, -x- 32 CPUs, -+- 64 CPUs. —— Mean and ---- maximum/minimum values taken over all processors of a simulation

on performance, as also evident from Fig. 8. The reason for the degradation of the floating-point performance lies in the increasing fraction of bookkeeping and communication tasks, which are typically non-floating point operations, when more blocks are employed. This means that in the same amount of time, fewer floating-point operations can be performed. The average vector length is primarily diminished by the increasingly smaller block sizes when they are split further. However, while the floating-point performance in Fig. 9(a) is largest for 16 CPUs and smallest for 64 CPUs for a given number of blocks, the vector lengths behave oppositely. Here the largest vector lengths occur for 64 CPUs, while the vector lengths for 32 and 16 processors are significantly smaller and approximately the same for all block counts. The reason for this inverse behaviour can again be found in the block distribution, which is visualised in Fig. 13. The mechanisms are most evident for the simulations with the lowest number of blocks, $n_b = 68$ blocks in this case. Whereas in all three cases with 16, 32 and 64 processors the dimensions of the individual blocks are the same, the properties of their distribution onto the CPUs is varying significantly. For the highest number of 64 processors, each CPU holds one only one block, with the exception of four CPUs containing two blocks each. This leads to a very uneven load distribution. In contrast, for 16 processors, most of the CPUs process four blocks, and some even five. It is clearly visible that this leads to a much more homogeneous load balancing. As discussed above, the load balancing has a direct influence on the floating-point rate. The maximum floating-point performance for a given block count is reached on the CPU with the highest amount of work. This processor conducts floating-point operations during most of the simulation and only interrupts this task for small periods to exchange boundary information between blocks. This is in contrast to CPUs with less work load, which spend a considerable fraction of time communicating and idling to wait for processors which are still processing data. With this observation the properties of Fig. 9(a) are readily explainable. The maximum floating-point rate is similar in all cases, since they all contain a CPU with considerably more cells than the others, which performs floating-point operations during most of the time. The mean and minimum floating-point rates are determined by variations in the cell count and the minimum number of cells on a processor, respectively. The fewer processors are employed to distribute the blocks on them, the smaller the variations in work load on the individual CPUs. Consequently, the average floating-point rate lies closer to the maximum. As expected and evident from Fig. 13, this is the case for 16 processors, whereas for 64 processors most of the CPUs hold an amount of work which lies closer to the minimum. Thus the mean floating-point rate lies closer to the curve of the minimum. The minimum floating-point rate is reached on the CPU with the lowest amount of work. This means that the ratio of the maximum to the minimum cell count in Table 2 is a suitable metric for the ratio of the maximum to the minimum floating-point rate. At 64 processors, the unfavourable load-balancing yields considerably higher cell-count ratios as for fewer CPUs. Therefore, its minimum floating-point rate is much lower. Note that for 102 blocks, the curves of the maximum floating-point rate exhibit a dip for all three cases. At the lowest number of processors it is most distinct. The reason for this behaviour is not evident from our benchmarking data and seems to be a peculiarity of the block setup with 102 blocks. We believe that only a detailed analysis of the MPI communication could bring further insight into this matter.

The minimum, mean and maximum of the average vector length in Fig. 9(b) are very similar for both 16 and 32 processors, since their block distributions yield about the same load-balancing properties for all cases (see Fig. 13). Especially the ratio of the largest number of cells on a processor to the smallest in Table 2 is a good indicator with direct influence on the vector length. For all simulations with 16 and 32 CPUs, this ratio is comparable for 16 and 32 CPUs. On the other hand, this measure is considerably higher for 64 CPUs, and consequently the minimum, mean and especially the maximum of the average vector length are considerably higher. Due to the more inhomogeneous load-balancing in this case, the difference between the minimum and maximum vector length is significantly larger than for a lower number of processors. Furthermore, for 16 and 32 processors, the mean vector length stays just above to their curves of the minimum vector length. In the case of 64 CPUs, however, the mean vector length lies more in the middle between the minimum and maximum vector lengths. For the higher block counts, it approaches the curve of the maximum vector length. Note that the minimum vector length for 64 processors is about as high as the maximum vector length for 16 and 32 CPUs.

After the detailed investigation of the impact of the block and processor count on the performance and related measures, we will consider the overall effect of block splitting on the NEC SX-8 in an aggregate performance chart. In Fig. 10, the performance of all computations on the NEC SX-8 is displayed with the usual symbols with both linear and logarithmic axes. The performance development with 34 blocks was already discussed in Sect. 5.1. After a linear initial parallel scaling for up to 4 processors, the performance stagnates at around 16 processors due to an unfavourable load balancing, and a higher level of parallelism does not yield any notable performance improvements. An alleviation of this problem was found in the refinement of the domain decomposition by splitting the mesh blocks. The three sets of computations with 16, 32 and 64 processors each with a varied block number were studied above in more detail. For 16 processors, the performance is at best only slightly increased by block splitting, but can also be severely degraded by it: already at a moderate number of blocks, the performance is actually lower than with the un-split block configuration (34 blocks). At 32 processors, where the load balancing is much worse, block splitting exhibits a generally more positive effect. The performance almost doubles when going to 34 to 68 blocks, and continually degrades for a higher number of blocks. At the highest investigated block count, 340 blocks, the performance is approximately the same as with 34 blocks. At 64 processors, a computation with 34 blocks is technically not possible. However, it is remarkable that already with 68 blocks and thus bad load balancing (cf. plot ⑯ in Fig. 13), the performance is optimum, and a higher number of blocks only degrades the performance.

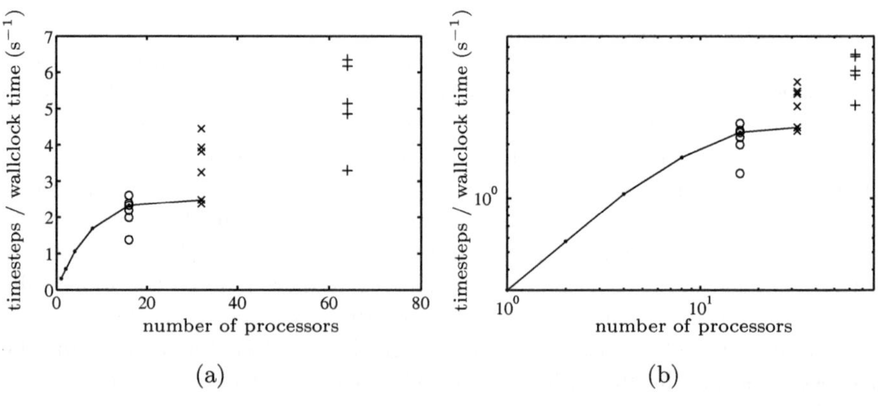

Fig. 10. Effect of block splitting on parallel performance of NEC SX-8. —— 34 blocks with varying number of processors; ○ 16 CPUs, × 32 CPUs and + 64 CPUs with a varying number of blocks ranging from 68 to 340. See Fig. 8(a) for the dependence of the performance on the block count for the three CPU configurations. (**a**) Linear axes, (**b**) logarithmic axes

When considering the hull curve through the maximum values for each processor count, it is notable that the performance continues to scale rather well for a higher number of CPUs, and the performance stagnation due to inhomogeneous load balancing can be readily overcome by block splitting. However, in contrast to the Cray XT3, the block count exhibits a relatively strong effect on performance, and a too copious use of block splitting can even deteriorate the performance, especially at a low number of processors. We therefore conclude that block splitting is also a viable approach to overcome load balancing issues on the NEC SX-8. In contrast to scalar machines, it is however advisable to keep the number of blocks as low as possible. A factor of two to the unbalanced block count causing performance stagnation appears to be a good choice. When following this recommendation, the performance scales well on the NEC SX-8 to a relatively high number of processors.

5.4 Putting Everything Together: The Big Picture

In this section, we compare the benchmarking data obtained on the Cray XT3 to the one gathered on the NEC SX-8, especially in view of the performance improvements that are possible with block splitting. In Fig. 11(a) and (b), we display this information in two plots with linear and logarithmic axes.

Using the original configuration with 34 blocks, all three machines (NEC SX-5, NEC SX-8 and Cray XT3) quickly run into performance stagnation after a short initial almost-linear scaling region. While the linear scaling region is slightly longer on the Cray XT3 than on the two vector machines, its stagnating performance level (with 16 CPUs) barely reaches the result of the NEC SX-5 on a single processor. While complete performance saturation cannot be reached on the NEC SX-5 due to its too small queue and machine size, its "extrapolated" stagnation performance is almost an order of magnitude higher, and the NEC SX-8 is approximately twice as fast on top of that.

For our past simulations, the performance on the two vector machines, especially on the NEC SX-8, has been sufficient for an adequate simulation turnover time, while the Cray XT3 result is clearly unsatisfactory. By refining the domain decomposition through block splitting, the load balancing is drastically improved, and the simulation performance using 32 CPUs on the Cray jumps up by a factor of about four. At this level, it is just about competitive with the NEC SX-5 using 4 CPUs and no block splitting (which would not yield notable performance improvements here anyway). Furthermore, at this setting, the Cray XT3 output equals approximately the result with one or two NEC SX-8 processors (also without block splitting).

Since the block splitting strategy has the general benefit of extending the linear parallel scaling range, an increase of the number of CPUs comes along with considerable performance improvements. On the Cray XT3, the performance using 512 processors roughly equals the SX-8 output with 8 CPUs. The 512 CPUs on the Cray correspond to about one third of the machine installed at CSCS, which can be considered the maximum allocation that is realistic

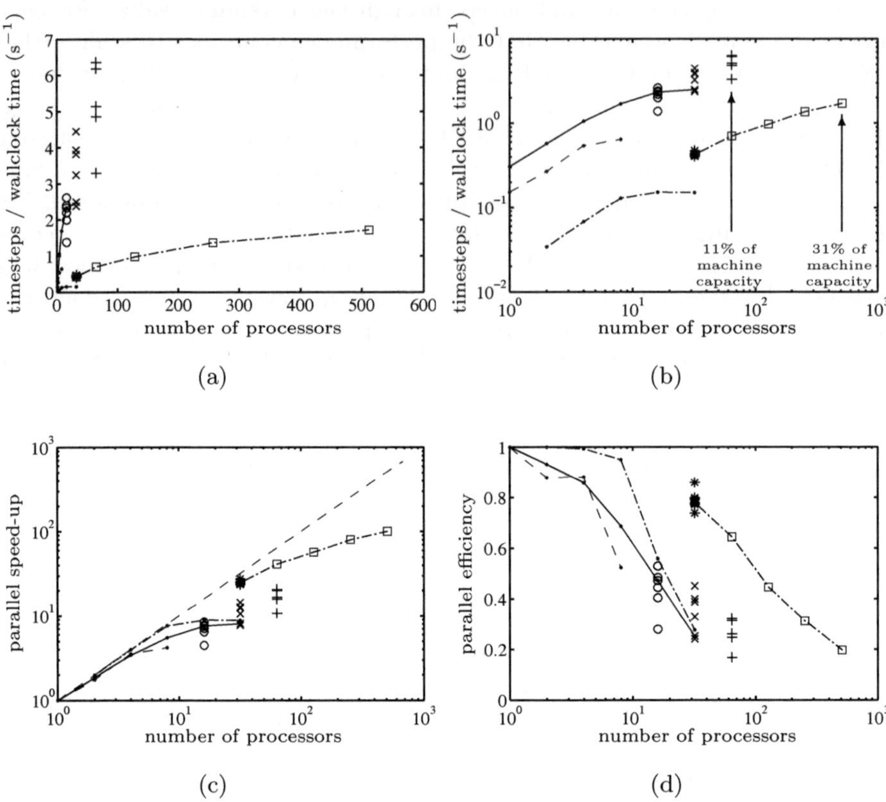

Fig. 11. Aggregate chart for varying number of processors. (**a**) Performance measure (linear axes), (**b**) performance measure (logarithmic axes), (**c**) parallel speed-up, (**d**) parallel efficiency. —— NEC SX-8 with 34 blocks; ∘ 16 CPUs, × 32 CPUs and + 64 CPUs on NEC SX-8 with a varying number of blocks ranging from 68 to 340. - - - - NEC SX-5 with 34 blocks, —·— Cray XT3 with 34 blocks, ∗ Cray XT3 with a varying number of blocks ranging from 68 to 680 and 32 CPUs, —⊡— Cray XT3 with 680 blocks, - - - - ideal scaling in (**c**)

in everyday operation, and thus this result marks the maximum performance of this test case on the Cray XT3. Since a similar performance is achievable on the NEC SX-8 using only one out of 72 nodes (slightly more than 1% of the machine capacity), we conclude that calculations on the NEC SX-8 are much more efficient than on the Cray XT3 for our code and test case. Further increases of the CPU count on the NEC SX-8, still without block splitting, yield notable performance improvements to about 50% above the maximum Cray XT3 performance. However, when block splitting is employed on the NEC SX-8, the full potential of the machine is uncovered with the given simulation setup, and the performance approximately doubles for 32 processors, when increasing from 34 to 68 blocks. The maximally observed performance, with 68 blocks on 64 processors of the SX-8 (8 nodes, or 11% of the machine capacity), surpasses the result of the Cray (using 512 CPUs, or about 31% of the machine) by a factor of approximately 2.5. This maximum performance is achieved with a parallel speed-up of about 20, resulting in a parallel efficiency of about 30%, see Figs. 11(c) and (d). Note that the allocation of more than 8 nodes of the NEC SX-8 is readily done in everyday operation, and its parallel scaling chart is still far from saturation for this number of CPUs, as evident from Fig. 11(c).

It is also instructive to consider the parallel efficiencies of the different machines in Fig. 11(d). For the initial block configuration with 34 blocks, the efficiencies of the two vector computers fall rather quickly due to load-balancing problems for more than 4 CPUs. For 32 processors, the efficiency lies just little above 20%. In contrast, the efficiency of the Cray XT3 remains close to the ideal value for up to 16 CPUs, but then it also sinks rapidly to about the same value of 20%. On both machines, the efficiency rises considerably by improvements in load-balancing due to block splitting. But whereas the parallel efficiency only doubles to about 45% on the NEC SX-8 with 32 processors and 68 blocks, more than the three-fold efficiency of about 75% is obtained on the Cray XT3. When further increasing the number of processors from 32, the parallel efficiencies of both the Cray XT3 and the NEC SX-8 decrease again due to communication overhead, albeit at a lower rate than it was caused by the load-balancing deficiencies before. Whereas the slope of the falling efficiency gets steeper with an increasing number of NEC SX-8 processors, the decrease of the Cray XT3 efficiency occurs roughly in a straight line in the semi-logarithmic diagram. For 512 CPUs on the Cray XT3, the efficiency reaches the level of 20%, which was obtained on 32 processors without block-splitting. However, while an unfavourable load-balancing is the reason for the low efficiency with 32 processors, the large communication overhead is responsible for the low value with 512 processors. The parallel efficiency on the NEC SX-8 with 64 processors is slightly higher (about 30%), but extrapolation suggests that an efficiency of 20% would be reached for slightly more than 100 CPUs.

The different behaviour of the performance and the parallel efficiency on the Cray XT3 and the NEC SX-8 for a varying number of blocks was already discussed in Sect. 5.3, when considering the effects of block splitting on the NEC SX-8. Therefore, we will review them only quickly in Fig. 12, where both quantities are shown together.

As clearly visible, the splitting of the blocks is generally beneficial on both machines. However, whereas the maximum performance and parallel efficiency are reached on the Cray XT3 for a relatively high block count (340 blocks), both the performance and parallel efficiency decrease considerably on the NEC SX-8 after an initial jump from 34 to 68 blocks. As typical for a scalar machine, the parallel efficiency of the Cray XT3 is considerably higher (75%–85%) than that of the NEC SX-8 vector computer, which lies in the range of 30%–55% (see Fig. 12(b)). On the other hand, the performance

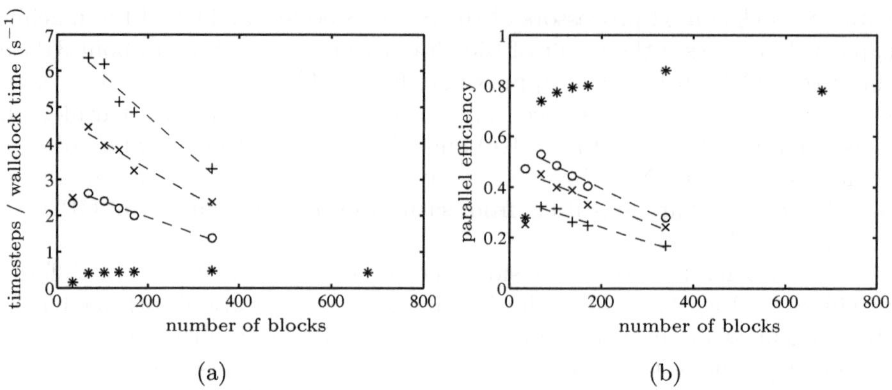

Fig. 12. Aggregate chart for varying number of blocks. (**a**) Performance measure, (**b**) parallel efficiency. ∘ 16 CPUs, × 32 CPUs and + 64 CPUs on NEC SX-8, ---- linear fit through falling parts of the curves. ∗ Cray XT3 with 32 CPUs

measure in Fig. 12(a) (which is of more practical interest) is much better on the NEC SX-8, due to its higher CPU performance and vector capabilities, which are favourable for the given application.

6 Conclusions

We conducted a comparative performance assessment with the code NSMB on different high-performance computing platforms at CSCS in Manno and HLRS in Stuttgart using a typical computational fluid dynamics simulation case involving time-resolved turbulent flow. The investigation was centred around the question if and how it is possible to achieve a similar performance on the new massively-parallel Cray XT3 at CSCS as obtained on the NEC SX-5 and SX-8 vector machines at CSCS and HLRS, respectively.

While for the given test case the processor performance of the mentioned vector computers is sufficient for low simulation turnover times even at a low number of processors, the Cray CPUs are considerably slower. Therefore, correspondingly more CPUs have to be employed on the Cray to compensate. However, this is usually not easily feasible due to the block-parallel nature of the simulation code in combination with the coarse-grained domain decomposition of our typical flow cases. While the total block count is a strict upper limit for the number of CPUs in a parallel simulation, a severe degradation of the load-balancing renders parallel simulations with a block-to-CPU number ratio of less than 4 very inefficient.

An alleviation of this problem can be found in the block-splitting technique, were the total number of blocks is artificially increased by splitting them with an existing utility programme. The finer granularity of the domain sub-partitions allows for a more homogeneous distribution of the work to the individual processors, which leads to a more efficient parallelisation and a drastically improved performance. While a twofold increase of the blocks already caused a more than threefold simulation speedup on the Cray XT3, a further splitting of the blocks did not yield considerable performance improvements. The optimum simulation performance for the given test case was reached for a CPU-to-block ratio of 10. On the NEC SX-8, the block splitting technique generally also yields favourable results. However, it is advisable to keep the total number of blocks on this machine as low as possible to avoid an undue degradation of the average vector length and thus a diminished parallel performance. Our findings on the NEC SX-8 are supported with a detailed analysis of the dependence of the simulation performance, floating-point rate and vector length on the processor and block counts, specifically in view of the properties of the specific load distribution.

The most important benefit from block splitting can be seen in the extension of the almost-linear parallel scaling range to a considerably higher number of processors. Given the low number of blocks in our typical simulations, block splitting seems to be the only possibility allowing for an efficient use of massively-parallel supercomputers with a simulation performance that is comparable to simulations with a low number of CPUs on vector machines. However, when employing an increased number of CPUs on the NEC SX-8, corresponding to a similar allocation of the total machine size than on the Cray XT3, the SX-8 performance considerably exceeds the capabilities of the XT3. Furthermore, an increased number of sub-partitions comes along with an overhead in bookkeeping and communication. Also the complexity of the pre- and post-processing rises considerably.

We conclude that a supercomputer with a relatively low number of high-performance CPUs such as the NEC SX-8 is far better suited for our typical numerical simulations with this code than a massively-parallel machine with a large number of relatively low-performance CPUs such as the Cray XT3.

Our further work on this topic will cover the evaluation of the newly-installed IBM Power 5 at CSCS, to which we did not have access for this study. Its CPUs are somewhat more powerful than the processors of the Cray XT3, and it offers shared-memory access within one node of 16 processors, which reduces the communication overhead. Additionally, the NSMB simulation code was already optimised for the predecessor machine IBM SP4. However, recent performance data gathered with another of our codes dampens the performance expectations for this machine. The fact that the performance on one SP5 node equals not more than a single NEC SX-8 processor in this case suggests that the new IBM can hardly achieve a comparable performance as on the NEC SX-8, even with the block-splitting technique.

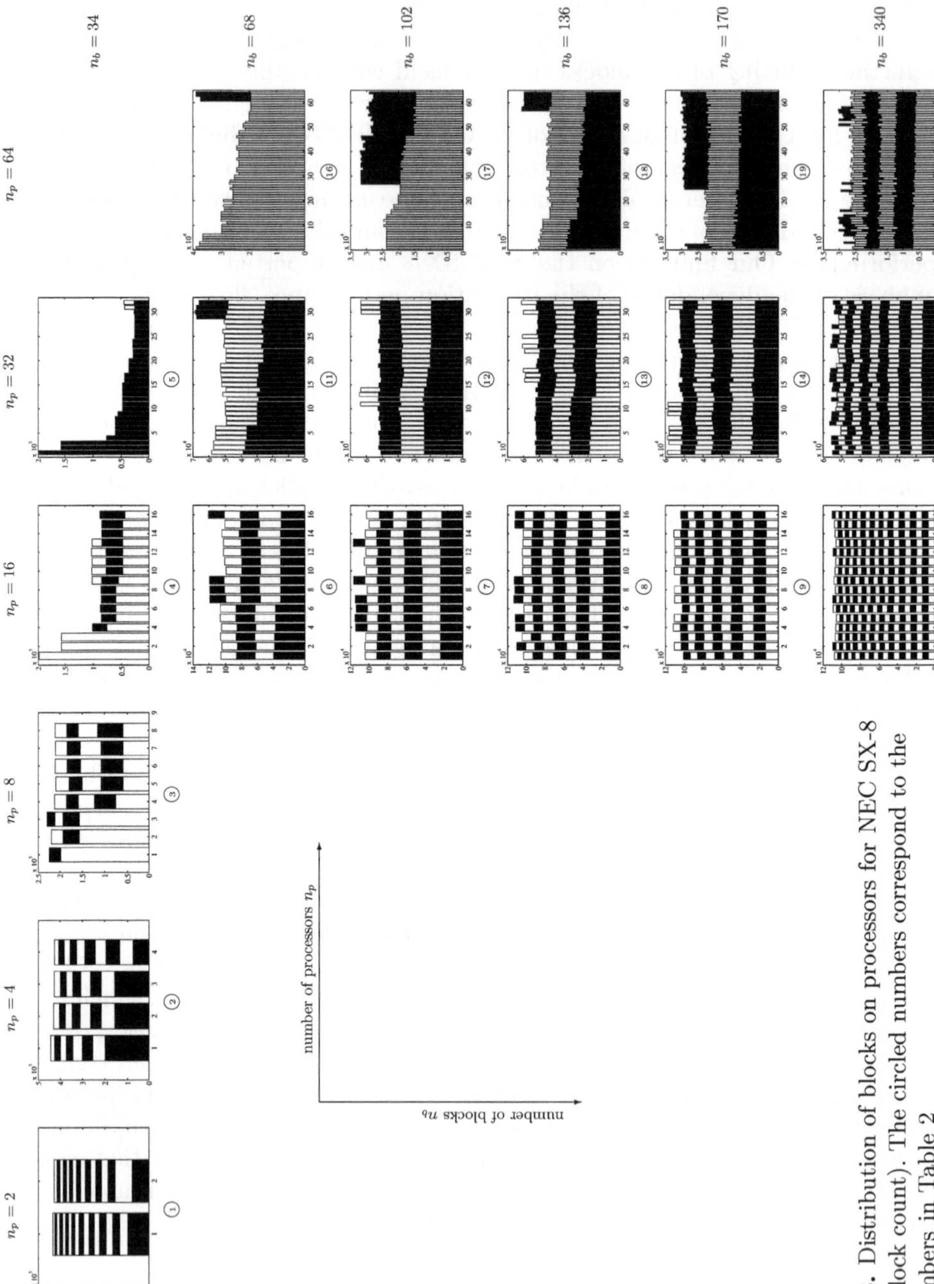

Fig. 13. Distribution of blocks on processors for NEC SX-8 (total block count). The circled numbers correspond to the run numbers in Table 2

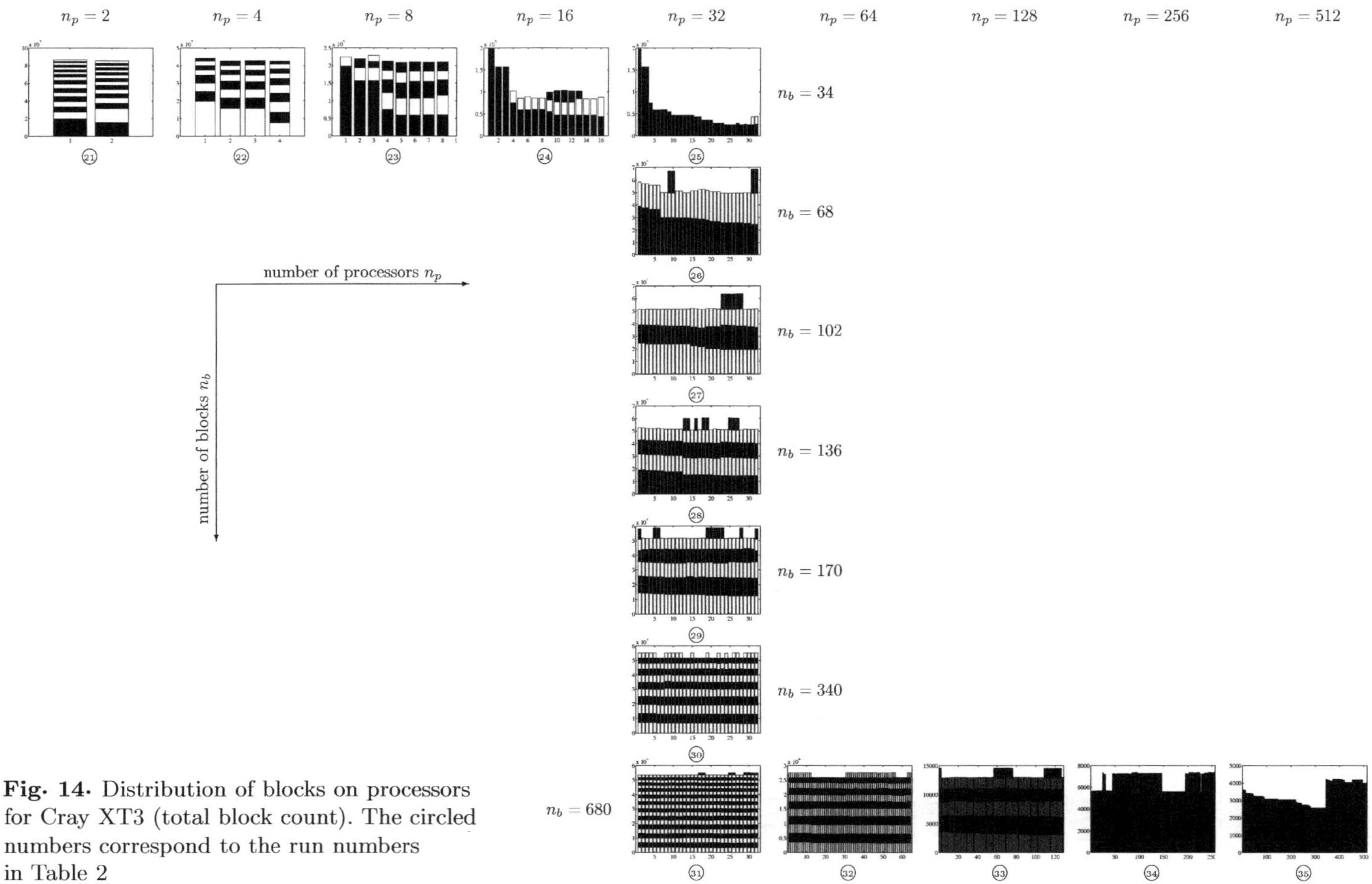

Fig. 14. Distribution of blocks on processors for Cray XT3 (total block count). The circled numbers correspond to the run numbers in Table 2

Acknowledgements

A part of this work was carried out under the HPC-EUROPA project (RII3-CT-2003-506079), with the support of the European Community – Research Infrastructure Action under the FP6 "Structuring the European Research Area" Programme. The hospitality of Prof. U. Rist and his group at the Institute of Aero and Gas Dynamics (University of Stuttgart) is greatly appreciated. We thank Stefan Haberhauer (NEC HPC Europe) and Peter Kunszt (CSCS) for fruitful discussions, as well as CSCS and HLRS staff for their support regarding our technical inquiries.

References

1. Vos, J. B., van Kemenade, V., Ytterström, A., and Rizzi, A. W., "Parallel NSMB: An Industrialized Aerospace Code for Complete Aircraft Simulations," *Proc. Parallel CFD Conference 1996*, edited by P. Schiano et al., North Holland, 1997.
2. Vos, J. B., Rizzi, A., Corjon, A., Chaput, E., and Soinne, E., "Recent Advances in Aerodynamics inside the NSMB (Navier Stokes Multi Block) consortium," *AIAA Paper 98-0225*, 1998.
3. Vos, J. B., Leyland, P., van Kemenade, V., Gacherieu, C., Duquesne, N., Lotstedt, P., Weber, C., Ytterström, A., and Saint Requier, C., *NSMB Handbook 4.5*.
4. Gacherieu, C., Collercandy, R., Larrieu, P., Soumillon, P., Tourette, L., and Viala, S., "Navier-Stokes calculations at Aerospace Matra Airbus for aircraft design," *Proc. ICAS, Harrogate, UK*, edited by G. I., Royal Aeronautical Society, London, UK, 2000.
5. Viala, S., Amant, S., and Tourette, L., "Recent achievements on Navier-Stokes methods for engine integration," *Proc. CEAS, Cambridge*, Royal Aeronautical Society, London, UK, 2002.
6. Mossi, M., *Simulation of benchmark and industrial unsteady compressible turbulent fluid flows*, Ph. D. thesis no. 1958, EPFL Lausanne, 1999.
7. Ziefle, J., Stolz, S., and Kleiser, L., "Large-Eddy Simulation of Separated Flow in a Channel with Streamwise-Periodic Constrictions," *17th AIAA Computational Fluid Dynamics Conference*, Toronto, Canada, June 6–9, 2005, AIAA Paper 2005-5353.
8. Ziefle, J. and Kleiser, L., "Large-Eddy Simulation of a Round Jet in Crossflow," *36th AIAA Fluid Dynamics Conference*, San Francisco, USA, June 5–8 2006, 2006, AIAA Paper 2006-3370.
9. TOP500 Team, "TOP500 Report for November 2006," Tech. rep., November 2006, also available as http://www.top500.org/list/2006/11/.
10. Ziefle, J. and Kleiser, L., "Large-Eddy Simulation of Film Cooling," 2007, in preparation.
11. Ytterström, A., "A Tool For Partitioning Structured Multiblock Meshes For Parallel Computational Mechanics," *The International Journal of Supercomputer Applications and High Performance Computing*, Vol. 11, 1997, pp. 336–343.

Applications III

Multiphysics Computational Fluid Dynamics

High Performance Computing Towards Silent Flows

E. Gröschel[1], D. König[2], S. Koh, W. Schröder, and M. Meinke

Institute of Aerodynamics, RWTH Aachen University,
Wüllnerstraße zw. 5 u. 7, 52062 Aachen, Germany,
[1] e.groeschel@aia.rwth-aachen.de,
[2] d.koenig@aia.rwth-aachen.de

Summary. The flow field and the acoustic field of various jet flows and a high-lift configuration consisting of a deployed slat and a main wing are numerically analyzed. The flow data, which are computed via large-eddy simulations (LES), provide the distributions being plugged in the source terms of the acoustic perturbation equations (APE) to compute the acoustic near field. The investigation emphasizes the core flow to have a major impact on the radiated jet noise. In particular the effect of heating the inner stream generates substantial noise to the sideline of the jet, whereas the Lamb vector is the dominant noise source for the downstream noise. Furthermore, the analysis of the airframe noise shows the interaction of the shear layer of the slat trailing edge and the slat gap flow to generate higher vorticity than the main airfoil trailing edge shear layer. Thus, the slat gap is the more dominant noise region for an airport approaching aircraft.

1 Introduction

In the recent years the emitted sound by aircraft has become a very contributing factor during the development process. This is due to the predicted growth of air-traffic as well as the stricter statutory provisions. The generated sound can be assigned to engine and airframe noise, respectively. The present paper deals with two specific noise sources, the jet noise and the slat noise.

Jet noise constitutes the major noise source for aircraft during take-off. In the last decade various studies [12, 25, 6, 5] focused on the computation of unheated and heated jets with emphasis on single jet configurations. Although extremely useful theories, experiments, and numerical solutions exist in the literature, the understanding of subsonic jet noise mechanisms is far from perfect. It is widely accepted that there exist two distinct mechanisms, one is associated with coherent structures radiating in the downstream direction and the other one is related to small scale turbulence structures contributing to the high frequency noise normal to the jet axis. Compared with single jets, coaxial jets with round nozzles can develop flow structures of very different topology, depending on environmental and initial conditions and, of course, on the temperature gradient between the inner or core stream and the bypass stream. Not much work has been done on such jet configurations and as such there are still many open questions [3]. For instance, how is the mixing process influenced by the development of the inner and outer shear layers What is the impact of the temperature distribution on the mixing and on the noise generation mechanisms The current investigation contrasts the flow field and acoustic results of a high Reynolds number cold single jet to a more realistic coaxial jet configuration including the nozzle geometry and a heated inner stream.

During the landing approach, when the engines are near idle condition, the airframe noise becomes important. The main contributor to airframe noise are high-lift devices, like slats and flaps, and the landing gear. The paper focuses here on the noise generated by a deployed slat.

The present study applies a hybrid method to predict the noise from turbulent jets and a deployed slat. It is based on a two-step approach using a large-eddy simulation (LES) for the flow field and approximate solutions of the acoustic perturbation equations (APE) [10] for the acoustic field. The LES comprises the neighborhood of the dominant noise sources such as the potential cores and the spreading shear layers for the jet noise and the slat cove region for the airframe noise. In a subsequent step, the sound field is calculated for the near field, which covers a much larger area than the LES source domain. Compared to direct methods the hybrid approach possess the potential to be more efficient in many aeroacoustical problems since it exploits the different length scales of the flow field and the acoustic field. To be more precise, in subsonic flows the characteristic acoustic length scale is definitely larger than that of the flow field. Furthermore, the discretization scheme of the acoustic solver is designed to mimic the physics of the wave operator.

The paper is organized as follows. The governing equations and the numerical procedure of the LES/APE method are described in Sect. 2. The simulation parameters of the cold single jet and the heated coaxial jet are given in the first part of Sect. 3 followed by the description of the high-lift configuration. The results for the flow field and the acoustical field are discussed in detail in Sect. 4. In each section, the jet noise and the slat noise problem are discussed subsequently. Finally, in Sect. 5, the findings of the present study are summarized.

2 Numerical Methods

2.1 Large-Eddy Simulations

The computations of the flow fields are carried out by solving the unsteady compressible three-dimensinal Navier-Stokes equations with a monotone-integrated large-eddy simulation (MILES) [7]. The block-structured solver is optimized for vector computers and parallelized by using the Message Passing Interface (MPI). The numerical solution of the Navier-Stokes equations is based on an vertex centered finite-volume scheme, in which the convective fluxes are computed by a modified AUSM method with an accuracy is 2nd order. For the viscous terms a central discretization is applied also of 2nd order accuracy. Meinke et al. showed in [21] that the obtained spatial precision is sufficient compared to a sixth-order method. The temporal integration from time level n to $n+1$ is done by an explicit 5-stage Runge-Kutta technique, whereas the coefficients are optimized for maximum stability and lead to a 2nd order accurate time approximation. At low Mach number flows a preconditioning method in conjunction with a dual-time stepping scheme can be used [2]. Furthermore, a multi-grid technique is implemented to accelerate the convergence of the dual-time stepping procedure.

2.2 Acoustic Simulations

The set of acoustic perturbation equations (APE) used in the present simulations corresponds to the APE-4 formulation proposed in [10]. It is derived by rewriting the complete Navier-Stokes equations as

$$\frac{\partial p'}{\partial t} + \bar{c}^2 \boldsymbol{\nabla} \cdot \left(\bar{\rho} \boldsymbol{u}' + \bar{\boldsymbol{u}} \frac{p'}{\bar{c}^2} \right) = \bar{c}^2 q_c \tag{1}$$

$$\frac{\partial \boldsymbol{u}'}{\partial t} + \boldsymbol{\nabla} \left(\bar{\boldsymbol{u}} \cdot \boldsymbol{u}' \right) + \boldsymbol{\nabla} \left(\frac{p'}{\bar{\rho}} \right) = \boldsymbol{q}_m . \tag{2}$$

The right-hand side terms constitute the acoustic sources

$$q_c = -\boldsymbol{\nabla} \cdot (\rho' \boldsymbol{u}')' + \frac{\bar{\rho}}{c_p} \frac{\bar{D} s'}{Dt} \tag{3}$$

$$\boldsymbol{q}_m = -\left(\boldsymbol{\omega} \times \boldsymbol{u} \right)' + T' \boldsymbol{\nabla} \bar{s} - s' \boldsymbol{\nabla} \bar{T} - \left(\boldsymbol{\nabla} \frac{(u')^2}{2} \right)' + \left(\frac{\boldsymbol{\nabla} \cdot \underline{\tau}}{\rho} \right)' . \tag{4}$$

To obtain the APE system with the perturbation pressure as independent variable the second law of thermodynamics in the first-order formulation is used. The left-hand side constitutes a linear system describing linear wave propagation in mean flows with convection and refraction effects. The viscous effects are neglected in the acoustic simulations. That is, the last source term in the momentum equation is dropped.

The numerical algorithm to solve the APE-4 system is based on a 7-point finite-difference scheme using the well-known dispersion-relation preserving scheme (DRP) [24] for the spatial discretization including the metric terms on curvilinear grids. This scheme accurately resolves waves longer than 5.4 points per wave length (PPW). For the time integration an alternating 5-6 stage low-dispersion low-dissipation Runge-Kutta scheme [15] is implemented.

To eliminate spurious oscillations the solution is filtered using a 6th-order explicit commutative filter [23, 26] at every tenth iteration step. As the APE system does not describe convection of entropy and vorticity perturbations [10] the asymptotic radiation boundary condition by Tam and Webb [24] is sufficient to minimize reflections on the outer boundaries. On the inner boundaries between the different matching blocks covering the LES and the acoustic domain, where the transition of the inhomogeneous to the homogeneous acoustic equations takes place, a damping zone is formulated to suppress artificial noise generated by a discontinuity in the vorticity distribution [22].

3 Computational Setup

3.1 Jet

The quantities u_j and c_j are the jet nozzle exit velocity and sound speed, respectively, and T_j and T_∞ the temperature at the nozzle exit and in the ambient fluid. Unlike the single jet, the simulation parameters of the coaxial jet have additional indices "p" and "s" indicating the primary and secondary stream. An isothermal turbulent single jet at $M_j = u_j/c_\infty = 0.9$ and $Re = 400,000$ is simulated. These parameters match with previous investigations performed by a direct noise computation via an acoustic LES by Bogey and Bailly [6] and a hybrid LES/Kirchhoff method by Uzun et al. [25]. The chosen Reynolds number can be regarded as a first step towards the simulation of real jet configurations. Since the flow parameters match those of various studies, a good database exists to validate our hybrid method for such high Reynolds number flows. The inflow condition at the virtual nozzle exit is given by a hyperbolic-tangent profile for the mean flow, which is seeded by random velocity fluctuations into the shear layers in form of a vortex ring [6] to provide turbulent fluctuations. Instantaneous LES data are sampled over a period of $\bar{T} = 3000 \cdot \Delta t \cdot u_j/R = 300.0$ corrsponding to approximately 6 times the time interval an acoustic wave needs to travel through the computational domain. Since the source data is cyclically fed into the acoustic simulation a modifed Hanning windowing [20] has been performed to avoid spurious noise generated by discontinuities in the source term distribution. More details on the computational set up can be found in Koh et al.[17]
The flow parameters of the coaxial jet comprises a velocity ratio of the secondary and primary jet exit velocity of $\lambda = u_{js}/u_{jp} = 0.9$, a Mach number 0.9

for the secondary and 0.877 for the primary stream, and a temperature ratio of $T_{js}/T_{jp} = 0.37$. An overview of the main parameter specifications is given in Table 1. To reduce the computational costs the inner part of the nozzle was not included in the simulation, but a precursor RANS simulation was set up to generate the inflow profiles for the LES. For the coaxial jet instantaneous data are sampled over a period of $\bar{T}_s = 2000 \cdot \Delta t \cdot c_\infty / r_s = 83$. This period corresponds to roughly three times the time interval an acoustic wave needs to propagate through the computational domain. As in the single jet computation, the source terms are cyclically inserted into the acoustic simulation. The grid topology and in particular the shape of the short cowl nozzle are shown in Fig.1. The computational grid has about $22 \cdot 10^6$ grid points.

3.2 High-Lift Configuration

Large-Eddy Simulation

The computational mesh consists of 32 blocks with a total amount of 55 million grid points. The extent in the spanwise direction is 2.1% of the clean chord length and is resolved with 65 points. Figures 2 and 3 depict the mesh near the airfoil and in the slat cove area, respectively. To assure a sufficient resolution in the near surface region of $\Delta x^+ \approx 100$, $\Delta y^+ \approx 1$, and $\Delta y^+ \approx 22$ [1] the analytical solution of a flate plate was used during the grid generation process to approximate the needed step sizes.
On the far-field boundaries of the computational domain boundary conditions based on the theory of characteristics are applied. A sponge layer following Israeli et al. [16] is imposed on these boundaries to avoid spurious reflections, which would extremely influence the acoustic analysis. On the walls an adiabatic no-slip boundary condition is applied and in the spanwise direction periodic boundary conditions are used.

Table 1. Flow properties coaxial jet.

Jet flow conditions of the (p)rimary and (s)econdary stream				
notation	dimension	SJ	CJ	parameter
Ma_p		-	0.9	Mach number primary jet
Ma_s		0.9	0.9	Mach number secondary jet
Ma_{ac}		0.9	1.4	Acoustic Mach number ($\frac{U_p}{c_\infty}$)
T_p	K	-	775	Static temperature primary jet
T_{tp}	K	-	879.9	Total temperature primary jet
T_s	K	288.	288.	Static temperature secondary jet
T_{ts}	K	335	335.	Total temperature secondary jet
T_∞	K	288.	288.	Ambient temperature
Re		$4 \cdot 10^5$	$2 \cdot 10^6$	Reynolds number

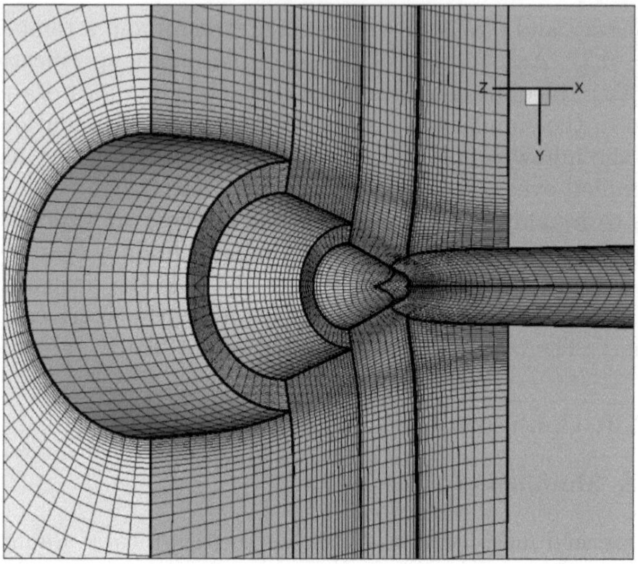

Fig. 1. The grid topology close to the nozzle tip is "bowl" shaped, i.e., grid lines from the primary nozzle exit end on the opposite side of the primary nozzle. Every second grid point is shown.

The computation is performed for a Mach number of $Ma = 0.16$ at an angle of attack of $\alpha = 13°$. The Reynolds number is set to $Re = 1.4 \cdot 10^6$. The inital conditions were obtained from a two-dimensinal RANS simulation.

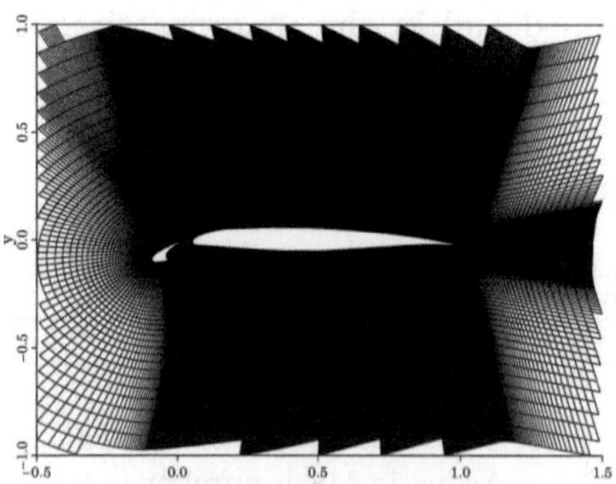

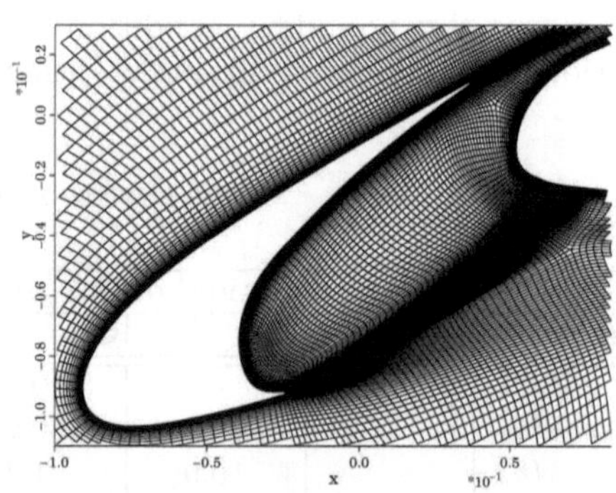

Fig. 2. LES grid of the high-lift configuration. Every 2nd grid point is depicted.

Fig. 3. LES grid in the slat cove area of the high-lift configuration. Every 2nd grid point is depicted.

Acoustic Simulation

The acoustic analysis is done by a two-dimensional approach. That is, the spanwise extent of the computational domain of the LES can be limited since especially at low Mach number flows the turbulent length scales are significantly smaller then the acoustic length scales and as such the noise sources can be considered compact. This treatment tends to result in somewhat over-predicted sound pressure levels which are corrected following the method described by Ewert et al. in [11].

The acoustic mesh for the APE solution has a total number of 1.8 million points, which are distributed over 24 blocks. Figure 4 shows a section of the used grid. The maximum grid spacing in the whole domain is chose to resolve 8 kHz as the highest frequency.

The acoustic solver uses the mean flow field obtained by averaging the unsteady LES data and the time dependent perturbed Lamb vector $(\omega \times u)'$, which is also computed from the LES results, as input data. A total amount of 2750 samples are used which describe a non-dimensional time periode of $T \approx 25$, non-dimensionalized with the clean chord length and the speed of sound c_∞.

To be in agreement with the large-eddy simulation the Mach number, the angle of attack and the Reynolds number are set to $Ma = 0.16$, $\alpha = 13°$ and $Re = 1.4 \cdot 10^6$, respectively.

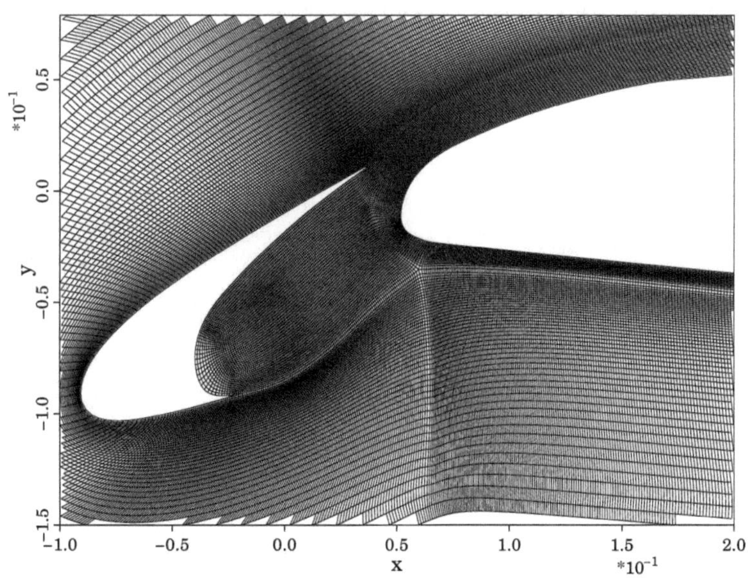

Fig. 4. APE grid of the high-lift configuration. Every 2nd grid point is depicted.

4 Results and Discussion

The results of the present study are divided into two parts. First, the flow field and the acoustic field of the cold single jet and the heated coaxial jet will be discussed concerning the mean flow properties, turbulent statistics and acoustic signature in the near field. To relate the findings of the coaxial jet to the single jet, the flow field and acoustic field of which has been validated in current studies [17] against the experimental results by [27] and numerical results by Bogey and Bailly [6], comparisons to the flow field and acoustic near field properties of the single jet computation are drawn. This part also comprises a discussion on the results of the acoustic near field concerning the impact by the additional source terms of the APE system, which are related to heating effects. The second part describes in detail the airframe noise generated by the deployed slat and the main wing element. Acoustic near field solutions are discussed on the basis of the LES solution alone and the hybrid LES/APE results.

4.1 Jet

Large-Eddy Simulation

In the following the flow field of the single jet is briefly discussed to show that the relevant properties of the high Reynolds number jet are well computed when compared with jets at the same flow condition taken from the literature. In Fig. 5 the half-width radius shows an excellent agreement with the LES by Bogey and Bailly [6] and the experiments by Zaman [27] indicating a potential core length of approximately 10.2 radii. The jet evolves downstream of the

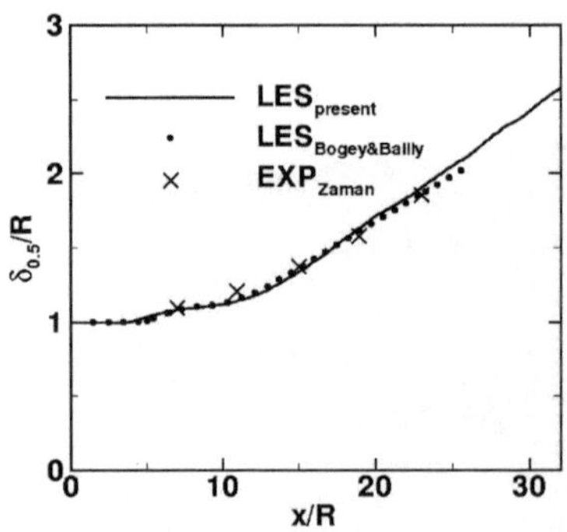

Fig. 5. Jet half-width radius in comparison with numerical [6] and experimental results [27].

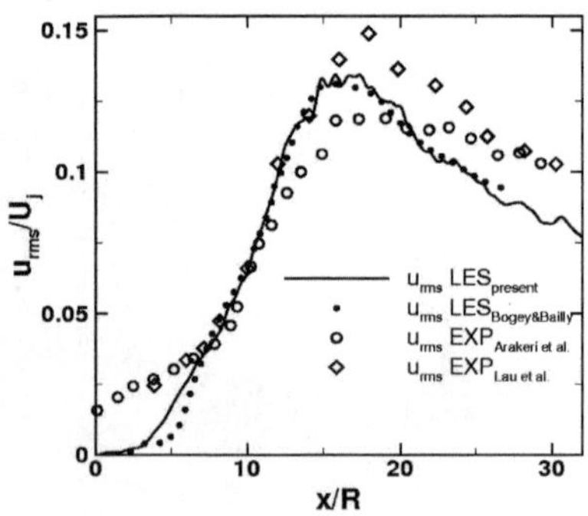

Fig. 6. Reynolds stresses $\sqrt{\overline{u'u'}/u_j^2}$ normalized by the nozzle exit velocity in comparison with numerical [6] and experimental [4, 19] results.

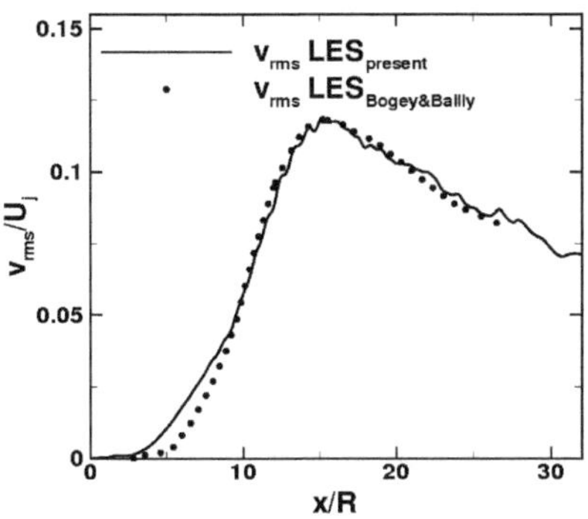

Fig. 7. Reynolds stresses $\sqrt{\overline{v'v'}/u_j^2}$ normalized by the nozzle exit velocity in comparison with numerical [6] results.

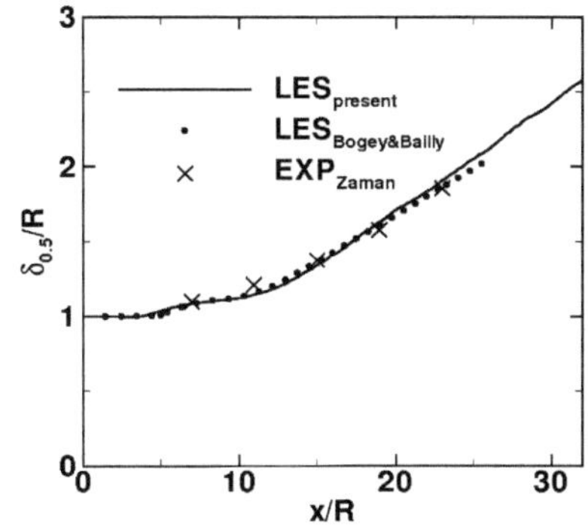

Fig. 8. Reynolds stresses $\sqrt{\overline{u'u'}/u_j^2}$ normalized by the nozzle exit velocity over jet half-width radius at $x/R = 22$ in comparison with numerical [6] results.

potential core according to experimental findings showing the quality of the lateral boundary conditions to allow a correct jet spreading. Furthermore, in Figs. 6 and 7 the turbulent intensities $\sqrt{\overline{u'u'}/u_j^2}$ and $\sqrt{\overline{v'v'}/u_j^2}$ along the center line rise rapidly after an initially laminar region to a maximum peak near the end of the potential core and decrease further downstream. The obtained values are in good agreement with those computed by Bogey and Bailly [6] and the experimental results by Arakeri et al. [4] and Lau et al. [19]. The self-similarity of the jet in Fig. 8 is well preserved. From these findings it seems appropriate to use the present LES results for jet noise analyses, which are performed in the next subsection.

The flow field analysis of the coxial jet starts with Fig. 9 showing instantaneous density contours with mapped on mean velocity field. Small vortical and slender ring-like structures are generated directly at the nozzle lip. Further downstream, these structures start to stretch and become unstable, eventually breaking into smaller structures. The degree of mixing in the shear layers

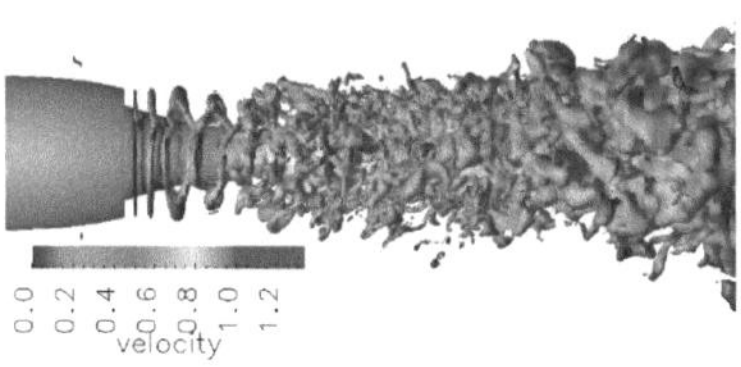

Fig. 9. Instantaneous density contours with mapped on velocity field.

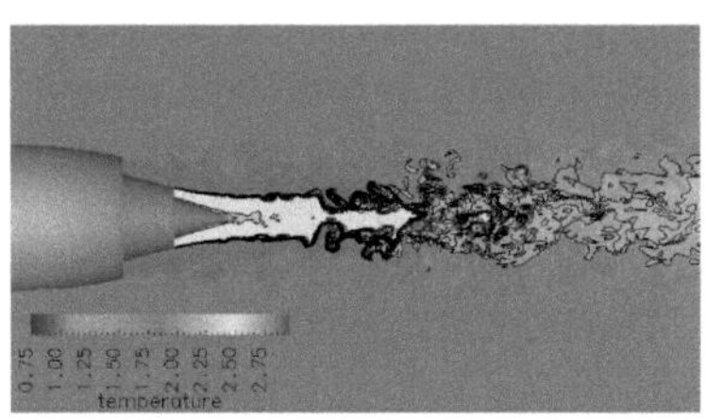

Fig. 10. Instantaneous temperature contours ($z/R_s = 0$ plane).

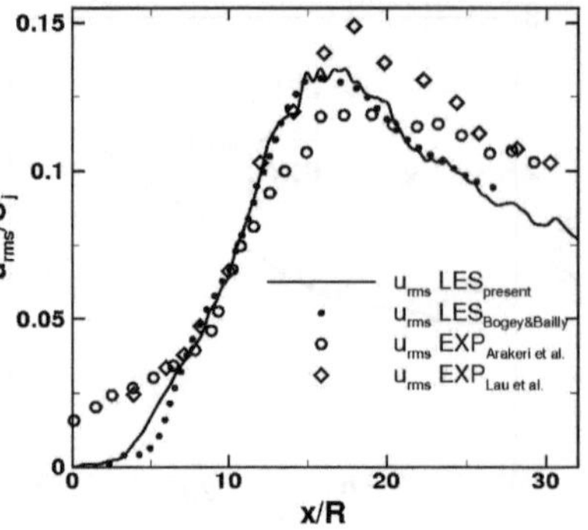

Fig. 11. Mean flow development of coaxial jet in parallel planes perpendicular to the jet axis in comparison with experimental results.

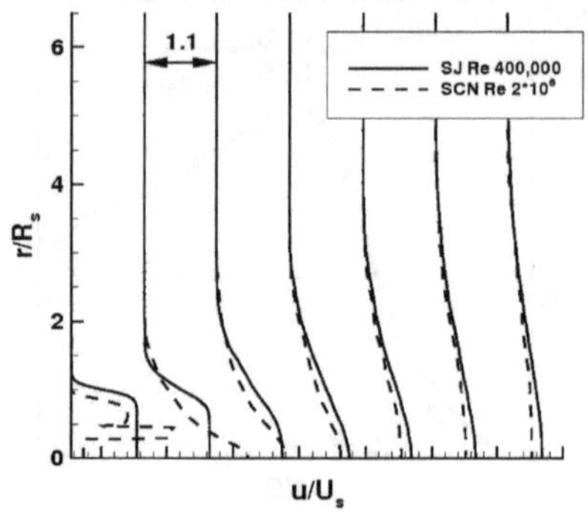

Fig. 12. Axial velocity profiles for cold single jet and heated coaxial jet.

between the inner and outer stream, the so-called primary mixing region, is generally very high. This is especially noticeable in Fig. 10 with the growing shear layer instability separating the two streams. Spatially growing vortical structures generated in the outer shear layer seem to affect the inner shear layer instabilities further downstream. This finally leads to the collapse and break-up near the end of the inner core region.

Figure 11 shows mean flow velocity profiles based on the secondary jet exit velocity of the coaxial jet at different axial cross sections ranging from $x/R_S = 0.0596$ to $x/R_s = 14.5335$ and comparisons to experimental results. A good agreement, in particular in the near nozzle region, is obtained, however, the numerical jet breaks-up earlier than in the experiments resulting in a faster mean velocity decay on the center line downstream of the potential core.

The following three Figs. 12 to 14 compare mean velocity, mean density, and Reynolds stress profiles of the coaxial jet to the single jet in planes normal to the jet axis and equally distributed in the streamwise direction from $x/R_s = 1$ to $x/R_s = 21$. In the initial coaxial jet exit region the mixing of the primary shear layer takes place. During the mixing process, the edges of the initially sharp density profile are smoothed. Further downstream the secondary jet shear layers start to break up causing a rapid exchange and mixing of the fluid in the inner core. This can be seen by the fast decay of the mean density profile in Fig. 13.

During this process, the two initially separated streams merge and show at $x/R_s = 5$ a velocity profile with only one inflection point roughly at $r/R_s = 0.5$. Unlike the density profile, the mean axial velocity profile decreases only slowly downstream of the primary potential core. In the self-similar region the velocity decay and the spreading of the single and the coaxial jet is similar.

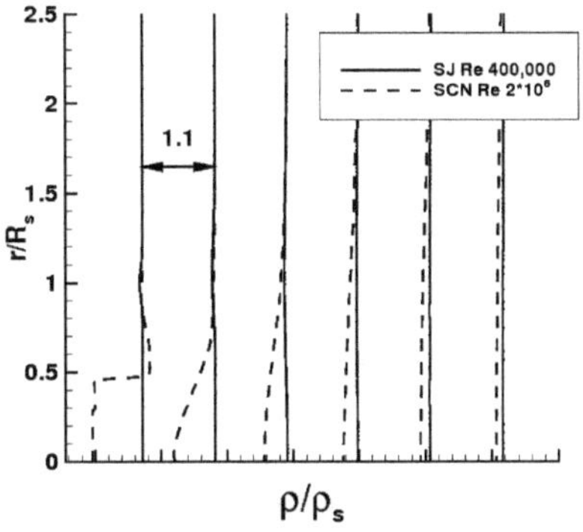

Fig. 13. Density profiles for cold single jet and heated coaxial jet.

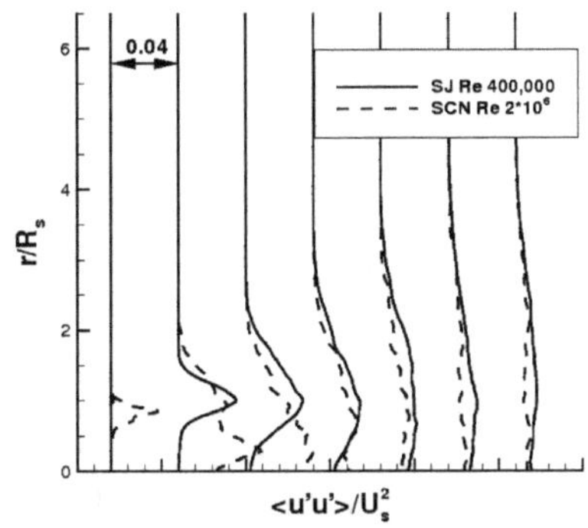

Fig. 14. Reynolds stresses profiles for cold single jet and heated coaxial jet.

The break-up process enhances the mixing process yielding higher levels of turbulent kinetic energy on the center line. The axial velocity fluctuations of the coaxial jet starts to increase at $x/R_s = 1$ in the outer shear layer and reach at $x/R_s = 9$ high levels on the center line, while the single jet axial fluctuations start to develop not before $x/r_s = 5$ and primarily in the shear layer but not on the center line. This difference is caused by the density and entropy gradient, which is the driving force of this process. This is confirmed by the mean density profiles. These profiles are redistributed beginning at $x/r_s = 1$ until they take on a uniform shape at approx. $x/r_s = 9$. When this process is almost finished the decay of the mean axial velocity profile sets in. This redistribution evolves much slower over several radii in the downstream direction.

Acoustic Simulation

The presentation of the jet noise results is organized as follows. First, the main characteristics of the acoustic field of the single jet from previous noise [13],[17] computations are summarized, by which the present hybrid method has been successfully validated against. Then, the acoustic fields for the single and coaxial jet are discussed. Finally, the impact of different source terms on the acoustic near field is presented.

Unlike the direct acoustic approach by an LES or a DNS, the hybrid methods based on an acoustic analogy allows to separate different contributions to the noise field. These noise mechanisms are encoded in the source terms of the acoustic analogy and can be simulated separately exploiting the linearity of the wave operator. Previous investigations of the single jet noise demonstrated the fluctuating Lamb vector to be the main source term for cold jet noise problems. An acoustic simulation with the Lamb vector only was performed and the sound field at the same points was computed and compared with the solution containing the complete source term.

The overall acoustic field of the single and coaxial jet is shown in Figs. 15 and 16 by instantaneous pressure contours in the near field, i.e., outside the source region, and contours of the Lamb vector in the acoustic source region. The acoustic field is dominated by long pressure waves of low frequency radiating in the downstream direction. The dashed line in Fig. 15 indicates the measurement points at a distance of 15 radii from the jet axis based on the outer jet radius at which the acoustic data have been sampled. Fig. 17 shows the acoustic near field signature generated by the Lamb vector only in comparison with an, in terms of number of grid points, highly resolved LES and the direct noise computation by Bogey and Bailly. The downstream noise is well captured by the LES/APE method and is consistent with the highly resolved LES results. The increasing deviation of the overall sound pressure level at obtuse angles with respect to the jet axis is due to missing contributions from nonlinear and entropy source terms. A detailed investigation can be found in Koh et al.[17]. Note that the results by Bogey and Bailly are 2 to 3 dB too high compared to the present LES and LES/APE distributions. Since different grids (Cartesian grids by Bogey and Bailly and boundary fitted grids in the present simulation) and different numerical methods for the compressible flow field have been used resulting resulting in varying boundary conditions, e.g.,the resolution of the initial momentum thickness, differences in the sensitive acoustic field are to be expected. The findings of the hybrid LES/Kirchhoff approach by Uzun et al. [25] do also compare favorably with the present solutions.

The comparison between the near field noise signature generated by the Lamb vector only of the single and the coaxial jet at the same measurement line shows almost the same characteristic slope and a similar peak value location along the jet axis. This is suprising, since the flow field development of both jets including mean flow and turbulent intensities differed strongly.

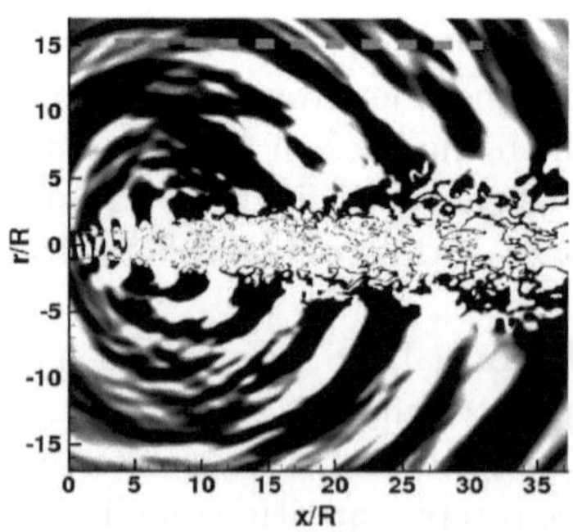

Fig. 15. Pressure contours of the single jet by LES/APE generated by the Lamb vector only. Dashed line indicates location of observer points to compute the acoustic near field.

Fig. 16. Pressure contours outside the source domain and the y-component of the Lamb vector inside the source domain of the coaxial jet.

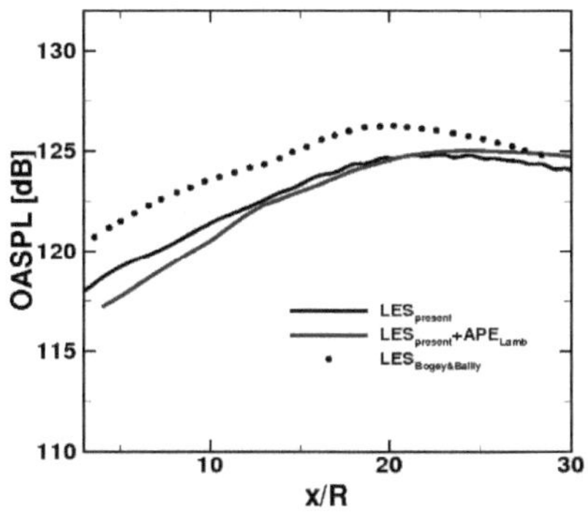

Fig. 17. Overall sound pressure level (OASPL) in dB for $r/R = 15$. Comparison with data from Bogey and Bailly [6].

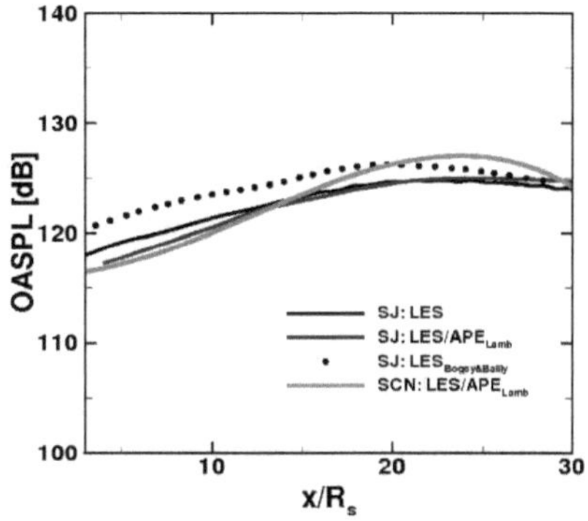

Fig. 18. Comparison of the acoustic field between the single jet and the coaxial jet generated by the Lamb vector only. Comparison with data from Bogey and Bailly [6].

Finally, Figs. 19 and 20 show the predicted far field directivity at 60 radii from the jet axis by the Lamb vector only and by the Lamb vector and the entropy source terms, respectively, in comparison with numerical and experimental results at the same flow condition. To obtain the far field noise signature, the near field results have been scaled to the far field by the 1/r-law assuming the center of directivity at $x/R_s = 4$. The acoustic results generated by the Lamb vector only match very well the experimental results at angles lower than 40 degree. At larger angles from the jet axis the OASPL falls off more rapidly.

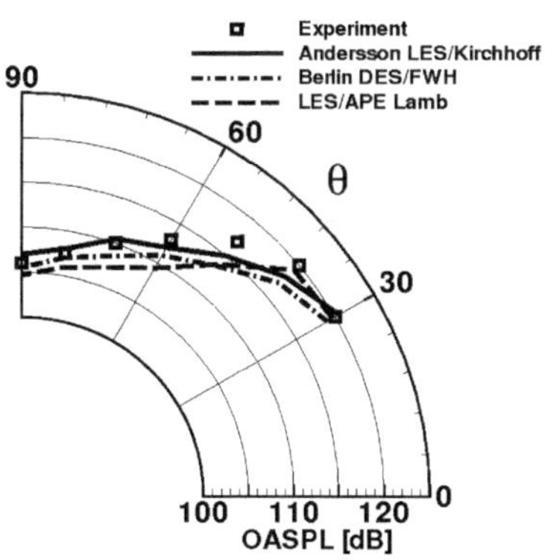

Fig. 19. Directivity at $r/R_s = 60$ generated by the Lamb vector only. Comparison with experimental and numerical results.

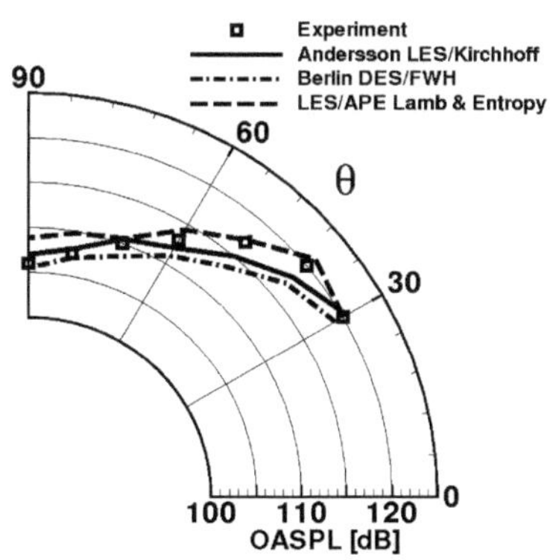

Fig. 20. Directivity at $r/R_s = 60$ generated by the Lamb vector and entropy sources. Comparison with experimental and numerical results..

This deviation is due to the missing contributions from the entropy sourece terms. When including those source terms in the computation, the LES/APE are in good agreement with the experimental results up to angles of 70 degree. That observation confirms previous studies [14] on the influence of different source terms. To be more precise, the Lamb vector radiates dominantly in the downstream direction, whereas the entropy sources radiate to obtuse angles from the jet axis.

4.2 High-Lift Configuration

Large-Eddy Simulation

The large-eddy simulation has been run for about 5 non-dimensional time units based on the freestream velocity and the clean chord length. During this time a fully developed turbulent flow field was obtained. Subsequently, samples for the statistical analysis and also to compute the aeroacoustic source terms were recorded. The sampling time interval was chosen to be approximatly 0.0015 time units. A total of 4000 data sets using 7 Terabyte of disk space have been collected which cover an overall time of approximatly 6 non-dimensional time units. The maximum obtained floating point operations per second (FLOPS) amounts 6.7 GFLOPS, the average value was 5.9 GFLOPS. An average vectorization ratio of 99.6% was achieved with a mean vector length of 247.4.
First of all, the quality of the results should be assessed on the basis of the proper mesh resolution near the walls. Figures 21 to 24 depict the determined values of the grid resolution and shows that the flate plate approximation yields satisfactory results. However, due to the accelerated and decelerated flow on the suction and pressure side, repectively, the grid resolution departs somewhat from the approximated values. In the slat cove region the resolution reaches everywhere the required values for large-eddy simulations of wall bounded flows ($\Delta x^+ \approx 100$, $\Delta y^+ \approx 2$, and $\Delta y^+ \approx 20$ [1]).

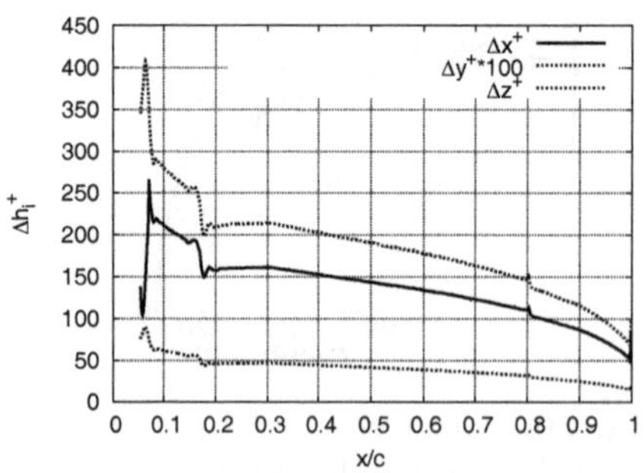

Fig. 21. Grid resolution near the wall: Suction side of the main wing.

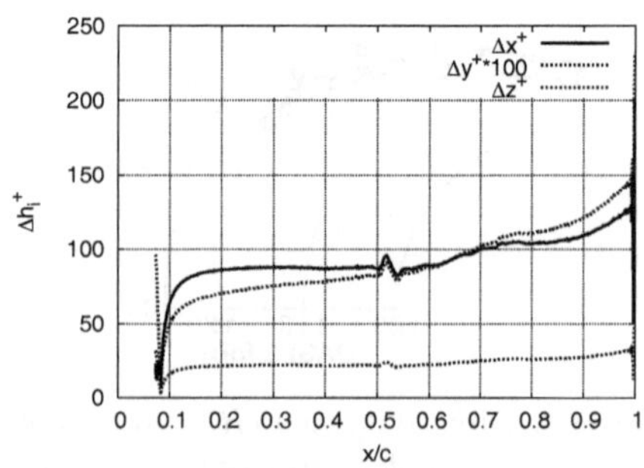

Fig. 22. Grid resolution near the wall: Pressure side of the main wing.

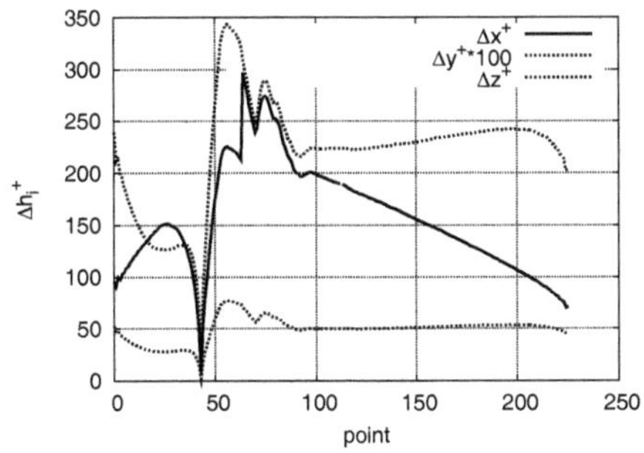

Fig. 23. Grid resolution near the wall: Suction side of the slat.

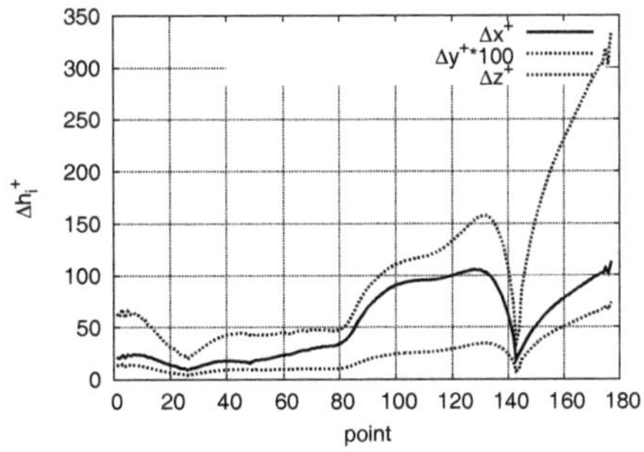

Fig. 24. Grid resolution near the wall: Slat cove.

The Mach number distribution and some selected streamlines of the time and spanwise averaged flow field is presented in Fig. 25. Apart form the two stagnation points one can see the area with the highest velocity on the suction side short downstream of the slat gap. Also recognizable is a large recirculation domain which fills the whole slat cove area. It is bounded by a shear layer which develops form the slat cusp and reattaches close to the end of the slat trailing edge.

The pressure coefficient c_p computed by the time averaged LES solution is compared in Fig. 26 with RANS results [9] and experimental data. The measurements were carried out at DLR Braunschweig in an anechoic wind tunnel with an open test section within the national project FREQUENZ. These experiments are compared to numerical solutions which mimic uniform freestream conditions. Therefore, even with the correction of the geometric angle of attack of 23° in the measurements to about 13° in the numerical solution no perfect match between the experimental and numerical data can be expected.

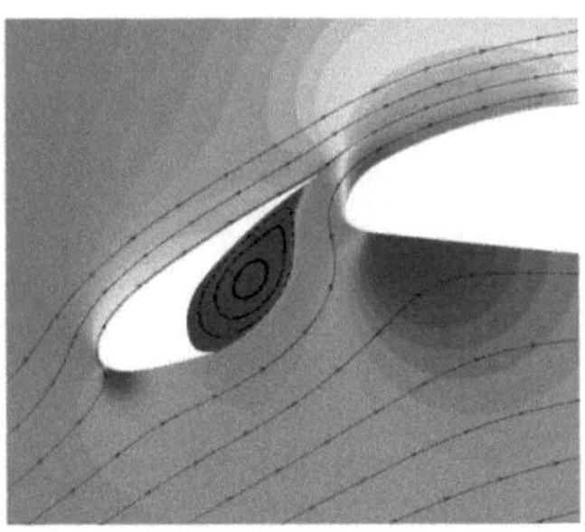

Fig. 25. Time and spanwise averaged Mach number distribution and some selected streamlines.

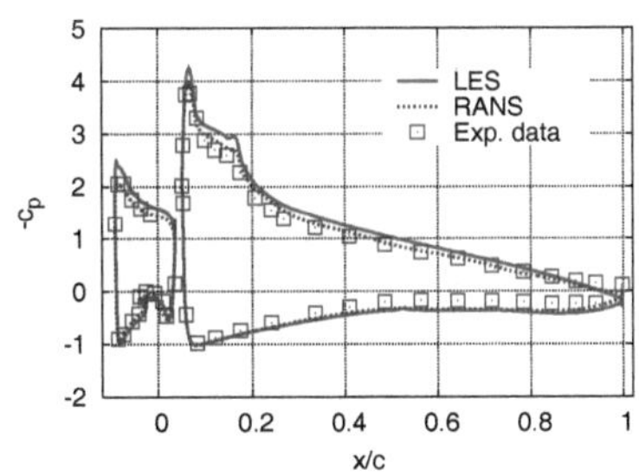

Fig. 26. Comparison of the c_p coefficient between LES, RANS [9] and experimental data [18].

Figures 27 to 29 show the turbulent vortex structures by means of λ_2 contours. The color mapped onto these contours represents the mach number. The shear layer between the recirculation area and the flow passing through the slat gap develops large vortical structures near the reattachment point. Most of these structures are convected through the slat gap while some vortices are trapped in the recirculation area and are moved upstream to the cusp. This behavior is in agreement with the findings of Choudhari et al. [8]. Furthermore, like the investigations in [8] the analysis of the unsteady data indicates a fluctuation of the reattachment point. On the suction side of the slat, shortly downstream of the leading edge, the generation of the vortical structures in Fig. 27 visualizes the transition of the boundary layer. This turbulent boundary layer passes over the slat trailing edge and interacts with the vortical structures convected through the slat gap. Figure 29 illustrates some more pronounced vortices being generated in the reattachment region and whose axes are aligned with the streamwise direction. They can be considered some kind of Görtler vortices created by the concave deflection of the flow.

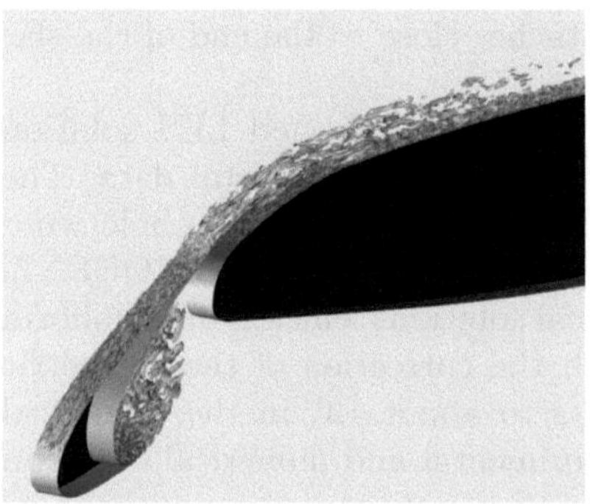

Fig. 27. λ_2 contours in the slat region.

Fig. 28. λ_2 contours in the slat region.

Fig. 29. λ_2 contours in the slat gap area.

Fig. 30. Time and spanwise averaged turbulent kinetic energy in the slat cove region.

The distribution of the time and spanwise averaged turbulent kinetic energy $k = \frac{1}{2}\left(u'^2 + v'^2 + w'^2\right)$ is depicted in Fig. 30. One can clearly identify the shear layer and the slat trailing edge wake. The peak values occur, in agreement with [8], in the reattachment area. This corresponds to the strong vortical structures in this area evidenced in Fig. 28.

Acoustic Simulation

A snapshot of the distribution of the acoustic sources by means of the perturbed Lamb vector $(\omega \times u)'$ is shown in Figs. 31 and 32. The strongest acoustic sources are caused by the normal component of the Lamb vector. The peak value occurs on the suction side downstream of the slat trailing edge, whereas somewhat smaller values are determined near the main wing trailing edge.

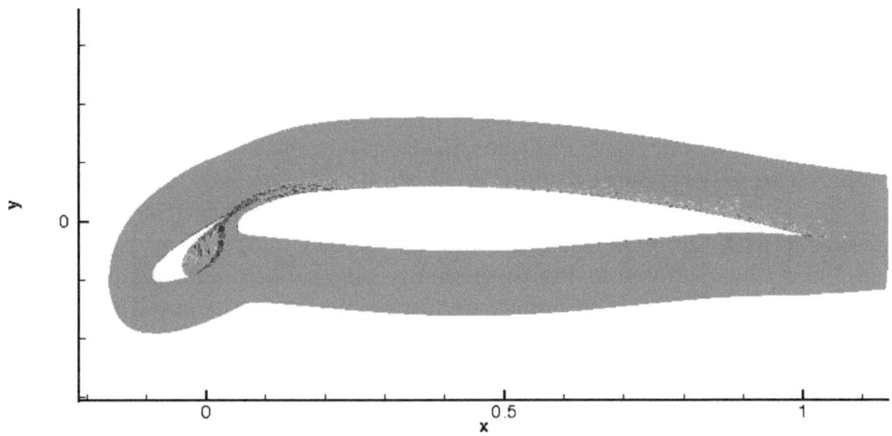

Fig. 31. Snapshot of the x-component of the Lamb Vector.

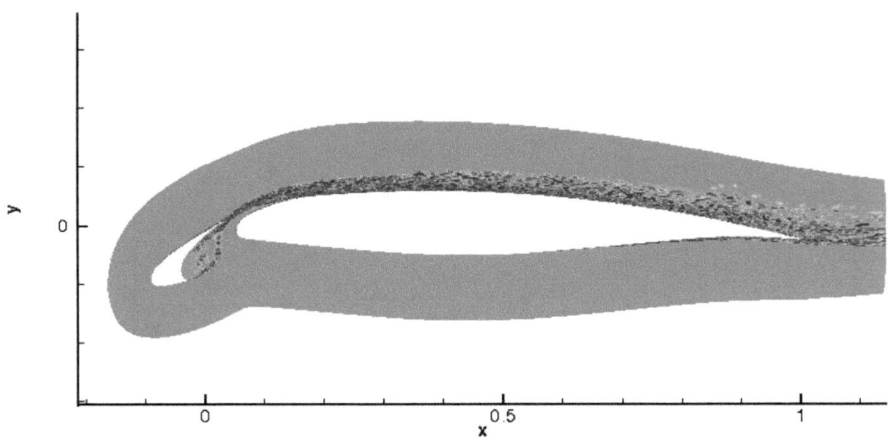

Fig. 32. Snapshot of the y-component of the Lamb Vector.

Fig. 33. Pressure contours based on the LES/APE solution.

Fig. 34. Pressure contours based on the LES solution.

Figures 33 and 34 illustrate a snapshot of the pressure fluctuations based on the APE and the LES solution. Especially in the APE solution the interaction between the noise of the main wing and that of the slat is obvious. A closer look reveals that the slat sources are dominant compared to the main airfoil trailing edge sources. It is clear that the LES mesh is not able to resolve the high frequency waves in some distance from the airfoil.
The power spectral density (PSD) for an observer point at x=-1.02 and y=1.76 compared to experimental results are shown in Fig. 35 [18]. The magnitude and the decay of the PSD at increasing Strouhal number (Sr) is in good agreement with the experimental findings. A clear correlation of the tonal components is not possible due to the limited period of time available for the Fast Fourier Transformation which in turn comes from the small number of input data.
The directivities of the slat gap noise source and the main airfoil trailing edge source are shown in Fig. 36 on a circle at radius $R = 1.5$ centered near the trailing edge of the slat. The following geometric source definitions were used.

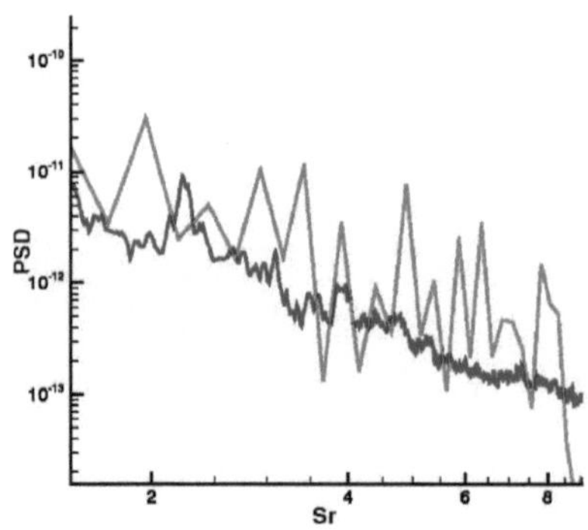

Fig. 35. Power spectral density for a point at x=-1.02 and y=1.76.

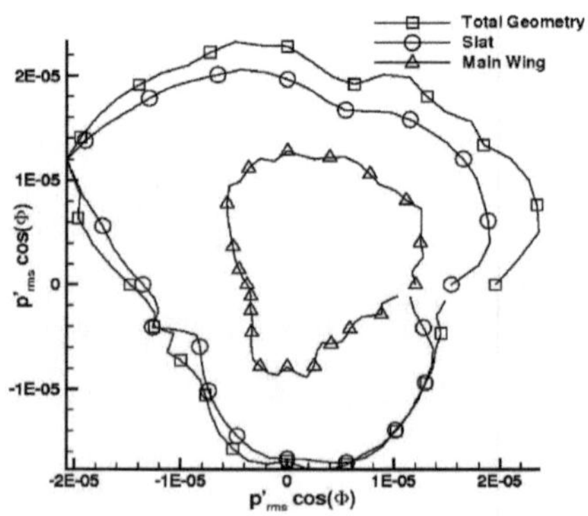

Fig. 36. Directivities for a circle with $R = 1.5$ based on the APE solution.

The slat source covers the part from the leading edge of the slat through 40% chord of the main wing. The remaining part belongs to the main wing trailing edge source. An embedded boundary formulation is used to ensure that no artificial noise is generated [22]. It is evident that the sources located near the slat cause a stronger contribution to the total sound field than the main wing trailing edge sources. This behavior corresponds to the distribution of the Lamb vector.

5 Conclusion

In the present paper we successfully computed the dominant aeroacoustic noise sources of aircraft during take-off and landing, that is, the jet noise and the slat noise by means of a hybrid LES/APE method. The flow parameters were chosen to match current industrial requirements such as nozzle geometry, high Reynolds numbers, heating effects etc. The flow field and acoustic field were computed in good agreement with experimental results showing the correct noise generation mechanisms to be determined.
The dominant source term in the APE formulation for the cold single jet has been shown to be the Lamb vector, while for the coaxial jets additional source terms of the APE-4 system due to heating effects must be taken into account. These source terms are generated by temperature and entropy fluctuations and by heat release effects and radiate at obtuse angles to the far field. The comparison between the single and coaxial jets revealed differences in the flow field development, however, the characteristics of the acoustic near field signature was hardly changed. The present investigation shows that the noise levels in the near field of the jet are not directly connected to the statistics of the Reynolds stresses.
The analysis of the slat noise study shows the interaction of the shear layer of the slat trailing edge and slat gap flow to generate higher vorticity than the main airfoil trailing edge shear layer. Thus, the slat gap is the dominant noise source region. The results of the large-eddy simulation are in good agreement with data from the literature. The acoustic analysis shows the correlation between the areas of high vorticity, especially somewhat downstream of the slat trailing edge and the main wing trailing edge, and the emitted sound.

Acknowledgments

The jet noise investigation, was funded by the Deutsche Forschungsgemeinschaft and the Centre National de la Recherche Scientifique (DFG-CNRS) in the framework of the subproject "Noise Prediction for a Turbulent Jet" of the research group 508 “Noise Generation in Turbulent flows”. The slat noise study was funded by the national project FREQUENZ. The APE solutions were computed with the DLR PIANO code the development of which is part of the cooperation between DLR Braunschweig and the Institute of Aerodynamics of RWTH Aachen University.

References

1. *LESFOIL: Large Eddy Simulation of Flow Around a High Lift Airfoil*, chapter Contribution by ONERA. Springer, 2003.
2. N. Alkishriwi, W. Schröder, and M. Meinke. A large-eddy simulation method for low mach number flows using preconditioning and multigrid. *Computers and Fluids*, 35(10):1126–1136, 2006.
3. N. Andersson, L.-E. Eriksson, and L. Davidson. Les prediction of flow and acoustcial field of a coaxial jet. Paper 2005-2884, AIAA, 2005.
4. V. Arakeri, A. Krothapalli, V. Siddavaram, M. Alkislar, and L. Lourenco. On the use of microjets to suppress turbulence in a mach 0.9 axissymmetric jet. *J. Fluid Mech.*, 490:75–98, 2003.
5. D. J. Bodony and S. K. Lele. Jet noise predicition of cold and hot subsonic jets using large-eddy simulation. CP 2004-3022, AIAA, 2004.
6. C. Bogey, C.and Bailly. Computation of a high reynolds number jet and its radiated noise using large eddy simulation based on explicit filtering. *Computers and Fluids*, 35:1344–1358, 2006.
7. J. P. Boris, F. F. Grinstein, E. S. Oran, and R. L. Kolbe. New insights into large eddy simulation. *Fluid Dynamics Research*, 10:199–228, 1992.
8. M. M. Choudhari and M. R. Khorrami. Slat cove unsteadiness: Effect of 3d flow structures. In *44st AIAA Aerospace Sciences Meeting and Exhibit.* AIAA Paper 2006-0211, 2006.
9. M. Elmnefi. Private communication. Institute of Aerodynamics, RWTH Aachen University, 2006.
10. R. Ewert and W. Schröder. Acoustic pertubation equations based on flow decomposition via source filtering. *J. Comput. Phys.*, 188:365–398, 2003.
11. R. Ewert, Q. Zhang, W. Schröder, and J. Delfs. Computation of trailing edge noise of a 3d lifting airfoil in turbulent subsonic flow. AIAA Paper 2003-3114, 2003.
12. J. B. Freund. Noise sources in a low-reynolds-number turbulent jet at mach 0.9. *J. Fluid Mech.*, 438:277 – 305, 2001.
13. E. Gröschel, M. Meinke, and W. Schröder. Noise prediction for a turbulent jet using an les/caa method. Paper 2005-3039, AIAA, 2005.
14. E. Gröschel, M. Meinke, and W. Schröder. Noise generation mechanisms in single and coaxial jets. Paper 2006-2592, AIAA, 2006.
15. F. Q. Hu, M. Y. Hussaini, and J. L. Manthey. Low-dissipation and low-dispersion runge-kutta schemes for computational acoustics. *J. Comput. Phys.*, 124(1):177–191, 1996.
16. M. Israeli and S. A. Orszag. Approximation of radiation boundary conditions. *Journal of Computational Physics*, 41:115–135, 1981.
17. S. Koh, E. Gröschel, M. Meinke, and W. Schröder. Numerical analysis of sound sources in high reynolds number single jets. Paper 2007-3591, AIAA, 2007.
18. A. Kolb. Private communication. FREQUENZ, 2006.
19. J. Lau, P. Morris, and M. Fisher. Measurements in subsonic and supersonic free jets using a laser velocimeter. *J. Fluid Mech.*, 193(1):1–27, 1979.

20. D. Lockard. An efficient, two-dimensional implementation of the ffowcs williams and hawkings equation. *J. Sound Vibr.*, 229(4):897–911, 2000.
21. M. Meinke, W. Schröder, E. Krause, and T. Rister. A comparison of second- and sixth-order methods for large-eddy simulations. *Computers and Fluids*, 31:695–718, 2002.
22. W. Schröder and R. Ewert. LES-CAA Coupling. In *LES for Acoustics*. Cambridge University Press, 2005.
23. J. S. Shang. High-order compact-difference schemes for time dependent maxwell equations. *J. Comput. Phys.*, 153:312–333, 1999.
24. C. K. W. Tam and J. C. Webb. Dispersion-relation-preserving finite difference schemes for computational acoustics. *J. Comput. Phys.*, 107(2):262–281, 1993.
25. A. Uzun, A. S. Lyrintzis, and G. A. Blaisdell. Coupling of integral acoustics methods with les for jet noise prediction. Pap. 2004-0517, AIAA, 2004.
26. O. V. Vasilyev, T. S. Lund, and P. Moin. A general class of commutative filters for les in complex geometries. *J. Comput. Phys.*, 146:82–104, 1998.
27. K. B. M. Q. Zaman. Flow field and near and far sound field of a subsonic jet. *Journal of Sound and Vibration*, 106(1):1–16, 1986.

Fluid-Structure Interaction: Simulation of a Tidal Current Turbine

Felix Lippold and Ivana Buntić Ogor[1]

Universität Stuttgart, Institute of Fluid Mechanics and Hydraulic Machinery
lippold,buntic@ihs.uni-stuttgart.de

Summary. Current trends in the development of new technologies for renewable energy systems show the importance of tidal and ocean current exploitation. But this also means to enter new fields of application and to develop new types of turbines. Latest measurements at economically interesting sites show strong fluctuations in flow and attack angles towards the turbine. In order to examine the dynamical behaviour of the long and thin structure of the turbine blades, coupled simulations considering fluid flow and structural behaviour need to be performed. For this purpose the parallel Navier-Stokes code FENFLOSS, developped at the IHS, is coupled with the commercial FEM-Code ABAQUS. Since the CFD domain has to be modelled in a certain way, the grid size tends to be in the range of about some million grid points. Hence, to solve the coupled problem in an acceptable timeframe the unsteady CFD calculations have to run on more than one CPU. Whereas, the structural grid is quite compact and, does not request that much computational power. Furthermore, the computational grid, distributed on different CPUs, has to be adapted after each deformation step. This also involves additional computational effort.

1 Basic equations

In order to simulate the flow of an incompressible fluid the momentum equations and mass conservation equation are derived in an infinitesimal control volume. Including the turbulent fluctuations yields the incompressible Reynolds-averaged Navier-Stokes equations, see Ferziger et al.[FP02]. Considering the velocity of the grid nodes U_{G} due to the mesh deformation results in the Arbitrary-Langrange-Euler (ALE) formulation, see Hughes [HU81].

$$\frac{\partial U_i}{\partial x_i} = 0 \tag{1}$$

$$\frac{\partial U_i}{\partial t} + (U_j - U_{\mathrm{G}})\frac{\partial U_i}{\partial x_j} = -\frac{1}{\rho}\frac{\partial p}{\partial x_i} + \frac{\partial}{\partial x_j}\left[\nu\left(\frac{\partial U_i}{\partial x_j} + \frac{\partial U_j}{\partial x_i}\right) - \underbrace{\overline{u_i' u_j'}}_{\text{Reynolds Stresses}}\right]. \tag{2}$$

The Reynolds Stresses are usually modelled following Boussinesq's vortex viscosity principle. To model the resulting turbulent viscosity, for most engineering problems k-ε and k-ω-models combined with logarithmic wall functions or Low-Reynolds formulations are applied.

The discretization of the momentum equations using a Petrov-Galerkin Finite Element approach, see Zienkiewicz [ZI89] and Gresho et al. [GR99], yields a non-linear system of equations. In FENFLOSS uses a point iteration to solve this problem numerically. For each iteration the equations are linearized and then smoothed by an ILU-preconditioned iterative BICGStab(2) solver, see van der Vorst [VVO92]. The three velocity components can be solved coupled or decoupled followed by a modified UZAWA pressure correction, see Ruprecht [RU89]. Working on parallel architectures, MPI is applied in the preconditioner and the matrix-vector and scalar products, see Maihoefer [MAI02].

The discretised structural equations with mass, damping and stiffness matrices $\boldsymbol{M}$, $\boldsymbol{D}$, and $\boldsymbol{K}$, load vector $\boldsymbol{f}$, and displacments $\boldsymbol{u}$ can be written as

$$\boldsymbol{M}\ddot{\boldsymbol{u}} + \boldsymbol{D}\dot{\boldsymbol{u}} + \boldsymbol{K}\boldsymbol{u} = \boldsymbol{f} \quad , \tag{3}$$

see Zienkiewicz [ZI89].

2 Dynamic mesh approach

The first mesh update method discussed here uses an interpolation between the nodal distance between moving and fixed boundaries to compute the new nodal position after a displacement step of the moving boundary. The most simple approach is to use a linear interpolation value $0 \leq \kappa \leq 1$. Here we are using a modification of the parametre $\kappa = \frac{|\boldsymbol{s}|}{|\boldsymbol{r}|+|\boldsymbol{s}|}$ proposed by Kjellgren and Hyvärinen [KH98].

$$\tilde{\kappa} = \begin{cases} 0, & \kappa < \delta \\ \frac{1}{2}(\cos(1 - (\frac{\kappa-\delta}{1-2\delta}) \cdot \pi) + 1, & \delta \leq \kappa \leq 1 - \delta \\ 1, & 1 - \delta < \kappa \leq 1 \end{cases} \tag{4}$$

This parametre is found from the nearest distance to the moving boundary and the distance to the fixed boundary in the opposite direction.

To use this approach for parallel computations the boundaries have to be available on all processors. Since a graph based domain decomposition is used here this is not implicitely given. Hence, the boundary exchange has to be implemented additionally. Usually the number of boundary nodes is considerably small, i.e. the additional communication time and overhead is negligible.

Another approach is based on a pseudo-structural approach using the lineal springs introduced by Batina [BAT89]. In order to stabilise the element angles this method is combined with the torsional springs approach by Farhat et. al. [FAR98-1, DEG02] for two- and three-dimensional problems. Since we use quadrilateral and hexahedral elements the given formulation for the torsional springs is enhanced for these element types.

Here, a complete matrix is built to solve the elasto-static problem of the dynamic mesh. The stiffness matrix built for the grid smoothing has the same graph as the matrix of the implicit CFD-problem. Hence, the the whole cfd-solver structure including the memory can be reused and the matrix graph has to be computed only once for both fields, the fluid and the structure. This means that there is almost no extra memory needed for the moving mesh. Furthermore, the parallelised and optimised, preconditioned BICGStab(2) performs good on cache and vector CPUs for the solution of the flow problem, which brings a good performance for the solution of the dynamic mesh equations, as well. Nevertheless, the stiffness matrix has to be computed and solved for every mesh update step, i.e. usually every time step. The overall computation time for a three-dimensional grid shows that the percentage of computing time needed for the mesh update compared to the total time is independent of the machine and number of CPUs used, see Lippold [LI06]. This means that the parallel performance is of the same quality as the one of the flow solver.

Regarding the usage of the torsional springs two issues have to be addressed. Computing the torsional part of the element matrices includes a bunch of nested loops costing quite a considerable amount of computational time. Because of that, the respective routine requires approximately twice the time than the lineal part does. Due to the fact that the lineal springs use the edge length to determine the stiffness but the torsional springs use the element area or volume, respectively, it is important to notice that the entries contributed to the matrix may differ within some scales. Hence, using the additional torsional springs the numbers shown above are changing for these two reasons. Furthermore, it yields higher values on the off-diagonal elements of the matrix which reduces its condition number. Hence, the smoother needs more iterations to reduce the residual to a given value. This leads to additional computational time. Furthermore the element quality might be unsatisfying if the matrix entries coming from the torsional springs are dominant. Meaning that the smoothed grid fulfills the angle criterion but not the nodal distance. In order to reduce this disadvantage, the matrix entries of the torsional springs have to be scaled to the same size as the contribution of the lineal springs.

3 Coupling fluid and structure

Seen from the physical point of view a fluid-structure interaction is a two field problem. But numerically, the problem has three fields, where the third field is the fluid grid that has to be updated after every deformation step of the structure, to propagate the movement of the fluid boundaries, the wetted structure, into the flow domain.

The solution of the numerical problem can be arraged in a monolithic scheme, which solves the structural and flow equations in one step. This method is applied for strongly coupled problems, e.g. for modal analyses. Another scheme, suitable for loosely and intermediately coupled simulations, is the separated solution of flow and structural equations with two independent codes. In the separated scheme well proven codes and models can be used. However, a data exchange between the codes has to be arranged, including the interpolation between the two computational meshes. This is the reason why for engineering problems usually the second approach is employed. In order to account for the coupling and to avoid unstable simulations, some coupling schemes have been developped, e.g. see Farhat et. al. [FAR98-2]. Figure 1 shows three of the most used schemes.

Using the simple coupling scheme results in the flow chart shown in Fig. 2. The left part shows the solution of the flow problem as it is implemented in FENFLOSS. On the right side the structural solver and the grid deformation is sketched. The moving grid algorithm may be implemented either directly in FENFLOSS or as a stand alone software called during the coupling step. Here, an extended interpolation method, which is implemented in FENFLOSS, see Lippold [LI06], is used.

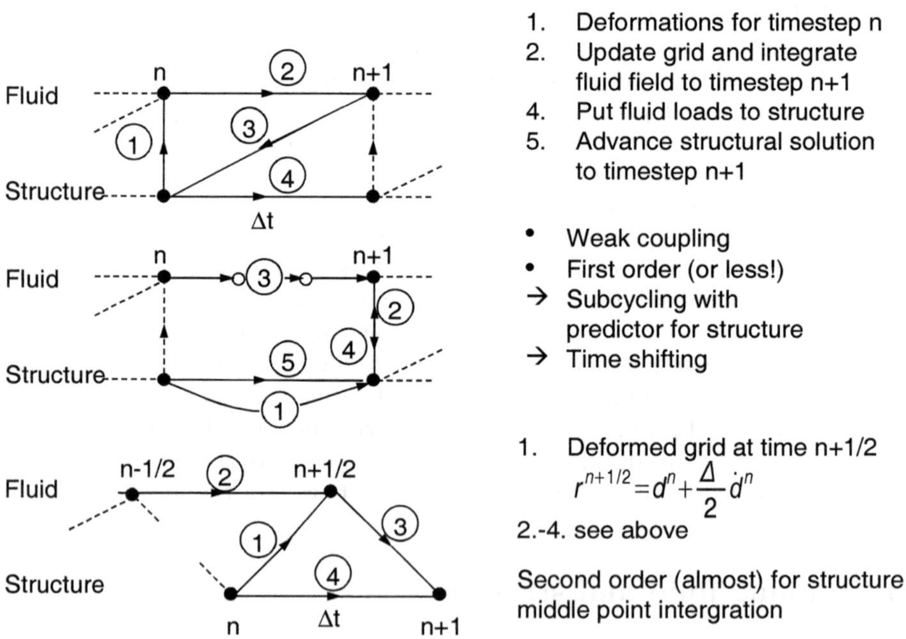

Fig. 1. Exchange schemes for loosely coupled problems

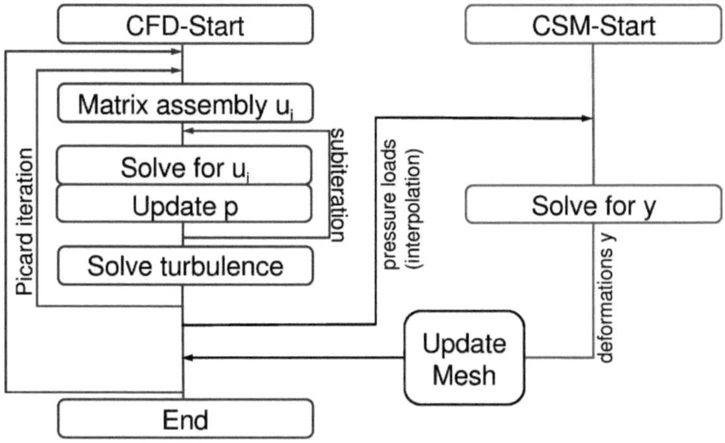

Fig. 2. Coupling scheme

4 Results

First results are obtained with an inclined (10°) NACA0012 wing in 3D. The wing is clamped at one end and free at the other end. The fluid used for these simulations is water with a density of $\rho = 1000$ kg/m^3 at a flow velocity of $v_\infty = 10.0$m/s. The interpolation method, presented above, shows a good performance and yields a good mesh quality for this application.

The fluid grid contains about 100000 nodes. Hence, two processors are sufficient to compute the flow within an acceptable time-frame. For the structural model a grid consisting of one domain with 1500 nodes and linear deformation elements is used.

Figure 3 shows the original and the deformed shape including the surface pressure on the wing in static equilibrium of the fluid-structure system. High pressure is marked with red and low pressure with blue shadings.

Fig. 3. Original and deformed structure in equilibrium state.

Furthermore, simulation results for the flow around the tidal current turbine runner, see Fig. 4, show a good agreement with available measurements of a reduced runner model. The computational grid used for these simulations consists of 2 Million nodes. Constant flow velocities were used at the domain boundaries.

Acknowledgements

Most results and developements were obtained under the HPC-EUROPA project (RII3-CT-2003-506079), with the support of the European Community - Research Infrastructure Action under the FP6 "Structuring the European Research Area" Programme as well as the InGrid-Project (www.ingrid-info.de) funded by the German government.

5 Further investigations and outlook

The next steps that will be taken is the modelling of the real structure of a tidal current turbine runner and the investigation of its static and dynamic behaviour. Furthermore, unsteady flow conditions at the runner inlet will be applied. The vibrations will be studied using unsteady two-way flow-structure simulations.

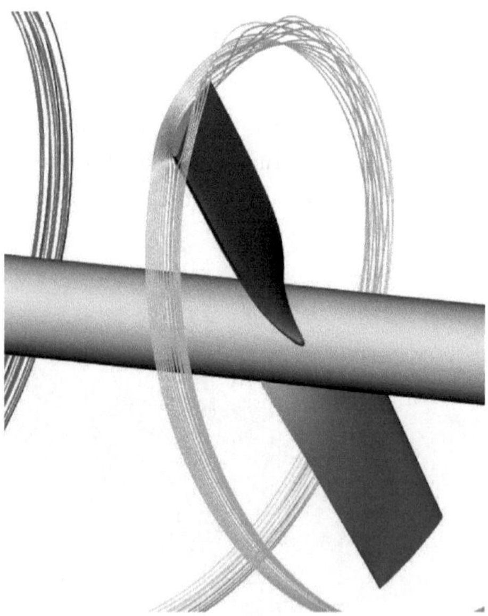

Fig. 4. Pressure distribution (blue -25000 Pa, red 12000 Pa) and streamlines.

References

[FP02] Ferziger, J.H., Perić, M.: Computational Methods for Fluid Dynamics (third Ed.). Springer (2002).

[HU81] Hughes, T.J.R., Liu, W.K., Zimmermann, T.K.: Lagrangian-Eulerian Finite Element Formulation for Viscous Flows. Computer Methods in Applied Mechanics and Engineering. **29**, 329-349 (1981)

[ZI89] Zienkiewicz, O.C., Taylor, R.L.: The Finite Element Method (Vol. I). McGraw-Hill (1989)

[GR99] Gresho, P.M., Sani, R.L.: Incompressible Flow and the Finite Element Method (Vol. I). John Wiley & Sons (1999)

[MAI02] Maihoefer, M.: Effiziente Verfahren zur Berechnung dreidimensionaler Stroemungen mit nichtpassenden Gittern (PhD-Thesis). University of Stuttgart, (2002)

[RU89] Ruprecht, A.: Finite Elemente zur Berechnung dreidimensionaler turbulenter Stroemungen in komplexen Geometrien (PhD-Thesis). University of Stuttgart (1989)

[VVO92] van der Vorst, H.A.: BI-CGSTAB: A fast and smoothly converging variant of BI-CG for the solution of nonsymmetric linear systems. SIAM Journal of Scientific Stat. Computing, **13**, 631-644, (1992)

[BAT89] Batina, J.T.: Unsteady Euler airfoil solutions using unstructured dynamic meshes. AIAA Paper No. 89-0115, AIAA 27th Aerospace Sciences Meeting, Reno, Nevada (1989)

[FAR98-1] Farhat, C., Degand, C., Koobus, B., Lesoinne, M.: Torsional springs for two-dimensional dynamic unstructred fluid meshes. Computational Methods in Applied Mechanics and Engineering, **163**, 231-245 (1998)

[DEG02] Degand, C., Farhat, C.: A three-dimensional torsional spring analogy method for unstructured dynamic meshes. Computers and Structures, **80**, 305-316 (2002)

[FAR98-2] Farhat, C., Lesoinne, M., le Tallec, P.: Load and motion transfer algorithms for fluid-structure interaction problems with non-matching interfaces. Computer Methods in Applied Mechanics and Engineering, **157**, 95-114 (1998)

[LI06] Lippold, F., Fluid-structure interaction in an axial fan. HPC-Europa report (2006)

[KH98] Kjellgren, P., Hyvärinen, J.: An Arbitrary Langrangian-Eulerian Finite Element Method. Computational Mechanics, **21**, 81-90 (1998)

Coupled Problems in Computational Modeling of the Respiratory System

Lena Wiechert[1], Timon Rabczuk[1,2], Michael Gee[1], Robert Metzke[1] and Wolfgang A. Wall[1]

[1] Chair of Computational Mechanics, Technical University of Munich, Boltzmannstraße 15, D-85747 Garching, Germany, {wiechert,gee,metzke,wall}@lnm.mw.tum.de

[2] Mechanical Engineering Department, University of Canterbury, Private Bag 4800, Christchurch 8140, New Zealand, timon.rabczuk@canterbury.ac.nz

Summary. Biomechanical simulations are an important field of application for high performance computing due to the complexity and largeness of involved problems. This paper is concerned with coupled problems in the human respiratory system with emphasis on mechanical ventilation. In this context we focus on the modeling aspects of pulmonary alveoli and the lower airways. Our alveolar model is based on artificially generated random geometries and takes into account realistic tissue behavior as well as interfacial phenomena. On the part of the structural solver a smoothed aggregation algebraic multigrid method is used. For the first four generations of the bronchial tree, a geometry based on human computer tomography scans obtained from in-vivo experiments is employed. With the help of fluid-structure interaction simulations, flow patterns and airway wall stresses for normal breathing and mechanical ventilation of the healthy and diseased lung are investigated.

1 Introduction

Mechanical ventilation of the human lung plays a significant role in medicine, especially in case of patients with acute lung diseases such as ALI (acute lung injury) and ARDS (acute respiratory distress syndrome) where it is known to be a vital supportive therapy. Improper methods of ventilation, however, can cause mechanical overstraining of parenchymal tissue resulting in additional inflammatory injuries. This complication is commonly called ventilator-induced lung injury (VILI) and is responsible for a significant increase in mortality rate.

Since up to now it is unclear how to improve ventilation strategies in order to prevent VILI and thereby minimize mortality, we want to bring a little more light into some of the involved phenomena.

The work described in this paper was done on the basis of our in-house research finite element program BACI covering a wide range of applications in computational mechanics, like e.g. multi-field and multi-scale problems, structural and fluid dynamics, material modeling and finite element technology. The code is parallelized using MPI and runs on a variety of platforms, on single processor systems as well as on clusters.

In the following, we will focus on the modeling aspects of the respiratory system, thereby revealing a possible field of application for high performance computing. Since it is computationally not feasible to study the entire pulmonary system, investigations are restricted to parts of it.

In the second chapter of this paper, our model of pulmonary alveoli accounting for realistic tissue behavior and interfacial phenomena will be introduced. We will present a smoothed aggregation algebraic multigrid method used in our simulations of artificially generated random geometries. The subsequent chapter deals with fluid-structure interaction (FSI) simulations of unsteady incompressible airflow in an airway model of the first four generations based on computer tomography (CT) scans. After a short overview on the governing equations, our computational models and solution techniques, we will present some results of our studies regarding healthy and diseased lungs under normal breathing and mechanical ventilation.

2 Modeling of Pulmonary Alveoli

VILI mainly occurs at the alveolar level of the lungs in terms of primary mechanical and secondary inflammatory injuries. Primary injuries are consequences of alveolar overexpansion or frequent recruitment and derecruitment inducing high shear stresses. Since mechanical stimulation of cells can result in the release of proinflammatory mediators – a phenomenon commonly called mechanotransduction – secondary inflammatory injuries directly follow and can spread over to other organs resulting in multi-organ failure. According to [1] it is therefore crucial to understand alveolar dynamic behavior in order to investigate mechanisms of VILI.

2.1 Labyrinthine Algorithm

Since currently no real geometries of pulmonary alveoli are available for simulations due to the low resolution of conventional imaging techniques, we are interested in finding ways to artificially generate them. For that purpose the labyrinthine algorithm presented in [2] is extended for the application to complex geometries.

A labyrinthine algorithm enables the generation of random pathways through an a priori defined assemblage of base cells. All cells have to be connected to a given starting cell and should be passed only once except in case of branching cells. If a cell is affiliated in the course of the labyrinth creation, it is stored in a queue. In every step, the cell located in front of the queue is set active and thus can create a path to a new cell by randomly choosing one of its neighboring cells that are not already passed. Afterwards, the active cell and the just affiliated new cell are moved to the end of the queue. If, however, the active cell has no unpassed neighbors in the beginning or at the end of the step, it is deleted from the queue. The procedure is repeated until the queue is worn out.

Generated labyrinths should contain no detours for the sake of preserving minimal overall pathlength, a feature of great importance regarding effective gas transport in the lungs. In [2], this is accomplished by introducing the concept of priority directions. Provided that cell geometries are simple, this method works well. In case of tetrakaidecahedral cells, however, it can be shown that the conventional labyrinthine algorithm fails in preserving optimal mean pathlength due to possible diagonal connection of cells.

Hence a new approach was developed enforcing generation of optimal random labyrinths while allowing the starting cell to be arbitrarily located within the given alveolar ensemble. In this connection it has to be explicitly verified that the pathway to the randomly chosen new cell via the currently active cell is an optimal one. If no shorter pathlength via other cells in the queue exists, then the new cell can be connected to the active cell. Otherwise another possible new cell has to be selected from the range of neighbors. In case that either no other neighboring cell is at hand or no other adjacent cell can be affiliated optimally, the subsequent cell in the queue is set active. For a more detailed presentation we refer to [3].

Based on the presented generalized labyrinthine algorithm arbitrarily large alveolar ensembles can be generated and structurally meshed with hexahedral finite elements. An example of a created labyrinthine pathway and the corresponding alveolar geometry is depicted in Fig. 1.

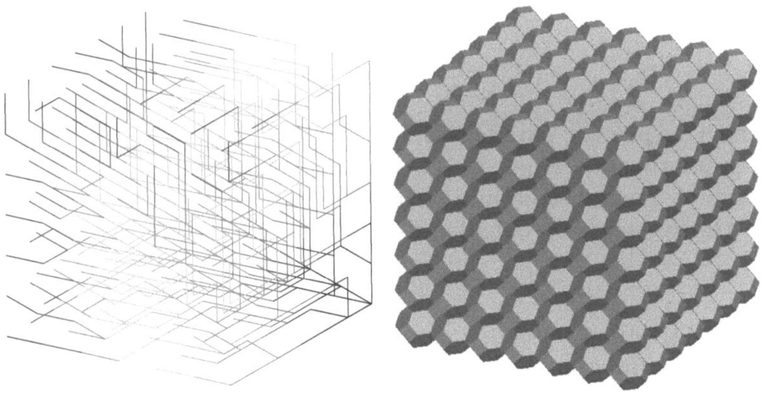

Fig. 1. Left: Connections of centers of tetrakaidecahedral cells based on labyrinthine algorithm. Colours characterize the distance to the starting cell. Right: Corresponding artificial geometry of an alveolar ensemble

2.2 Organization and Modeling of Alveolar Parenchymal Tissue

Alveolar tissue is composed of interstitial cells and the so-called extracellular matrix (ECM) consisting of connective tissue fiber networks and an amorphous ground substance made up mainly of proteoglycans. Since the contribution of interstitial cells to parenchymal mechanics seems to be marginal according to [4], we focus on modeling the behavior of the ECM.

Under the assumption of hyperelasticity, we can describe the mechanical behavior of the ECM via a potential, the so-called strain-energy density function

$$S = 2\frac{\partial W}{\partial C} \tag{1}$$

with $\boldsymbol{S}$, $\boldsymbol{C}$ and W being the second Piola-Kirchhoff stress tensor, the right Cauchy-Green strain tensor and the strain-energy density function, respectively.

The strain-energy function used in the following is based on [5], [6] and [7]. W is composed of functions for the matrix including ground substance and elastin fibers and for the collagen fiber families, each fulfilling the principles of objectivity and material symmetry as well as the requirements of polyconvexity and the stress-free reference state.

For the ground substance, a modification of the isotropic neo-Hookean material model is used

$$W_{gs} = c\left(\frac{I_1}{I_3^{\frac{1}{3}}} - 3\right) \qquad c > 0 \tag{2}$$

with c representing a shear-modulus-like parameter and I_1 and I_3 being the first and third invariants of the right Cauchy-Green tensor.

Additionally, a penalty function is employed for the enforcement of incompressibility

$$W_{pen} = \epsilon\left(I_3^{\gamma} + \frac{1}{I_3^{\gamma}} - 2\right) \qquad \epsilon > 0, \quad \gamma > 1 \tag{3}$$

with ϵ and γ as penalty parameters.

The orientation of collagen fibers varies according to an orientation density distribution. A general structural tensor $\boldsymbol{H}$ is introduced (exemplarily for the case of a transversely isotropic material with preferred direction $\boldsymbol{e}_3$)

$$\boldsymbol{H} = \kappa \boldsymbol{I} + (1 - 3\kappa)\, \boldsymbol{e}_3 \otimes \boldsymbol{e}_3. \tag{4}$$

In this connection κ represents a parameter derived from the orientation density distribution function $\rho(\theta)$

$$\kappa = \frac{1}{4}\int_0^{\Pi} \rho(\theta) sin^3(\theta) d\theta. \tag{5}$$

Fiber orientation in alveolar tissue seems to be rather random, hence lung parenchyma can be treated as a homogeneous, isotropic continuum following

[8]. In that case κ is equal to $\frac{1}{3}$.
A new invariant K of the right Cauchy-Green tensor is defined by

$$K = tr\left(\boldsymbol{CH}\right). \tag{6}$$

The strain-energy function of the non-linear collagen fiber network then reads

$$W_{fib} = \begin{cases} \frac{k_1}{2k_2}\left[exp\left[k_2\left(K-1\right)^2\right]-1\right] & \text{for } K \geq 1 \\ 0 & \text{for } K < 1 \end{cases} \tag{7}$$

with $k_1 \geq 0$ as a stress-like parameter and $k_2 > 0$ as a dimensionless parameter.

Finally, our strain-energy density function takes the following form

$$W = W_{gs} + W_{fib} + W_{pen}. \tag{8}$$

Unfortunately only very few experimental data are published regarding the mechanical behavior of alveolar tissue. To the authors' knowledge, no material parameters for single alveolar walls are derivable since up to now only parenchyma was tested (see for example [9], [10], [11]). For that purpose, we fitted the material model to experimental data published in [12] for lung tissue sheets.

2.3 Modeling of Interfacial Phenomena due to Surfactant

Pulmonary alveoli are covered by a thin, continuous liquid lining with a monomolecular layer of surface active agents (the so-called surfactant) on top of it. It is widely believed that the resulting interfacial phenomena contribute significantly to the lungs' retraction force. That is why taking into account surface stresses appearing in the liquid lining of alveoli is of significant importance.

For our model of pulmonary alveoli we are not primarily interested in the liquid lining itself but rather in its influence on the overall mechanical behavior. Therefore we do not model the aquateous hypophase and the surfactant layer explicitly but consider the resulting surface phenomena in the interfacial structural finite element (FE) nodes of the alveolar walls by enriching them with corresponding internal force and tangent stiffness terms (cf. Fig. 2).

The infinitesimal internal work done by the surface stress γ reads

$$dW_{surf} = \gamma(S)\mathrm{d}S \tag{9}$$

with dS being the infinitesimal change in interfacial area. Consequently we obtain for the overall work

$$W_{surf} = \int_{S_0}^{S} \gamma(S^*)\mathrm{d}S^*. \tag{10}$$

The variation of the overall work with respect to the nodal displacements $\mathbf{d}$ then takes the following form

$$\delta W_{surf} = \left(\frac{\partial}{\partial \mathbf{d}} \left(\int_{S_0}^{S} \gamma(S^*) \mathrm{d}S^* \right) \right)^T \delta \mathbf{d} = \left(\frac{\partial}{\partial S} \left(\int_{S_0}^{S} \gamma(S^*) \mathrm{d}S^* \right) \frac{\partial S}{\partial \mathbf{d}} \right)^T \delta \mathbf{d}. \tag{11}$$

Using

$$\frac{\mathrm{d}}{\mathrm{d}x} \int_a^x f(t) \mathrm{d}t = f(x) \tag{12}$$

yields

$$\delta W_{surf} = \gamma(S) \left(\frac{\partial S}{\partial \mathbf{d}} \right)^T \delta \mathbf{d} = \mathbf{f}_{surf}^T \delta \mathbf{d}. \tag{13}$$

with the internal force vector

$$\mathbf{f}_{surf} = \gamma(S) \frac{\partial S}{\partial \mathbf{d}}. \tag{14}$$

The consistent tangent stiffness matrix derived by linearization of (14) therefore reads

$$\mathbf{A}_{surf} = \gamma(S) \frac{\partial}{\partial \mathbf{d}} \left(\frac{\partial S}{\partial \mathbf{d}} \right)^T + \frac{\partial \gamma(S)}{\partial \mathbf{d}} \left(\frac{\partial S}{\partial \mathbf{d}} \right)^T. \tag{15}$$

For details refer also to [13] where, however, an additional surface stress element was introduced in contrast to the above mentioned concept of enriching the interfacial structural nodes.

Unlike e.g. water with its constant surface tension, surfactant exhibits a dynamically varying surface stress γ depending on the interfacial concentration of surfactant molecules. We use the adsorption-limited surfactant model developed in [14] to capture this dynamic behavior.

It is noteworthy that no scaling techniques as in [15], where a single explicit function for γ is used, are necessary, since the employed surfactant model itself delivers the corresponding surface stress depending on both input parameter and dynamic data.

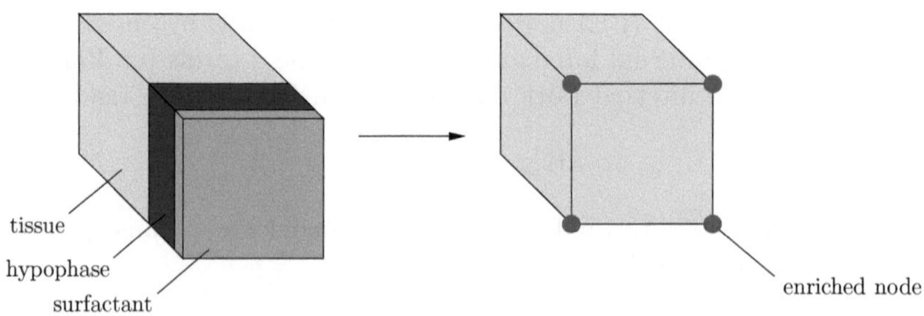

Fig. 2. Left: Actual configuration. Right: Simplified FE model

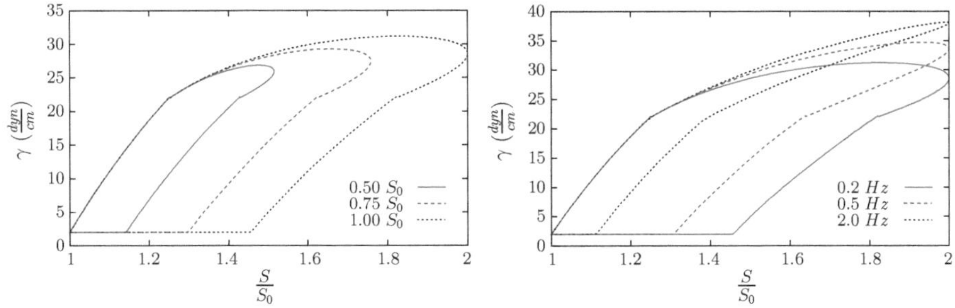

Fig. 3. Dynamic behavior of surfactant model for different sinusoidal amplitudes and different frequencies

For illustrative purposes, we have plotted the course of γ for different frequencies and amplitudes of area change if interfacial area is changed sinusoidally in Fig. 3. First results of simulations with single alveolar models are presented in Fig. 4. Clearly, interfacial phenomena play a significant role for the overall mechanical behavior of pulmonary alveoli as can be seen in the differences of overall displacements during sinusoidal ventilation. The comparison of the results for surfactant and water demonstrates the efficiency of surfactant in decreasing the surface tension of the aquateous hypophase, thereby reducing work of breathing and stabilizing alveoli at low lung volumes. Since interfacial phenomena play a more distinct role for geometries exhibiting a larger curvature, the changes in stiffness are more pronounced for the

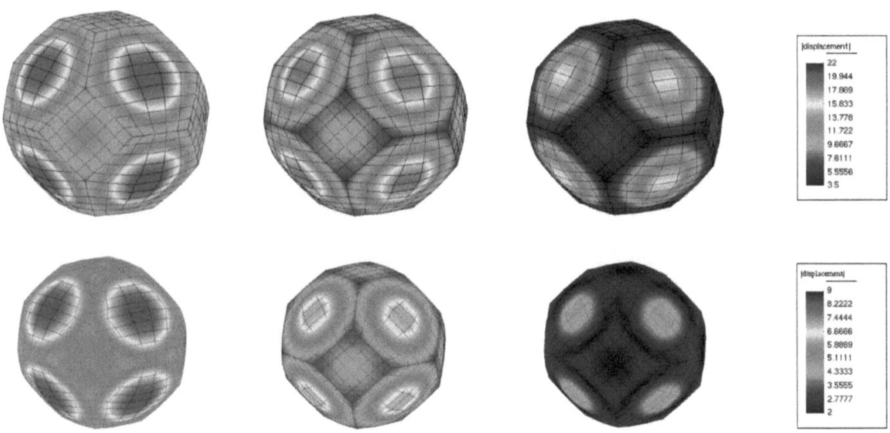

Fig. 4. Absolute displacements for sinusoidal ventilation of single alveoli with different interfacial configurations for two characteristic geometric sizes. Top: Characteristic geometric size comparable to human alveoli. Bottom: Characteristic geometric size comparable to small animals like e.g. hamsters. Left: No interfacial phenomena. Middle: Dynamic surfactant model. Right: Water with constant surface tension

smaller alveoli shown in the bottom of Fig. 4. Therefore differences between species have to be taken into account when e.g. comparing experimental data.

2.4 Structural Dynamics Solver

Since large alveolar ensembles are analyzed in our studies, an efficient solver is sorely needed. In this section, we give an overview of multigrid as well as a brief introduction to smoothed aggregation multigrid (SA), which we use in parallel versions in order to solve the resulting systems of equations.

Multigrid Overview

Multigrid methods are among the most efficient iterative algorithms for solving the linear system, $\mathbf{Ad} = \mathbf{f}$, associated with elliptic partial differential equations. The basic idea is to damp errors by utilizing multiple resolutions in the iterative scheme. High-energy (or oscillatory) components are efficiently reduced through a simple smoothing procedure, while the low-energy (or smooth) components are tackled using an auxiliary lower resolution version of the problem (coarse grid). The idea is applied recursively on the next coarser level. An example multigrid iteration is given in Algorithm 1 to solve

$$\mathbf{A}_1\mathbf{d}_1 = \mathbf{f}_1. \tag{16}$$

The two operators needed to fully specify the multigrid method are the relaxation procedures, $\mathbf{R}_k$, $k = 1, \ldots, N_{levels}$, and the grid transfers, $\mathbf{P}_k$, $k = 2, \ldots, N_{levels}$. Note that $\mathbf{P}_k$ is an interpolation operator that transfers grid information from level $k+1$ to level k. The coarse grid discretization operator $\mathbf{A}_{k+1}$ $(k \geq 1)$ can be specified by the Galerkin product

$$\mathbf{A}_{k+1} = \mathbf{P}_k^T\mathbf{A}_k\mathbf{P}_k. \tag{17}$$

Algorithm 1 Multigrid V-cycle consisting of N_{levels} grids to solve $\mathbf{A}_1\mathbf{d}_1 = \mathbf{f}_1$.

{Solve $\mathbf{A}_k\mathbf{d}_k = \mathbf{f}_k$}
procedure multilevel($\mathbf{A}_k, \mathbf{f}_k, \mathbf{d}_k, k$)
if $(k \neq N_{levels})$ **then**
    $\mathbf{d}_k = \mathbf{R}_k(\mathbf{A}_k, \mathbf{f}_k, \mathbf{d}_k)$;
    $\mathbf{r}_k = \mathbf{f}_k - \mathbf{A}_k\mathbf{d}_k$;
    $\mathbf{A}_{k+1} = \mathbf{P}_k^T\mathbf{A}_k\mathbf{P}_k$;
    $\mathbf{d}_{k+1} = 0$;
    multilevel($\mathbf{A}_{k+1}, \mathbf{P}_k^T\mathbf{r}_k, \mathbf{d}_{k+1}, k+1$);
    $\mathbf{d}_k = \mathbf{d}_k + \mathbf{P}_k\mathbf{d}_{k+1}$;
    $\mathbf{d}_k = \mathbf{R}_k(\mathbf{A}_k, \mathbf{f}_k, \mathbf{d}_k)$;
else
    $\mathbf{d}_k = \mathbf{A}_k^{-1}\mathbf{f}_k$;
end if

The key to fast convergence is the complementary nature of these two operators. That is, errors not reduced by $\mathbf{R}_k$ must be well interpolated by $\mathbf{P}_k$.

Even though constructing multigrid methods via algebraic concepts presents certain challenges, algebraic multigrid (AMG) can be used for several problem classes without requiring a major effort for each application. Here, we focus on a strategy to determine the $\mathbf{P}_k$'s based on algebraic principles. It is assumed that $\mathbf{A}_1$ and $\mathbf{f}_1$ are given.

Smoothed Aggregation Multigrid

We briefly describe a special type of algebraic multigrid called smoothed aggregation multigrid. For a more detailed description, see [16] and [17]. Specifically, we focus on the construction of smoothed aggregation interpolation operators $\mathbf{P}_k$ $(k \geq 1)$.

The interpolation $\mathbf{P}_k$ is defined as a product of a given prolongator smoother $\mathbf{S}_k$ and a tentative prolongator $\hat{\mathbf{P}}_k$

$$\mathbf{P}_k = \mathbf{S}_k \hat{\mathbf{P}}_k \quad , \quad k = 1, ..., N_{levels} - 1 \; . \tag{18}$$

The basic idea of the tentative prolongator is that it must accurately interpolate certain near null space (kernel) components of the discrete operator $\mathbf{A}_k$. Once constructed, the tentative prolongator is then improved by the prolongator smoother in a way that reduces the energy or smoothes the basis functions associated with the tentative prolongator. Constructing $\hat{\mathbf{P}}_k$ consists of deriving its sparsity pattern and then specifying its nonzero values. The sparsity pattern is determined by decomposing the set of degrees of freedoms associated with $\mathbf{A}_k$ into a set of so called *aggregates* $\mathcal{A}_i^k$, such that

$$\bigcup_{i=1}^{N_{k+1}} \mathcal{A}_i^k = \{1, ..., N_k\} \; , \; \mathcal{A}_i^k \cap \mathcal{A}_j^k = 0 \; , \; 1 \leq i < j \leq N_{k+1} \tag{19}$$

where N_k denotes the number of nodal blocks on level k.

The ideal ith aggregate $\mathcal{A}_i^k$ on level k would formally be defined by

$$\mathcal{A}_i^k = \{j_i\} \cup \{\mathbb{N}(j_i)\} \tag{20}$$

where j_i is a so called root nodal block in $\mathbf{A}_k$ and

$$\mathbb{N}(j) = \{n : ||(\mathbf{A}_k)_{jn}|| \neq 0 \quad \text{and} \quad n \neq j\} \tag{21}$$

is the neighborhood of nodal blocks, $(\mathbf{A}_k)'_{jn}s$, that share a nonzero off-diagonal block entry with node i. While ideal aggregates would only consist of a root nodal block and its immediate neighboring blocks, it is usually not possible to entirely decompose a problem into ideal aggregates. Instead, some aggregates which are a little larger or smaller than an ideal aggregate must be created.

For this paper, each nodal block contains m_k degrees of freedom where for simplicity we assume that the nodal block size m_k is constant throughout $\mathbf{A}_k$.

With N_k denoting the number of nodal blocks in the system on level k this results in $n_k = N_k m_k$ being the dimension of $\mathbf{A}_k$.

Aggregates $\mathcal{A}_i^k$ can be formed based on the connectivity and the strength of the connections in $\mathbf{A}_k$. For an overview of serial and parallel aggregation techniques, we refer to [18].

Although we speak of 'nodal blocks' and 'connectivity' in an analogy to finite element discretizations here, it shall be stressed that a node is a strictly algebraic entity consisting of a list of degrees of freedom. In fact, this analogy is only possible on the finest level; on coarser levels, $k > 1$, a node denotes a set of degrees of freedom associated with the coarse basis functions whose support contain the same aggregate on level $k-1$. Hence, each aggregate $\mathcal{A}_i^k$ on level k gives rise to one node on level $k+1$ and each degree of freedom (DOF) associated with that node is a coefficient of a particular basis function associated with $\mathcal{A}_i^k$.

Populating the sparsity structure of $\hat{\mathbf{P}}_k$ derived from aggregation with appropriate values is the second step. This is done using a matrix $\mathbf{B}_k$ which represents the near null space of $\mathbf{A}_k$. On the finest mesh, it is assumed that $\mathbf{B}_k$ is given and that it satisfies $\tilde{\mathbf{A}}_k \mathbf{B}_k = \mathbf{0}$, where $\tilde{\mathbf{A}}_k$ differs from $\mathbf{A}_k$ in that Dirichlet boundary conditions are replaced by natural boundary conditions. Tentative prolongators and a coarse representation of the near null space are constructed simultaneously and recursively to satisfy

$$\mathbf{B}_k = \hat{\mathbf{P}}_k \mathbf{B}_{k+1} \ , \ \hat{\mathbf{P}}_k^T \hat{\mathbf{P}}_k = \mathbf{I} \ , \ k = 1, ..., N_{levels} - 1 \ . \tag{22}$$

This guarantees exact interpolation of the near null space by the tentative prolongator. To do this, each nodal aggregate is assigned a set of columns of $\hat{\mathbf{P}}_k$ with a sparsity structure that is disjoint from all other columns. We define

$$I_k^m(i,j) = \begin{cases} 1 & \text{if } i = j, \ i \text{ a DOF in } l^{th} \text{ nodal block with } l \in \mathcal{A}_m^k \\ 0 & \text{otherwise} \end{cases} \tag{23}$$

to be the aggregatewise identity. Then,

$$\mathbf{B}_k^m = \mathbf{I}_k^m \mathbf{B}_k \ , \ m = 1, ..., N_{k+1} \tag{24}$$

is an aggregate-local block of the near nullspace. $\mathbf{B}_k$ is restricted to individual aggregates using (24) to form

$$\bar{\mathbf{B}}_k = \begin{pmatrix} \mathbf{B}_k^1 & & & \\ & \mathbf{B}_k^2 & & \\ & & \ddots & \\ & & & \mathbf{B}_k^{N_{k+1}} \end{pmatrix} , \tag{25}$$

and an aggregate-local orthonormalization problem

$$\mathbf{B}_k^i = \mathbf{Q}_k^i \mathbf{R}_k^i \ , \ i = 1, ..., N_{k+1} \tag{26}$$

is solved applying a QR algorithm. The resulting orthonormal basis forms the values of a block column of

$$\hat{\mathbf{P}}_k = \begin{pmatrix} \mathbf{Q}_k^1 & & & \\ & \mathbf{Q}_k^2 & & \\ & & \ddots & \\ & & & \mathbf{Q}_k^{N_{k+1}} \end{pmatrix}, \tag{27}$$

while the coefficients $\mathbf{R}^i$ define the coarse representation of the near null space

$$\mathbf{B}_{k+1} = \begin{pmatrix} \mathbf{R}_k^1 \\ \mathbf{R}_k^2 \\ \vdots \\ \mathbf{R}_k^{N_{k+1}} \end{pmatrix}. \tag{28}$$

The exact interpolation of the near null space, (22), is considered to be an essential property of an AMG grid transfer. It implies that error components in the near null space (which are not damped by conventional smoothers) are accurately approximated (and therefore eliminated) on coarse meshes. Unfortunately, (22) is not sufficient for an effective multigrid cycle. In addition, one needs to also bound the energy of the grid transfer basis functions. One way to do this, the tentative prolongator is improved via a prolongator smoother. The usual choice for the prolongator smoother is

$$\mathbf{S}_k = \left(\mathbf{I} - \frac{4}{3\lambda_k}\mathbf{D}^{-1}\mathbf{A}_k\right) \tag{29}$$

where $\mathbf{D} = \mathrm{diag}(\mathbf{A}_k)$ and λ_k is an upper bound on the spectral radius of the matrix on level k; $\rho\left(\mathbf{D}^{-1}\mathbf{A}_k\right) \leq \lambda_k$. This corresponds to a damped Jacobi smoothing procedure applied to each column of the tentative prolongator.

It can be easily shown, that (22) holds for the smoothed prolongator. In particular,

$$\begin{aligned} \mathbf{P}_k\mathbf{B}_{k+1} &= (\mathbf{I} - \omega\mathbf{D}^{-1}\mathbf{A}_k)\hat{\mathbf{P}}_k\mathbf{B}_{k+1} \\ &= (\mathbf{I} - \omega\mathbf{D}^{-1}\mathbf{A}_k)\mathbf{B}_k \\ &= \mathbf{B}_k \; as \; \mathbf{A}_k\mathbf{B}_k = \mathbf{0} \end{aligned} \tag{30}$$

where $\omega = 4/(3\lambda_k)$. It is emphasized, that once $\hat{\mathbf{P}}_k$ is chosen, the sparsity pattern of $\mathbf{P}_k$ is defined.

With $\mathbf{A}_1$, $\mathbf{B}_1$ and $\mathbf{b}_1$ given, the setup of the standard isotropic smoothed aggregation multigrid hierarchy can be performed using (19), (26), (22), (18) and finally (17). For a more detailed discussion on smoothed aggregation we refer to [16], [17], [18] and [19].

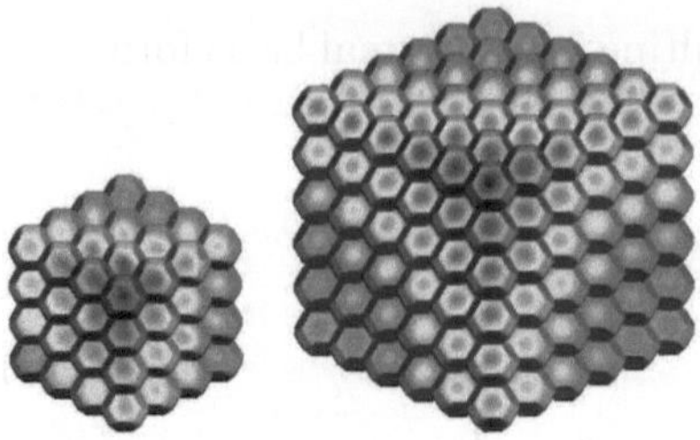

DOF	184,689	960,222
processors	16	32
CG iterations per solve	170	164
solver time per call	8.9 s	25.8 s
AMG setup time	1.5 s	2.2 s

Fig. 5. Left: Alveolar geometries for simulations. Right: Solver details

Application to pulmonary alveoli

We used a SA preconditioned conjugate gradient method [18] with four grids on two alveolar geometries depicted in Fig. 5. Chebychev smoothers were employed on the finer grids, whereas an LU decomposition was applied on the coarsest grid. Convergence was assumed when

$$\frac{||r||}{||r_0||} < 10^{-6} \tag{31}$$

with $||r||$ and $||r_0||$ as L^2-norm of the current and initial residuum, respectively. The number of DOF as well as details concerning solver and setup times and number of iterations per solve are summarized in Fig. 5. The time per solver call for both simulations is given in Fig. 6. It is noteworthy that $O(n)$ overall scalability is achieved with the presented approach for these examples with complex geometries.

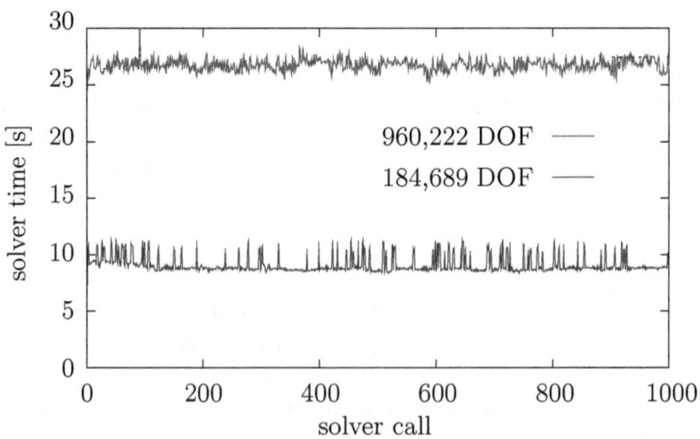

Fig. 6. Solution time per solver call

3 Fluid-Structure Interaction of Lower Airways

Currently appropriate boundary conditions for pulmonary alveoli are not yet known. To bridge the gap between the respiratory zone where VILI occurs and the ventilator where pressure and flow are known, it is essential to understand the airflow in the respiratory system. In a first step we have studied flow in a CT-based geometry of the first four generations of lower airways [20]. In a second step we also included flexibility of airway walls and investigated fluid-structure interaction effects [21]. The CT scans are obtained from in-vivo experiments of patients under normal breathing and mechanical ventilation.

3.1 Governing Equations

We assume an incompressible Newtonian fluid under transient flow conditions. The underlying governing equation is the Navier Stokes equation formulated on time dependent domains.

$$\left.\frac{\partial \boldsymbol{u}}{\partial t}\right|_{\boldsymbol{\chi}} + \left(\boldsymbol{u} - \boldsymbol{u}^G\right) \cdot \boldsymbol{\nabla} \boldsymbol{u} - 2\nu \boldsymbol{\nabla} \cdot \boldsymbol{\varepsilon}(\boldsymbol{u}) + \nabla p = \boldsymbol{f}^F \qquad \text{in } \Omega^{\mathrm{F}} \tag{32}$$

$$\boldsymbol{\nabla} \cdot \boldsymbol{u} = 0 \qquad \text{in } \Omega^{\mathrm{F}} \tag{33}$$

where $\boldsymbol{u}$ is the velocity vector, $\boldsymbol{u}^G$ is the grid velocity vector, p is the pressure and $\boldsymbol{f}^F$ is the body force vector. A superimposed F refers to the fluid domain and $\boldsymbol{\nabla}$ denotes the nabla operator. The parameter $\nu = \mu/\rho^F$ is the kinematic viscosity with viscosity μ and fluid density ρ^F. The kinematic pressure is represented by p where $\bar{p} = p\,\rho^F$ is the physical pressure within the fluid field. The balance of linear momentum (32) refers to a deforming arbitrary Lagrangean Eulerian (ALE) frame of reference denoted by $\boldsymbol{\chi}$ where the geometrical location of a mesh point is obtained from the unique mapping $\boldsymbol{x} = \varphi(\boldsymbol{\chi}, t)$.

The stress tensor of a Newtonian fluid is given by

$$\boldsymbol{\sigma}^F = -\bar{p}\boldsymbol{I} + 2\mu\boldsymbol{\varepsilon}(\boldsymbol{u}) \tag{34}$$

with the compatibility condition

$$\boldsymbol{\varepsilon}(\boldsymbol{u}) = \frac{1}{2}\left(\boldsymbol{\nabla}\boldsymbol{u} + \boldsymbol{\nabla}\boldsymbol{u}^T\right) \tag{35}$$

where $\boldsymbol{\varepsilon}$ is the rate of deformation tensor. The initial and boundary conditions are

$$\begin{aligned} \boldsymbol{u}(t=0) &= \boldsymbol{u}_0 && \text{in } \Omega^{\mathrm{F}} \\ \boldsymbol{u} &= \hat{\boldsymbol{u}} && \text{on } \Gamma_{\mathrm{D}}^{\mathrm{F}} \\ \boldsymbol{\sigma} \cdot \boldsymbol{n} &= \hat{\boldsymbol{h}}^F && \text{on } \Gamma_{\mathrm{N}}^{\mathrm{F}} \end{aligned} \tag{36}$$

where $\Gamma_{\mathrm{D}}^{\mathrm{F}}$ and $\Gamma_{\mathrm{N}}^{\mathrm{F}}$ denote the Dirichlet and Neumann partition of the fluid boundary, respectively, with normal $\boldsymbol{n}$, with $\Gamma_{\mathrm{D}}^{\mathrm{F}} \bigcap \Gamma_{\mathrm{N}}^{\mathrm{F}} = \emptyset$. $\hat{\boldsymbol{u}}$ and $\hat{\boldsymbol{h}}^F$ are the prescribed velocities and tractions.

The governing equation in the solid domain is the linear momentum equation given by

$$\rho^S \ddot{\boldsymbol{d}} = \boldsymbol{\nabla}_0 \cdot \boldsymbol{S} + \rho^S \boldsymbol{f}^S \qquad \text{in } \Omega^{\mathrm{S}} \tag{37}$$

where $\boldsymbol{d}$ are the displacements, the superimposed dot denotes material time derivatives and the superimposed S refers to the solid domain. ρ^S and $\boldsymbol{f}^S$ represent the density and body force, respectively.

The initial and boundary conditions are

$$\begin{aligned} \boldsymbol{d}(t=0) &= \boldsymbol{d}_0 && \text{in } \Omega^{\mathrm{S}} \\ \dot{\boldsymbol{d}}(t=0) &= \dot{\boldsymbol{d}}_0 && \text{in } \Omega^{\mathrm{S}} \\ \boldsymbol{d} &= \hat{\boldsymbol{d}} && \text{on } \Gamma_{\mathrm{D}}^{\mathrm{S}} \\ \boldsymbol{S} \cdot \boldsymbol{n} &= \hat{\boldsymbol{h}}^S && \text{on } \Gamma_{\mathrm{N}}^{\mathrm{S}}, \end{aligned} \tag{38}$$

where $\Gamma_{\mathrm{D}}^{\mathrm{S}}$ and $\Gamma_{\mathrm{N}}^{\mathrm{S}}$ denote the Dirichlet and Neumann partition of the structural boundary, respectively, with $\Gamma_{\mathrm{D}}^{\mathrm{S}} \bigcap \Gamma_{\mathrm{N}}^{\mathrm{S}} = \emptyset$. $\hat{\boldsymbol{d}}$ and $\hat{\boldsymbol{h}}^S$ are the prescribed displacements and tractions.

Within this paper, we will account for geometrical nonlinearities but we will assume the material to be linear elastic. Since we expect only small strains and due to lack of experimental data, this assumption seems to be fair for first studies.

3.2 Partitioned Solution Approach

A partitioned solution approach is used based on a domain decomposition that separates the fluid and the solid. The surface of the solid Γ^{S} acts hereby as a natural coupling interface Γ_{FSI} across which displacement and traction continuity at all discrete time steps has to be fulfilled:

$$\boldsymbol{d}_\Gamma(t) \cdot \boldsymbol{n} = \boldsymbol{r}_\Gamma(t) \cdot \boldsymbol{n} \qquad \text{and} \qquad \boldsymbol{u}_\Gamma(t) \cdot \boldsymbol{n} = \boldsymbol{u}_\Gamma^G(t) \cdot \boldsymbol{n} = \left.\frac{\partial \boldsymbol{r}_\Gamma(t)}{\partial t}\right|_{\boldsymbol{x}} \cdot \boldsymbol{n} \tag{39}$$

$$\boldsymbol{\sigma}_\Gamma^S(t) \cdot \boldsymbol{n} = \boldsymbol{\sigma}_\Gamma^F(t) \cdot \boldsymbol{n} \tag{40}$$

where $\boldsymbol{r}$ are the displacements of the fluid mesh and $\boldsymbol{n}$ is the unit normal on the interface. Satisfying the kinematic continuity leads to mass conservation at Γ_{FSI}, satisfying the dynamic continuity yields conservation of linear momentum, and energy conservation finally requires to simultaneously satisfy both continuity equations.

The algorithmic framework of the partitioned FSI analysis is discussed in detail elsewhere, cf. e.g. [22], [23], [24] and[25].

3.3 COMPUTATIONAL MODEL

In the fluid domain, we used linear tetrahedral elements with GLS stabilization. The airways are discretized with 7-parameter triangular shell elements (cf. [26], [27], [28]). We refined the mesh from 110,000 up to 520,000 fluid elements and 50,000 to 295,000 shell elements, respectively, until the calculated mass flow rate was within a tolerance of 1%.

Time integration was done with a one-step-theta method with fixed-point iteration and $\theta = 2/3$. For the fluid, we employed a generalized minimal residual (GMRES) iterative solver with ILU-preconditioning.

We study normal breathing under moderate activity conditions with a tidal volume of $2l$ and a breathing cycle of $4s$, i.e. $2s$ inspiration and $2s$ expiration. Moreover, we consider mechanical ventilation where experimental data from the respirator is available, see Fig. 7. A pressure-time history can be applied at the outlets such that the desired tidal volume is obtained. For the case of normal breathing, the pressure-time history at the outlets is sinusoidal, negative at inspiration and positive at expiration as it occurs in “reality”. The advantage is that inspiration and expiration can be handled quite naturally within one computation. The difficulty is to calibrate the boundary conditions such that the desired tidal volume is obtained which is an iterative procedure.

To investigate airflow in the diseased lung, non-uniform boundary conditions are assumed. For that purpose, we set the pressure outlet boundary conditions consistently twice and three-times higher on the left lobe of the lung as compared to the right lobe. This should model a higher stiffness resulting from collapsed or highly damaged parts of lower airway generations.

3.4 Results

Normal Breathing – Healthy Lung

At inspiration the flow in the right bronchus exhibits a skew pattern towards the inner wall, whereas the left main bronchus shows an M-shape, see Fig. 8.

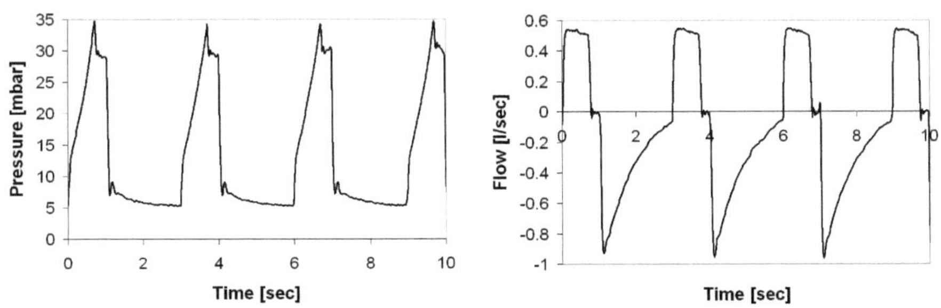

Fig. 7. Pressure-time and flow-time history of the respirator for the mechanically ventilated lung

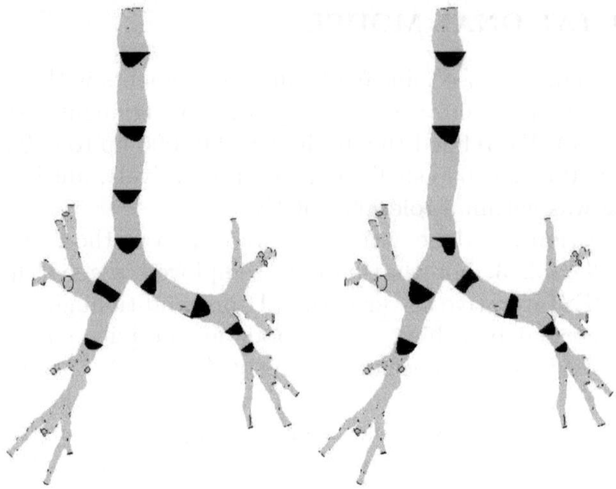

Fig. 8. Total flow structures at different cross sections for the healthy lung under normal beathing (left) and the diseased lung under mechanical ventilation (right)

The overall flow and stress distribution at inspiratory peak flow are shown in Fig. 9. The flow pattern is similar in the entire breathing cycle with more or less uniform secondary flow intensities except at the transition from inspiration to expiration. Stresses in the airway model are highest in the trachea as well as at bifurcation points. Due to the imposed boundary conditions,

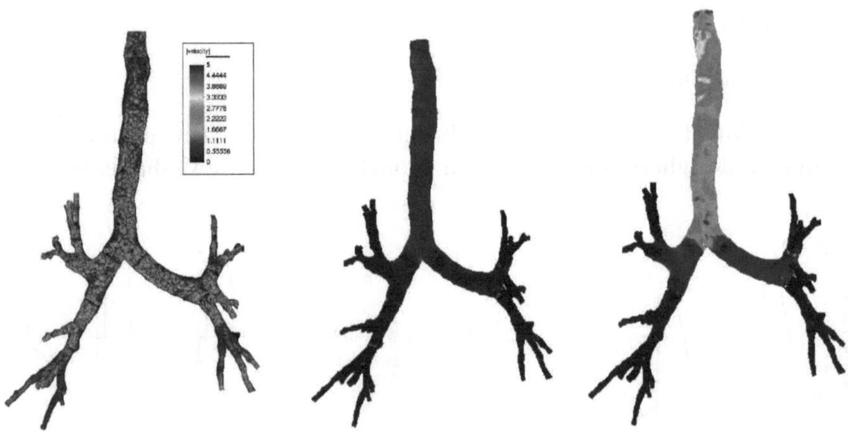

(a) normal breathing (b) normal breathing (c) mechanical ventilation

Fig. 9. Flow and principal tensile stress distribution in the airways of the healthy lung at inspiratory peak flow rate under normal breathing and mechanical ventilation

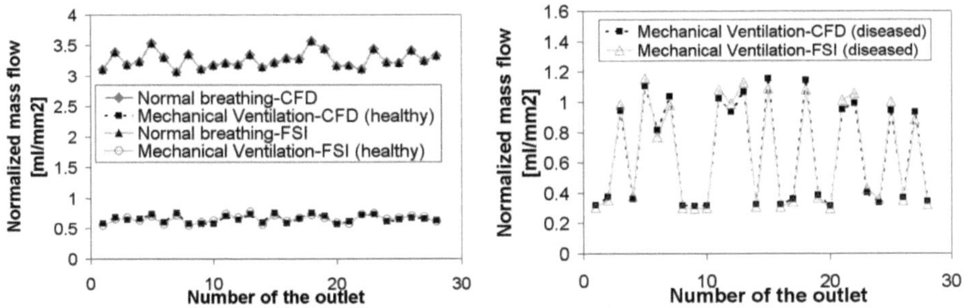

(a) healthy lung (b) diseased lung

Fig. 10. Normalized flow distribution at the outlets under normal breathing and mechanical ventilation for the healthy and diseased lung

velocity and stress distributions as well as normalized mass flow through the outlets are uniform as can be seen in Figs. 9 and 10.

Similar results are obtained for pure computational fluid dynamics (CFD) simulations where the airway walls are assumed to be rigid and not moving (cf. [20]). However, differences regarding secondary flow pattern can be observed between FSI and pure CFD simulations. The largest deviations occur in the fourth generation and range around 17% at selected cross sections.

Mechanical Ventilation – Healthy Lung

Airflow patterns under mechanical ventilation quantitatively differ from normal breathing because of the shorter inspiration time, the different pressure-time history curve and the smaller tidal volume. Despite the different breathing patterns, the principal flow structure is qualitatively quite similar in the trachea and the main bronchi. However, flow patterns after generation 2 are different particularly with respect to secondary flow. The stress distribution of the healthy lung under mechanical ventilation is shown on the right hand side of Fig. 9. Again, due to the imposed boundary conditions, stress distributions as well as normalized mass flow through the outlets are uniform as can be seen in Figs. 9 and 10.

Airflow during expiration differs significantly from inspiratory flow in contrast to normal breathing. At the end of the expiration, the pressure is set almost instantly to the positive endexpiratory pressure (PEEP) value of the ventilator. This results in a high peak flow rate right at the beginning of the expiration, see Fig. 7. The peak flow at this time is more than twice as high as the maximum peak flow rate under inspiration. The flow at the beginning of expiration is unsteady with a significant increase in secondary flow intensity. At the middle of the expiration cycle, airflow becomes quasi-steady and stresses in the airway walls as well as secondary flow intensities decrease again. The bulk of the inspirated tidal air volume is already expirated at that time.

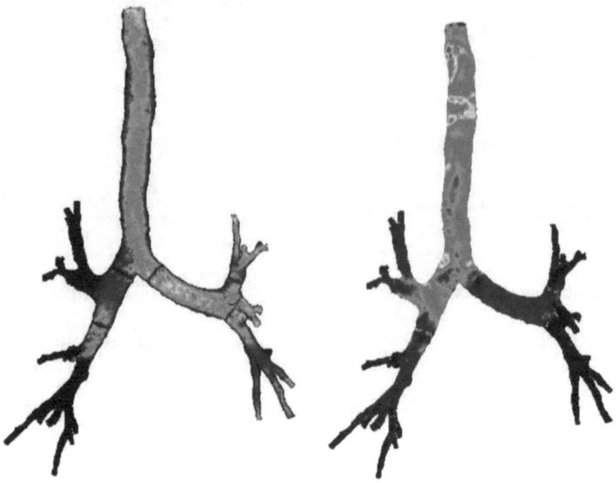

(a) velocity (b) stress

Fig. 11. Flow and principal tensile stress distribution in the airways of the diseased lung at inspiratory peak flow rate under mechanical ventilation

Mechanical Ventilation – Diseased Lung

Airflow structures obtained for diseased lungs differ significantly from those for healthy lungs in inspiration as well as in expiration. Flow and stress distributions are no longer uniform because of the different imposed pressure outlet boundary conditions, see Fig. 11. Only 25% of the tidal air volume enters the diseased part of the lung, i.e. the left lobe. The normalized mass flow calculated at every outlet of the airway model is shown in Fig. 10.

In Fig. 8 the differences in airflow structures of the healthy and diseased lung in terms of discrete velocity profiles during inspiratory flow are visualized.

The secondary flow structures are not only quite different from the healthy lung but they also deviate from the results for diseased lungs obtained in [20] where the airway walls were assumed to be rigid and nonmoving. Thus FSI-forces are significantly larger in simulations of the diseased lung and the influence of airway wall flexibility on the flow should therefore not be neglected.

In general, airway wall stresses are larger in the diseased compared to the healthy lung. Interestingly, stresses in the diseased lung are larger in the less ventilated parts due to the higher secondary flow intensities (especially close to the walls) found there. The highest stresses occur at the beginning of expiration. We have modified the expiration curves of the respirator and decreased the pressure less abruptly resulting in a significant reduction of airway wall stresses. This finding is especially interesting with respect to our long-term goal of proposing protective ventilation strategies allowing minimization of VILI.

4 Summary and Outlook

In the present paper, several aspects of coupled problems in the human respiratory system were addressed.

The introduced model for pulmonary alveoli comprises the generation of three-dimensional artificial geometries based on tetrakaidecahedral cells. For the sake of ensuring optimal mean pathlength – a feature of great importance regarding effective gas transport in the lungs – a labyrinthine algorithm for complex geometries is employed. A polyconvex hyperelastic material model incorporating general histologic information is applied to describe the behavior of parenchymal lung tissue. Surface stresses stemming from the alveolar liquid lining are considered by enriching interfacial structural nodes of the finite element model. For that purpose, a dynamic adsorption-limited surfactant model is applied. It could be shown that interfacial phenomena influence the overall mechanical behavior of alveoli significantly. Due to different sizes and curvatures of mammalian alveoli, the intensity of this effect is species dependent.

On the part of the structural solver, a smoothed aggregation algebraic multigrid method was applied. Remarkably, an $O(n)$ overall scalability could be proven for the application to our alveolar simulations.

The investigation of airflow in the bronchial tree is based on a human CT-scan airway model of the first four generations. For this purpose a partitioned FSI method for incompressible Newtonian fluids under transient flow conditions and geometrically nonlinear structures was applied. We studied airflow structures under normal breathing and mechanical ventilation in healthy and diseased lungs. Airflow under normal breathing conditions is steady except in the transition from inspiration to expiration. By contrast, airflow under mechanical ventilation is unsteady during the whole breathing cycle due to the given respirator settings. We found that results obtained with FSI and pure CFD simulations are qualitatively similar in case of the healthy lung whereas significant differences can be shown for the diseased lung. Apart from that, stresses are larger in the diseased lung and can be influenced by the choice of ventilation parameters.

The lungs are highly heterogeneous structures comprising multiple spatial length scales. Since it is neither reasonable nor computationally feasible to simulate the lung on the whole, investigations are restricted to certain interesting parts of it. Modeling the interplay between the different scales is essential in gaining insight into the lungs' behavior on both the micro- and the macroscale. In this context, coupling our bronchial and alveolar model and thus deriving appropriate boundary conditions for both models plays an essential role. Due to the limitations of mathematical homogenization and sequential multi-scale methods particularly in the case of nonlinear behavior of complex structures, an integrated scale coupling as depicted in Fig. 12 is desired, see e.g. [29]. This will be subject of future investigations.

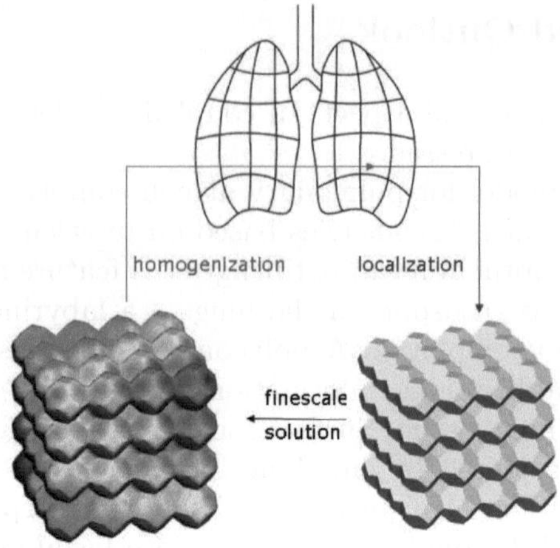

Fig. 12. Schematic description of multi-scale analyses based on integrated scale coupling

Despite the fact that we do not intend to compute an overall and fully resolved model of the lung each part that is involved in our investigations is a challenging area and asks for the best that HPC nowadays can offer.

Acknowledgement

Support by the German Science Foundation / Deutsche Forschungsgemeinschaft (DFG) is gratefully acknowledged. We also would like to thank our medical partners, i.e. the Guttmann workgroup (J. Guttmann, C. Stahl and K. Möller) at University Hospital Freiburg (Division of Clinical Respiratory Physiology), the Uhlig workgroup (S. Uhlig and C. Martin) at University Hospital Aachen (Institute for Pharmacology and Toxicology) and the Kauczor and Meinzer workgroups (H. Kauczor, M. Puderbach, S. Ley and I. Wegener) at German Cancer Research Center (Division of Radiology / Division of Medical and Biological Informatics).

References

1. J. DiRocco, D. Carney, and G. Nieman. The mechanism of ventilator-induced lung injury: Role of dynamic alveolar mechanics. In *Yearbook of Intensive Care and Emergency Medicine.* 2005.
2. H. Kitaoka, S. Tamura, and R. Takaki. A three-dimensional model of the human pulmonary acinus. *J. Appl. Physiol.*, 88(6):2260–2268, Jun 2000.
3. L. Wiechert and W.A. Wall. An artificial morphology for the mammalian pulmonary acinus. *in preparation*, 2007.

4. H. Yuan, E. P. Ingenito, and B. Suki. Dynamic properties of lung parenchyma: mechanical contributions of fiber network and interstitial cells. *J. Appl. Physiol.*, 83(5):1420–31; discussion 1418–9, Nov 1997.
5. G. A. Holzapfel, T. C. Gasser, and R. W. Ogden. Comparison of a multi-layer structural model for arterial walls with a fung-type model, and issues of material stability. *J. Biomech. Eng.*, 126(2):264–275, Apr 2004.
6. D. Balzani, P. Neff, J. Schröder, and G. A. Holzapfel. A polyconvex framework for soft biological tissues. adjustment to experimental data. *International Journal of Solids and Structures*, 43(20):6052–6070, 2006.
7. T. C. Gasser, R. W. Ogden, and G. A. Holzapfel. Hyperelastic modelling of arterial layers with distributed collagen fibre orientations. *Journal of the Royal Society Interface*, 3(6):15–35, 2006.
8. Y. C. B. Fung. Elasticity of Soft Biological Tissues in Simple Elongation. *Am. J. Physiol.*, 213:1532–1544, 1967.
9. T. Sugihara, C. J. Martin, and J. Hildebrandt. Length-tension properties of alveolar wall in man. *J. Appl. Physiol.*, 30(6):874–878, Jun 1971.
10. F. G. Hoppin Jr, G. C. Lee, and S. V. Dawson. Properties of lung parenchyma in distortion. *J. Appl. Physiol.*, 39(5):742–751, November 1975.
11. R. A. Jamal, P. J. Roughley, and M. S. Ludwig. Effect of Glycosaminoglycan Degradation on Lung Tissue Viscoelasticity. *Am. J. Physiol. Lung Cell. Mol. Physiol.*, 280:L306–L315, 2001.
12. S. A. F. Cavalcante, S. Ito, K. Brewer, H. Sakai, A. M. Alencar, M. P. Almeida, J. S. Andrade, A. Majumdar, E. P. Ingenito, and B. Suki. Mechanical Interactions between Collagen and Proteoglycans: Implications for the Stability of Lung Tissue. *J. Appl. Physiol.*, 98:672–9, 2005.
13. R. Kowe, R. C. Schroter, F. L. Matthews, and D. Hitchings. Analysis of elastic and surface tension effects in the lung alveolus using finite element methods. *J. Biomech.*, 19(7):541–549, 1986.
14. D. R. Otis, E. P. Ingenito, R. D. Kamm, and M. Johnson. Dynamic surface tension of surfactant TA: experiments and theory. *J. Appl. Physiol.*, 77(6):2681–2688, Dec 1994.
15. M. Kojic, I. Vlastelica, B. Stojanovic, V. Rankovic, and A. Tsuda. Stress integration procedures for a biaxial isotropic material model of biological membranes and for hysteretic models of muscle fibres and surfactant. *International Journal for Numerical Methods in Engineering*, 68(8):893–909, 2006.
16. P. Vaněk, J. Mandel, and M. Brezina. Algebraic multigrid based on smoothed aggregation for second and fourth order problems. *Computing*, 56:179–196, 1996.
17. P. Vaněk, M. Brezina, and J. Mandel. Convergence of algebraic multigrid based on smoothed aggregation. *Numer. Math.*, 88(3):559–579, 2001.
18. M.W. Gee, C.M. Siefert, J.J. Hu, R.S. Tuminaro, and M.G. Sala. Ml 5.0 smoothed aggregation user's guide. *SAND2006-2649, Sandia National Laboratories*, 2006.
19. M.W. Gee, J.J. Hu, and R.S. Tuminaro. A new smoothed aggregation multigrid method for anisotropic problems. *to appear*, 2006.
20. T. Rabczuk and W.A. Wall. Computational studies on lower airway flows of human and porcine lungs based on ct-scan geometries for normal breathing and mechanical ventilation. *in preparation*, 2007.
21. T. Rabczuk and W.A. Wall. Fluid-structure interaction studies in lower airways of healthy and diseased human lungs based on ct-scan geometries for normal breathing and mechanical ventilation. *in preparation*, 2007.

22. U. Küttler, C. Förster, and W.A. Wall. A solution for the incompressibility dilemma in partitioned fluid-structure interaction with pure dirichlet fluid domains. *Computational Mechanics*, 38:417–429, 2006.
23. C. Förster, W.A. Wall, and E. Ramm. Artificial added mass instabilities in sequential staggered coupling of nonlinear structures and incompressible flows. *Computer Methods in Applied Mechanics and Engineering*, 2007.
24. W.A. Wall. *Fluid-Struktur-Interaktion mit stabilisierten Finiten Elementen.* PhD thesis, Institut für Baustatik, Universität Stuttgart, 1999.
25. D.P. Mok. *Partitionierte Lösungsansätze in der Strukturdynamik und der Fluid-Struktur-Interaktion.* PhD thesis, Institut für Baustatik, Universität Stuttgart, 2001.
26. M. Bischoff. *Theorie und Numerik einer dreidimensionalen Schalenformulierung.* PhD thesis, Institut für Baustatik, University Stuttgart, 1999.
27. M. Bischoff and E. Ramm. Shear deformable shell elements for large strains and rotations. *International Journal for Numerical Methods in Engineering*, 1997.
28. M. Bischoff, W.A. Wall, K.-U. Bletzinger, and E. Ramm. Models and finite elements for thin-walled structures. In E. Stein, R. de Borst, and T.J.R. Hughes, editors, *Encyclopedia of Computational Mechanics - Volume 2: Solids, Structures and Coupled Problems.* John Wiley & Sons, 2004.
29. V. G. Kouznetsova. Computational homogenization for the multi-scale analysis of multi-phase materials. *PhD thesis, Technische Universiteit Eindhoven*, 2002.

FSI Simulations on Vector Systems – Development of a Linear Iterative Solver (BLIS)

Sunil R. Tiyyagura[1] and Malte von Scheven[2]

[1] High Performance Computing Center Stuttgart, University of Stuttgart, Nobelstrasse 19, 70569 Stuttgart, Germany.
sunil@hlrs.de

[2] Institute of Structural Mechanics, University of Stuttgart, Pfaffenwaldring 7, 70550 Stuttgart, Germany.
mvs@ibb.uni-stuttgart.de

Summary. This paper addresses the algorithmic and implementation issues associated with fluid structure interaction simulations, specially on vector architecture. Firstly, the fluid structure coupling algorithm is presented and then a newly developed parallel sparse linear solver is introduced and its performance discussed.

1 Introduction

In this paper we focus on the performance improvement of the fluid structure interaction simulations on vector systems. The work described here was done on the basis of the research finite element program *Computer Aided Research Analysis Tool* (CCARAT), that is developed and maintained at the Institute of Structural Mechanics of the University of Stuttgart. The research code CCARAT is a multipurpose finite element program covering a wide range of applications in computational mechanics, like e.g. multi-field and multi-scale problems, structural and fluid dynamics, shape and topology optimization, material modeling and finite element technology. The code is parallelized using MPI and runs on a variety of platforms.

The major time consuming portions of a finite element simulation are calculating the local element contributions to the globally assembled matrix and solving the assembled global system of equations. As much as 80% of the time in a very large scale simulation can be spent in the linear solver, specially if the problem to be solved is ill-conditioned. While the time taken in element calculation scales linearly with the size of the problem, often the time in the sparse solver does not. Major reason being the kind of preconditioning needed for a successful solution. In Sect. 2 of this paper the fluid structure coupling algorithm implemented in CCARAT is presented. Sect. 3 of this paper briefly analyses the performance of public domain solvers on vector architecture and then a newly developed parallel iterative solver (Block-based Linear Iterative Solver – BLIS) is introduced. In Sect. 4, a large-scale fluid-structure interaction example is presented. Sect. 5 discusses the performance of a pure fluid example and fluid structure interaction example on scalar and vector systems along with the scaling results of BLIS on the NEC SX-8.

2 Fluid structure interaction

Our partitioned fluid structure interaction environment is described in detail in Wall [1] or Wall et al. [2] and is therefore presented here in a comprising overview in figure 1. In this approach a non-overlapping partitioning is employed, where the physical fields fluid and structure are coupled at the interface Γ, i.e. the wetted structural surface. A third computational field Ω_M, the deforming fluid mesh, is introduced through an Arbitrary Lagrangian-Eulerian (ALE) description. Each individual field is solved by semi-discretization strategies with finite elements and implicit time stepping algorithms.

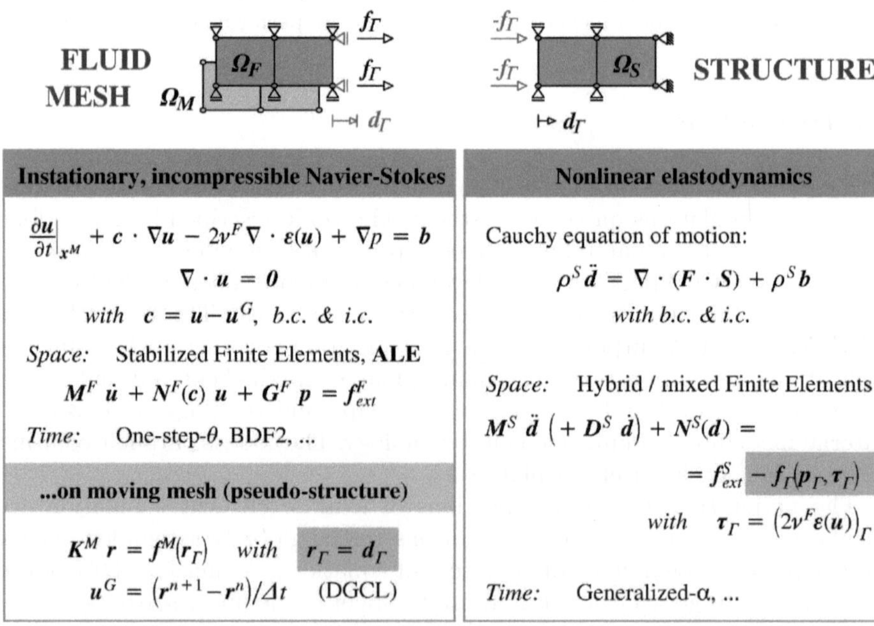

Fig. 1. Non-overlapping partitioned fluid structure interaction environment

Key requirement for the coupling schemes is to fulfill two coupling conditions: the kinematic and the dynamic continuity across the interface. Kinematic continuity requires that the position of structure and fluid boundary are equal at the interface, while dynamic continuity means that all tractions at the interface are in equilibrium:

$$\boldsymbol{d}_\Gamma(t) \cdot \boldsymbol{n} = \boldsymbol{r}_\Gamma(t) \cdot \boldsymbol{n} \quad \text{and} \quad \boldsymbol{u}_\Gamma(t) = \boldsymbol{u}_\Gamma^G(t) = f(\boldsymbol{r}_\Gamma(t)), \tag{1}$$

$$\boldsymbol{\sigma}_\Gamma^S(t) \cdot \boldsymbol{n} = \boldsymbol{\sigma}_\Gamma^F(t) \cdot \boldsymbol{n} \tag{2}$$

with $\boldsymbol{n}$ denoting the unit normal vector on the interface. Satisfying the kinematic continuity leads to mass conservation at Γ, satisfying the dynamic continuity leads to conservation of linear momentum, and energy conservation finally requires to simultaneously satisfy both continuity equations. In this paper (and in figure 1) only no-slip boundary conditions and sticking grids at the interface are considered.

3 Linear iterative solver

CCARAT uses Krylov subspace based sparse iterative solvers to solve the linearized structural and fluid equations described in Fig 1. Most public domain solvers like AZTEC [3], PETSc, Trilinos [4], etc. do not perform on vector architecture as well as they do on superscalar architectures. The main reason being their design considerations that primarily target performance on superscalar architectures thereby neglecting the following performance critical features of vector systems.

3.1 Vector length and indirect memory access

Average vector length is an important metric that has a huge effect on performance. In sparse linear algebra, the matrix object is sparse whereas the vectors are dense. So, any operations involving only the vectors, like the dot product, run with high performance on any architecture as they exploit spatial locality in memory. But, for any operations involving the sparse matrix object, like the matrix vector product (MVP), sparse storage formats play a crucial role in achieving good performance, specially on the vector architecture. This is extensively discussed in [5, 6].

The performance of sparse MVP on vector as well as on superscalar architectures is not limited by memory bandwidth, but by latencies. Due to sparse storage, the vector to be multiplied in a sparse MVP is accessed randomly (non-strided access). This introduces indirect memory access which is a memory latency bound operation. Blocking is employed on scalar as well as on vector architecture to reduce the amount of indirect memory access needed for the sparse MVP kernel using any storage format [7, 8]. The cost of accessing

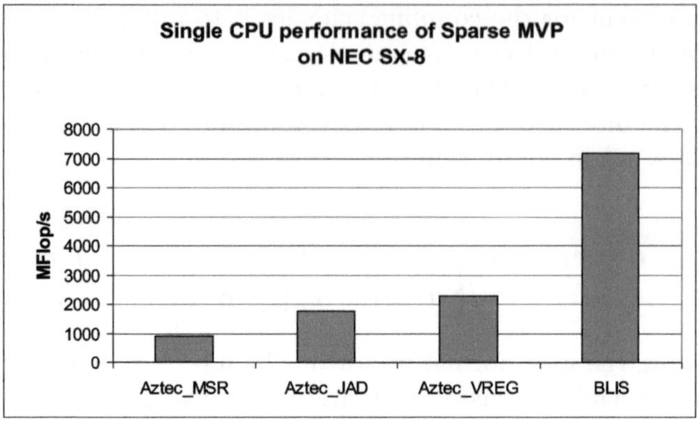

Fig. 2. Single CPU performance of Sparse MVP on NEC SX-8

the main memory is so high on superscalar systems when compared to vector systems (which usually have a superior memory subsystem performance) that this kernel runs at an order of magnitude faster on typical vector processors like the NEC SX-8 than commodity processors.

3.2 Block-based Linear Iterative Solver (BLIS)

In the sparse MVP kernel discussed so far, the major hurdle to performance is not memory bandwidth but the latencies involved due to indirect memory addressing. Block based computations exploit the fact that many FE problems typically have more than one physical variable to be solved per grid point. Thus, small blocks can be formed by grouping the equations at each grid point. Operating on such dense blocks considerably reduces the amount of indirect addressing required for sparse MVP [6]. This improves the performance of the kernel dramatically on vector machines [9] and also remarkably on superscalar architectures [10, 11]. A vectorized general parallel iterative solver (BLIS) targeting performance on vector architecture is under development. Block-based approach is adopted in BLIS primarily to reduce the penalty incurred due to indirect memory access on most hardware architectures. Some solvers already implement similar blocking approaches, but use BLAS routines when processing each block. This method will not work on vector architecture as the innermost loop is short when processing small blocks. So, explicitly unrolling the kernels is the key to achieve high sustained performance. This approach also has advantages on scalar architectures and is adopted in [7].

Available functionality in BLIS:

Presently, BLIS is working with finite element applications that have 3, 4 or 6 unknowns to be solved per grid point. JAD sparse storage format is used to store the dense blocks. This assures sufficient average vector length for operations done using the sparse matrix object (Preconditioning, Sparse MVP). The single CPU performance of sparse MVP, Fig. 2, with a matrix consisting of 4x4 dense blocks is around 7.2 GFlop/s (about 45% vector peak) on the NEC SX-8. The sustained performance in the whole solver is about 30% peak when the problem size is enough to fill the vector pipelines.

BLIS is based on MPI and includes well known Krylov subspace methods such as CG, BiCGSTAB and GMRES. Block scaling, block Jacobi, colored block symmetric Gauss-Seidel and block ILU(0) on subdomains are the available matrix preconditioners. Exchange of halos in sparse MVP can be done using MPI blocking, non-blocking or persistent communication.

Future work:

The restriction of block sizes will be solved by extending the solver to handle any number of unknowns. Blocking functionality will be provided in the solver in order to relieve the users from preparing blocked matrices in order to use the library. This makes adaptation of the library to an application easier. Reducing global synchronization at different places in Krylov subspace algorithms has to be extensively looked into for further improving scaling of the solver [12]. We also plan to implement domain decomposition based and multigrid preconditioning methods.

4 Numerical example

In this numerical example a simplified 2-dimensional representation of a cubic building with a flat membrane roof is studied. The building is situated in a horizontal flow with an initially exponential profile and a maximum velocity of 26.6 m/s. The fluid is Newtonian with dynamic viscosity $\nu_F = 0.1\ Ns/m^2$ and density $\rho_F = 1.25 Kg/m^3$.

In the following two different configurations are compared:

- a rigid roof, i.e. pure fluid simulation
- a flexible roof including fluid-structure interaction

For the second case the roof is assumed to be a very thin membrane ($t/l = 1/1000$) with Young's modulus $E_S = 1.0 \cdot 10^9\ N/m^2$, Poisson's ratio $\nu_S = 0.0$ and density $\rho_S = 1000.0\ Kg/m^3$.

The fluid domain is discretized by 25,650 GLS-stabilized Q1Q1 and the structure with 80 Q1 elements. The moving boundary of the fluid is considered via an ALE-Formulation only for the fluid subdomain situated above the membrane roof. Here 3,800 pseudo-structural elements are used to calculate the new mesh positions [2]. This discretization results in $\sim 85,000$ degrees of freedom for the complete system.

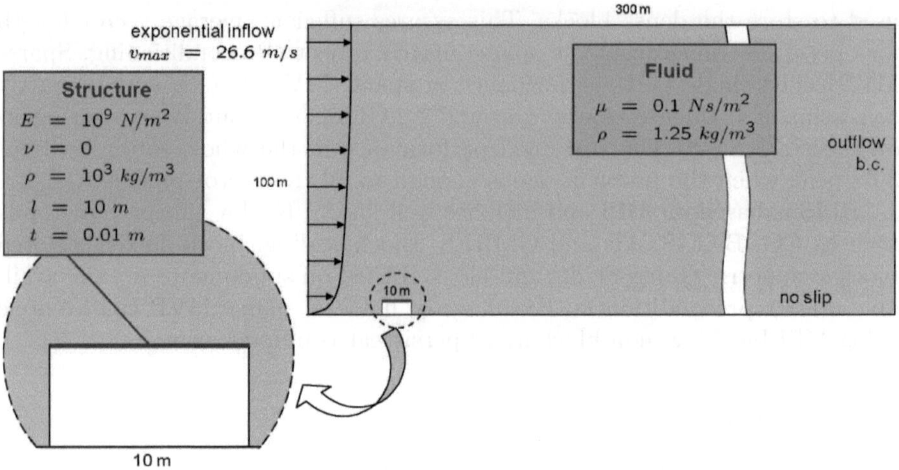

Fig. 3. Membrane Roof: Geometry and material parameters

The calculation was run for approx. 2,000 time steps with $\Delta t = 0.01\ s$, resulting in a simulation time of $\sim 20\ s$. For each timestep 4-6 iterations between fluid and structure field were needed to fulfill the coupling conditions. In the single fields 3-5 Newton iterations for fluid and 2-3 iterations for structure were necessary to solve the nonlinear problems.

The results for both the rigid and the flexible roof for $t = 9.0\ s$ are visualized in figure 4. For both simulations the pressure field clearly shows a large vortex, which emerges in the wake of the building and then moves slowly downstream. In addition, for the flexible roof smaller vortices are separating at the upstream edge of the building and traveling over the membrane roof.

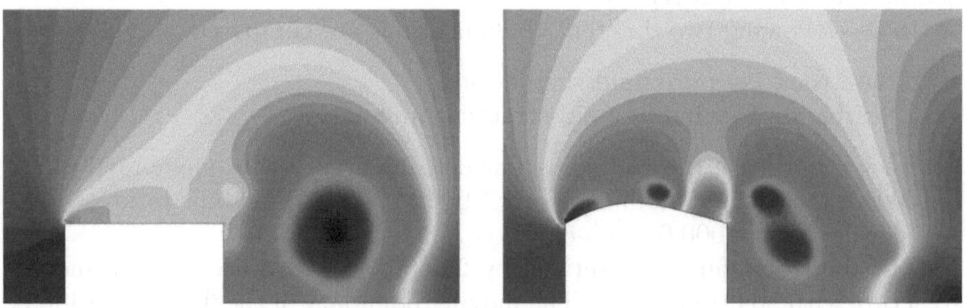

Fig. 4. Membrane Roof: Pressure field on deformed geometry (10-fold) for rigid roof (left) and flexible roof (right)

Fig. 5. Velocity in the midplane of the channel

These vortices, originating from the interaction between fluid and structure, cause the nonsymmetric deformation of the roof.

5 Performance

This section provides the performance analysis of finite element simulations on both scalar and vector architectures. Firstly, scaling of BLIS on NEC SX-8 is presented for a laminar flow problem with different discretizations. Then, performance of a pure fluid example and a FSI example is compared between two different hardware architectures. The machines tested are a cluster of NEC SX-8 SMPs and a cluster of Intel 3.2 GHz Xeon EM64T processors. The network interconnect available on NEC SX-8 is a proprietary multi-stage crossbar called IXS and Infiniband on the Xeon cluster. Vendor tuned MPI library is used on the SX-8 and Voltaire MPI library on the Xeon cluster.

5.1 Example used for scaling tests

In this example the laminar, unsteady 3-dimensional flow around a cylinder with a square cross-section is examined. The setup was introduced as a benchmark example by the DFG Priority Research Program "Flow Simulation on High Performance Computers" to compare different solution approaches of the Navier-Stokes equations[13]. The fluid is assumed to be incompressible Newtonian with a kinematic viscosity $\nu = 10^{-3}\ m^2/s$ and a density of $\rho = 1.0\ kg/m^3$. The rigid cylinder (cross-section: 0.1 m x 0.1m) is placed in a 2.5 m long channel with a square cross-section of 0.41 m by 0.41 m. On one side a parabolic inflow condition with the mean velocity $u_m = 2.25\ m/s$ is applied. No-slip boundary conditions are assumed on the four sides of the channel and on the cylinder.

5.2 BLIS scaling on NEC SX-8

Scaling of the solver on NEC SX-8 was tested for the above mentioned numerical example using stabilized 3D hexahedral fluid elements implemented in CCARAT. Table 1 lists all the six discretizations of the example used.

Figure 6 plots weak scaling of BLIS for different processor counts. Each curve represents performance using particular number of CPUs with varying problem size. All problems were run for 5 time steps where each non-linear time step needs about 3-5 newton iterations for convergence. The number of iterations needed for convergence in BLIS for each newton step varies largely between 200-2000 depending on the problem size (number of equations). The plots show the drop in sustained floating point performance of BLIS from over 6 GFlop/s to 3 GFlop/s depending on the number of processors used for each problem size.

The right plot of Fig. 6 explains the reason for this drop in performance in terms of drop in computation to communication ratio in BLIS. It has to be noted that major part of the communication with the increase in processor count is spent in MPI global reduction calls which need global syncronization. As the processor count increases, the performance curves climb slowly till the performance saturates. This behavior can be directly attributed to the time spent in communication which is clear from the right plot. These plots are hence important as they accentuate the problem with Krylov subspace algorithms where large problem sizes are needed to sustain high performance on large processor counts. This is a drawback for certain class of applications

Table 1. Different discretizations of the introduced example

Discretization	No.of elements	No. of nodes	No. of unknowns
1	33750	37760	151040
2	81200	88347	353388
3	157500	168584	674336
4	270000	285820	1143280
5	538612	563589	2254356
6	911250	946680	3786720

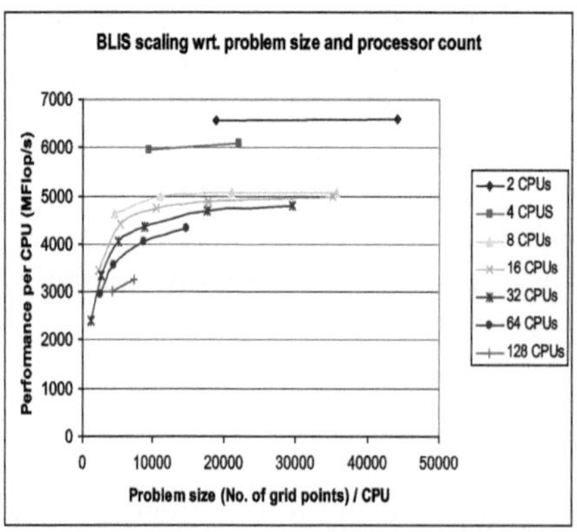

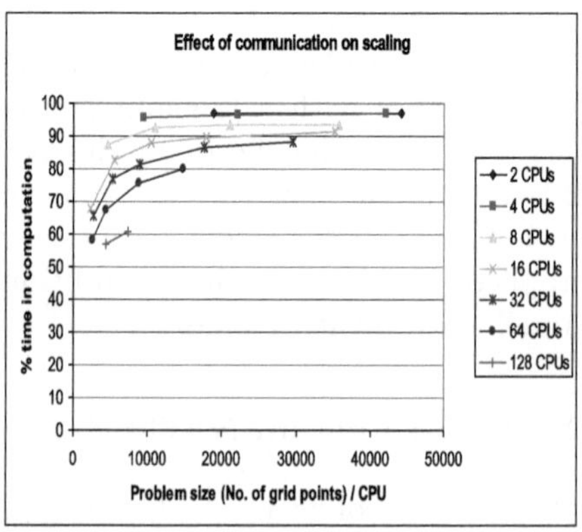

Fig. 6. Scaling of BLIS wrt. problem size on NEC SX-8 (left) Computation to communication ratio in BLIS on NEC SX-8 (right)

where the demand for HPC (High Performance Computing) is due to the largely transient nature of the problem. For instance, even though the problem size is moderate in some Fluid-Structure interaction examples, thousands of time steps are necessary to simulate the transient effects.

5.3 Performance comparison on scalar and vector machines

Here, performance is compared on 2 different hardware architectures between AZTEC and BLIS for a pure fluid example and for a fluid structure interaction (FSI) example. The peak floating-point performance of the Xeon processor is 6.4 GFlop/s and of the SX-8 is 16 GFlop/s.

The pure fluid example is run for 2 Newton iterations which needed 5 solver calls (linearization steps). In the FSI example the structural field is discretized using BRICK elements in CCARAT. So, block 3 and block 4 functionality in BLIS is used for this problem. It was run for 1 FSI time step which needed 21 solver calls (linearization steps). It can be noted from Tables 2 and 3 that the number of iterations needed for convergence in the solver vary between different preconditioners and also between different architectures for the same preconditioner. The reason for this variation between architectures is due to the difference in partitioning. Also the preconditioning in BLIS and AZTEC cannot be exactly compared as BLIS operates on blocks which normally results in superior preconditioning than point-based algorithms.

Even with all the above mentioned differences, the comparison is done on the basis of time to solution for the same problem on different systems and

Table 2. Performance comparison in solver between SX-8 and Xeon for a pure fluid example with 631504 equations on 8 CPUs

Machine	Solver	Precond.	BiCGSTAB iters. per solver call	MFlop/s per CPU	CPU time
SX-8	BLIS4	BJAC	65	4916	**110**
	BLIS4	BILU	125	1027	765
	AZTEC	ILU	48	144	3379
Xeon	BLIS4	BJAC	68	-	1000
	BLIS4	BILU	101	-	625
	AZTEC	ILU	59	-	**1000**

Table 3. Performance comparison in solver between SX-8 and Xeon for a fluid structure interaction example with 25168 fluid equations and 26352 structural equations

Machine	Solver	Precond.	CG iters. per solver call	MFlop/s per CPU	CPU time
SX-8	BLIS3,4	BJAC	597	6005	**66**
	AZTEC	ILU	507	609	564
Xeon	BLIS3,4	BILU	652	-	294
	AZTEC	ILU	518	-	**346**

using different algorithms. It is to be noted that the fastest preconditioner is taken on any particular architecture. The time to solution using BLIS is clearly much better on the SX-8 when compared to AZTEC. This is the main reason for developing a new general solver in the teraflop workbench. It is also interesting to note that the time for solving the linear systems (which is the most time consuming part of any unstructured finite element or finite volume simulation) is clearly less than a factor 5 on the SX-8 when compared to the Xeon cluster.

6 Summary

The fluid structure interaction framework has been presented. The reasons behind the dismal performance of most of the public domain sparse iterative solvers on vector machines were briefly stated. We then introduced the Block-based Linear Iterative Solver (BLIS) which is currently under development targeting performance on all architectures. Results show an order of magnitude performance improvement over other public domain libraries on the tested vector system. A moderate performance improvement is also measured on the scalar machines.

References

1. Wall, W.A.: Fluid-Struktur-Interaktion mit stabilisierten Finiten Elementen. phdthesis, Institut für Baustatik, Universität Stuttgart (1999)
2. Wall, W., Ramm, E.: Fluid-Structure Interaction based upon a Stabilized (ale) Finite Element Method. In: E. Oñate and S. Idelsohn (Eds.), Computational Mechanics, Proceedings of the Fourth World Congress on Computational Mechanics WCCM IV, Buenos Aires. (1998)
3. Tuminaro, R.S., Heroux, M., Hutchinson, S.A., Shadid, J.N.: Aztec user's guide: Version 2.1. Technical Report SAND99-8801J, Sandia National Laboratories (1999)
4. Heroux, M.A., Willenbring, J.M.: Trilinos users guide. Technical Report SAND2003-2952, Sandia National Laboratories (2003)
5. Saad, Y.: Iterative Methods for Sparse Linear Systems, Second Edition. SIAM, Philadelphia, PA (2003)
6. Tiyyagura, S.R., Küster, U., Borowski, S.: Performance improvement of sparse matrix vector product on vector machines. In Alexandrov, V., van Albada, D., Sloot, P., Dongarra, J., eds.: Proceedings of the Sixth International Conference on Computational Science (ICCS 2006). LNCS 3991, May 28-31, Reading, UK, Springer (2006)
7. Im, E.J., Yelick, K.A., Vuduc, R.: Sparsity: An optimization framework for sparse matrix kernels. International Journal of High Performance Computing Applications **(1)18** (2004) 135–158

8. Tiyyagura, S.R., Küster, U.: Linear iterative solver for NEC parallel vector systems. In Resch, M., Bönisch, T., Tiyyagura, S., Furui, T., Seo, Y., Bez, W., eds.: Proceedings of the Fourth Teraflop Workshop 2006, March 30-31, Stuttgart, Germany, Springer (2006)
9. Nakajima, K.: Parallel iterative solvers of geofem with selective blocking preconditioning for nonlinear contact problems on the earth simulator. GeoFEM 2003-005, RIST/Tokyo (2003)
10. Jones, M.T., Plassmann, P.E.: Blocksolve95 users manual: Scalable library software for the parallel solution of sparse linear systems. Technical Report ANL-95/48, Argonne National Laboratory (1995)
11. Tuminaro, R.S., Shadid, J.N., Hutchinson, S.A.: Parallel sparse matrix vector multiply software for matrices with data locality. Concurrency: Practice and Experience **(3)10** (1998) 229–247
12. Demmel, J., Heath, M., van der Vorst, H.: Parallel numerical linear algebra. Acta Numerica **2** (1993) 53–62
13. Schäfer, M., Turek, S.: Benchmark Computations of Laminar Flow Around a Cylinder. Notes on Numerical Fluid Mechanics **52** (1996) 547–566

Applications IV

Chemistry and Astrophysics

Simulations of Premixed Swirling Flames Using a Hybrid Finite-Volume/Transported PDF Approach

Stefan Lipp[1] and Ulrich Maas[2]

[1] University Karlsruhe, Institute for Technical Thermodynamics, lipp@itt.mach.uni-karlsruhe.de
[2] University Karlsruhe, Institute for Technical Thermodynamics maas@itt.mach.uni-karlsruhe.de

Abstract

The mathematical modeling of swirling flames is a difficult task due to the intense coupling between turbulent transport processes and chemical kinetics in particular for instationary processes like the combustion induced vortex breakdown. In this paper a mathematical model to describe the turbulence-chemistry interaction is presented. The described method consists of two parts. Chemical kinetics are taken into account with reduced chemical reaction mechanisms, which have been developed using the ILDM-Method ("Intrinsic Low-Dimensional Manifold"). The turbulence-chemistry interaction is described by solving the joint probability density function (PDF) of velocity and scalars. Simulations of test cases with simple geometries verify the developed model.

1 Introduction

In many industrial applications there is a high demand for reliable predictive models for turbulent swirling flows. While the calculation of non-reacting flows has become a standard task and can be handled using Reynolds averaged Navier-Stokes (RANS) or Large Eddy Simulation (LES) methods the modeling of reacting flows still is a challenging task due to the difficulties that arise from the strong non-linearity of the chemical source term which can not be modeled satisfactorily by using oversimplified closure methods.
PDF methods (probability density function) show a high capability for modeling turbulent reactive flows, because of the advantage of treating convection and finite rate non-linear chemistry exactly [1, 2]. Only the effect of molecular mixing has to be modeled [3]. In the literature different kinds of PDF approaches can be found. Some use stand-alone PDF methods in

which all flow properties are computed by a joint probability density function method [4, 5, 6, 7]. The transport equation for the joint probability density function that can be derived from the Navier-Stokes equations still contains unclosed terms that need to be modeled. These terms are the fluctuating pressure gradient and the terms describing the molecular transport. In contrast the above mentioned chemistry term, the body forces and the mean pressure gradient term already appear in closed form and need no more modeling assumtions.
Compared to RANS methods the structure of the equations appearing in the PDF context is remarkably different. The moment closure models (RANS) result in a set of partial differential equations, which can be solved numerically using finite-difference or finite-volume methods [8]. In contrast the transport equation for the PDF is a high-dimensional scalar transport equation. In general it has $7 + n_S$ dimensions which consist of three dimensions in space, three dimensions in velocity space, the time and the number of species n_S used for the description of the thermokinetic state. Due to this high dimensionality it is not feasible to solve the equation using finite-difference of finite-volume methods. For that reason Monte Carlo methods have been employed, which are widely used in computational physics to solve problems of high dimensionality, because the numerical effort increases only linearly with the number of dimensions.
Using the Monte Carlo method the PDF is represented by an ensemble of stochastic particles [9]. The transport equation for the PDF is transformed to a system of stochastic ordinary differential equations. This system is constructed in such a way that the particle properties, e.g. velocity, scalars, and turbulent frequency, represent the same PDF as in the turbulent flow.
In order to fulfill consistency of the modeled PDF, the mean velocity field derived from an ensemble of particles needs to satisfy the mass conservation equation [1]. This requires the pressure gradient to be calculated from a Poission equation. The available Monte Carlo methods cause strong bias determining the convective and diffusive terms in the momentum conservation equations. This leads to stability problems calculating the pressure gradient from the Poisson equation. To avoid these instabilities different methods to calculate the mean pressure gradient where used. One possibility is to couple the particle method with an ordinary finite-volume or finite-difference solver to optain the mean pressure field from the Navier-Stokes equations. These so called hybrid PDF/CFD methods are widely used by different authors for many types of flames [10, 11, 12, 13, 14, 15].
In the presented paper a hybrid scheme is used. The fields for mean pressure gradient and a turbulence charactaristic, e.g. the turbulent time scale, are derived solving the Reynolds averaged conservation equations for momentum, mass and energy for the flow field using a finite-volume method. The effect of turbulent fluctuations is modeled using a k-τ model [16]. Chemical kinetics are taken into account by using the ILDM method to get reduced chemical mechanisms [17, 18]. In the presented case the reduced mechanism describes

the reaction with three parameters which is on the one hand few enough to limit the simulation time to an acceptable extent and on the other hand sufficiently high to get a detailed description of the chemical reaction.
The test case for the developed model is a model combustion chamber investigated by serveral authors [19, 20, 21, 22]. With their data the results of the presented simulations are validated.

2 Numerical Model

As mentioned above a hybird CFD/PDF method is used in this work. In Fig. 1 a complete sketch of the solution precedure can be found. Before explaning the details of the implemented equations and discussing consistency and numerical matters the idea of the solution procedure shall be briefly overviewed.

The calulation starts with a CFD step in which the Navier-Stokes equations for the flow field are solved by a finite-volume method. The resulting mean pressure gradient together with the mean velocities and the turbulence characteristics is handed over to the PDF part. Here the joint probability density function of the scalars and the velocity is solved by a particle Monte Carlo method. The reaction progress is taken into account by reading from a lookup table based on a mechanism reduced with the ILDM method. As a result of this step the mean molar mass, the composition vector and the mean temperature field are returned to the CFD part. This internal iteration is performed until convergence is achieved.

2.1 CFD Model

The CFD code which is used to calculate the mean velocity and pressure field along with the turbulent kinetic energy and the turbulent time scale is called SPARC[3] and was developed by the Department of Fluid Machinery at Karlsruhe University. It solves the Favre-averaged compressible Navier Stokes equations using a Finite-Volume method on block structured non-uniform

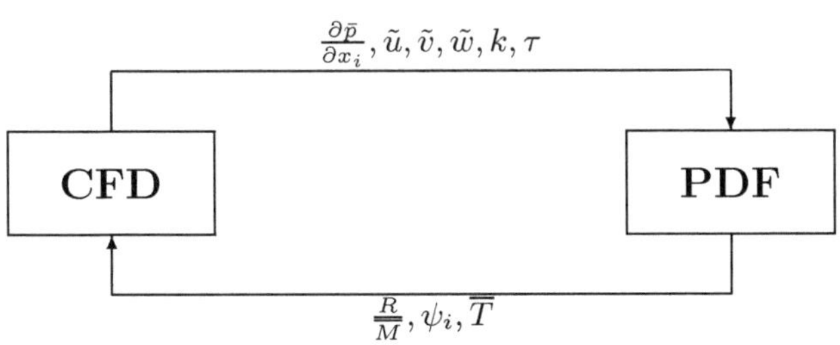

Fig. 1. Scheme of the coupling of CFD and PDF

[3] **S**tructured **Par**allel **R**esearch **C**ode

meshes. In this work a 2D axi-symmetric solution domain is used. Turbulence closure is provided using a two equation model solving a transport equation for the turbulent kinetic energy and a turbulent time scale [16].
In detail the equations read

$$\frac{\partial \bar{\rho}}{\partial t} + \frac{\partial \left(\bar{\rho}\tilde{u}_i\right)}{\partial x_i} = 0 \tag{1}$$

$$\frac{\partial \left(\bar{\rho}\tilde{u}_i\right)}{\partial t} + \frac{\partial}{\partial x_j}\left(\bar{\rho}\tilde{u}_i\tilde{u}_j + \overline{\rho u_i^{''} u_j^{''}} + \bar{\rho}\delta_{ij} - \tau_{ij}\right) = 0 \tag{2}$$

$$\frac{\left(\partial \bar{\rho}\tilde{e}\right)}{\partial t} + \frac{\partial}{\partial x_j}\left(\bar{\rho}\tilde{u}_j\tilde{e} + \tilde{u}_j\bar{p} + \overline{u_j^{''} p} + \overline{\rho u_j^{''} e^{''}} + \bar{q}_j - \overline{u_i \tau_{ij}}\right) = 0 \tag{3}$$

which are the conservation equations for mass, momentum and energy in Favre average manner, respectively. Modeling of the unclosed terms in the energy equation will not be described in detail any further but can be found for example in [8]. The unclosed cross correlation term in the momentum conservation equation is modeled using the Boussinesq approximation

$$\overline{\rho u_i^{''} u_j^{''}} = \bar{\rho}\mu_T \left(\frac{\partial \tilde{u}_i}{\partial x_j} + \frac{\partial \tilde{u}_j}{\partial x_i}\right) \tag{4}$$

with

$$\mu_T = C_\mu f_\mu k \tau \quad . \tag{5}$$

The parameter C_μ is an empirical constant with a value of $C_\mu = 0.09$ and f_μ accounts for the influence of walls. The turbulent kinetic energy k and the turbulent time scale τ are calculated from their transport equation which are [16]

$$\bar{\rho}\frac{\partial k}{\partial t} + \bar{\rho}\tilde{u}_j \frac{\partial k}{\partial x_j} = \tau_{ij}\frac{\partial u_i}{\partial x_j} - \bar{\rho}\frac{k}{\tau} + \frac{\partial}{\partial x_i}\left[\left(\mu + \frac{\mu_T}{\sigma_k}\right)\frac{\partial k}{\partial x_j}\right] \tag{6}$$

$$\begin{aligned} \bar{\rho}\frac{\partial \tau}{\partial t} + \bar{\rho}\tilde{u}_j \frac{\partial \tau}{\partial x_j} &= (1 - C_{\epsilon 1})\frac{\tau}{k}\tau_{ij}\frac{\partial u_i}{\partial x_j} + (C_{\epsilon 2} - 1)\bar{\rho} + \\ &\frac{\partial}{\partial x_j}\left[\left(\mu + \frac{\mu_T}{\sigma_{\tau 2}}\right)\frac{\partial k}{\partial x_j}\right] + \frac{2}{k}\left(\mu + \frac{\mu_T}{\sigma_{\tau 1}}\right)\frac{\partial k}{\partial x_k}\frac{\partial \tau}{\partial x_k} - \\ &\frac{2}{\tau}\left(\mu + \frac{\mu_T}{\sigma_{\tau 2}}\right)\frac{\partial \tau}{\partial x_k}\frac{\partial \tau}{\partial x_k} \quad . \end{aligned} \tag{7}$$

Here $C_{\epsilon 1} = 1.44$ and $\sigma_{\tau 1} = \sigma_{\tau 2} = 1.36$ are empirical model constants. The parameter $C_{\epsilon 2}$ is calculated from the turbulent Reynolds number $\mathrm{Re_t}$.

$$\mathrm{Re_t} = \frac{k\tau}{\mu} \tag{8}$$

$$C_{\epsilon 2} = 1.82\left[1 - \frac{2}{9}\exp\left(\left(-\mathrm{Re_t}/6\right)^2\right)\right] \tag{9}$$

2.2 Joint PDF Model

In the literature many different joint PDF models can be found, for example models for the joint PDF of velocity and composition [23, 24] or for the joint PDF of velocity, composition and turbulent frequency [25]. A good overview of the different models can be found in [12].
In most joint PDF approaches a turbulent (reactive) flow field is described by a one-time, one-point joint PDF of certain fluid properties. At this level chemical reactions are treated exactly without any modeling assumptions [1]. However, the effect of molecular mixing has to be modeled.
The state of the fluid at a given point in space and time can be fully described by the velocity vector $\mathbf{V} = (V_1, V_2, V_3)^T$ and the the composition vector $\mathbf{\Psi}$ containing the mass fractions of $n_S - 1$ species and the enthalpy h $\left(\mathbf{\Psi} = (\Psi_1, \Psi_2, \ldots, \Psi_{n_s-1}, h)^T\right)$. The probability density function is

$$f_{\mathbf{U}\phi}\left(\mathbf{V}, \mathbf{\Psi}; \mathbf{x}, t\right) \mathrm{d}\mathbf{V}\mathrm{d}\mathbf{\Psi} = \mathrm{Prob}\left(\mathbf{V} \le \mathbf{U} \le \mathbf{V} + \mathrm{d}\mathbf{V}, \mathbf{\Psi} \le \mathbf{\Phi} \le \mathbf{\Psi} + \mathrm{d}\mathbf{\Psi}\right) \quad (10)$$

and gives the probability that at one point in space and time one realization of the flow is within the interval

$$\mathbf{V} \le \mathbf{U} \le \mathbf{V} + \mathrm{d}\mathbf{V} \quad (11)$$

for its velocity vector and

$$\mathbf{\Psi} \le \mathbf{\Phi} \le \mathbf{\Psi} + \mathrm{d}\mathbf{\Psi} \quad (12)$$

for its composition vector.
According to [1] a transport equation for the joint PDF of velocity and composition can be derived. Under the assumption that the effect of pressure fluctuations on the fluid density is negligible the transport equation writes

$$\underbrace{\rho(\mathbf{\Psi})\frac{\partial \tilde{f}}{\partial t}}_{I} + \underbrace{\rho(\mathbf{\Psi})U_j\frac{\partial \tilde{f}}{\partial x_j}}_{II} + \underbrace{\left[\left(\rho(\mathbf{\Psi})g_j - \frac{\partial \langle p \rangle}{\partial x_j}\right)\right]\frac{\partial \tilde{f}}{\partial U_j}}_{III} + \underbrace{\frac{\partial}{\partial \Psi_\alpha}\left[\rho(\mathbf{\Psi})S_\alpha(\mathbf{\Psi})\tilde{f}\right]}_{IV}$$
$$= \underbrace{\frac{\partial}{\partial U_j}\left[\left\langle -\frac{\partial \tau_{ij}}{\partial x_i} + \frac{\partial p'}{\partial x_i}|\mathbf{U}, \mathbf{\Psi}\right\rangle \tilde{f}\right]}_{V} + \underbrace{\frac{\partial}{\partial \Psi_\alpha}\left[\left\langle -\frac{\partial J_i}{\partial x_i}|\mathbf{U}, \mathbf{\Psi}\right\rangle \tilde{f}\right]}_{VI}. \quad (13)$$

Term I describes the instationary change of the PDF, Term II its change by convection in physical space and Term III takes into account the influence of gravitiy and the mean pressure gradient on the PDF. Term IV includes the chemical source term which describes the change of the PDF in composition space due to chemical reactions. All terms on the left hand side of the equation appear in closed form, e.g. the chemical source term. In contrast the terms on the right hand side are unclosed and need further modeling. Many closing

assumptions for these two terms exist. In the following only the ones that are used in the present work shall be explained further.
Term V describes the influence of pressure fluctuations and viscous stresses on the PDF. Commonly a Langevin approach [26, 27] is used to close this term. In the presented case the SLM (Simplified Langevin Model) is used [1]. More sophisticated approaches that take into account the effect of non-isotropic turbulence or wall effects exist as well [26, 28]. But in the presented case of a swirling non-premixed free stream flame the closure of the term by the SLM is assumed to be adequate and was chosen because of its simplicity.
Term VI regards the effect of molecular diffusion within the fluid. This diffusion flattens the steep composition gradients which are created by the strong vortices in a turbulent flow. Several models have been proposed to close this term. The simplest model is the interaction by exchange with the mean model (IEM) [29, 30] which models the fact that fluctuations in the composition space relax to the mean. A more detailed model has been proposed by Curl [31] and modified by [32, 33] and is used in its modified form in the presented work. More recently new models based on Euclidian minimum spanning trees have been developed [34, 35] but are not yet implemented in this work.
As mentioned previously it is numerically unfeasable to solve the PDF transport equation with finite-volume or finite-difference methods because of its high dimensionality. Therefore a Monte Carlo method is used to solve the transport equation making use of the fact that the PDF of a fluid flow can be represented as a sum of δ-functions.

$$f^*_{\mathbf{U},\phi}(\mathbf{U},\mathbf{\Psi};\mathbf{x},t) = \sum_{i=1}^{N(t)} \delta\left(\mathbf{v}-\mathbf{u}^i\right)\delta\left(\phi-\mathbf{\Psi}^i\right)\delta\left(\mathbf{x}-\mathbf{x}^i\right) \tag{14}$$

Instead of the high dimensional PDF transport equation using a particle Monte Carlo method a set of (stochastic) ordinary differential equations are solved for each numerical particle discretizing the PDF. The evolution of the particle position $\mathbf{X}^*_\mathrm{i}$ reads

$$\frac{\mathrm{d}\mathbf{X}^*_\mathrm{i}}{\mathrm{d}t} = \mathbf{U}^*_\mathrm{i}(t) \tag{15}$$

in which $\mathbf{U}^*_\mathrm{i}$ is the velocity vector for each particle.
The evolution of the particles in the velocity space can be calculated according to the Simplified Langevin Model [1] by

$$\mathrm{d}U^*_i = -\frac{\partial\bar{p}}{\partial x_i}\mathrm{d}t - \left(\frac{1}{2}+\frac{3}{4}C_0\right)\left[U^*_i - \langle U_i\rangle\right]\frac{\mathrm{d}t}{\tau} + \sqrt{\frac{C_0 k}{\tau}}\mathrm{d}W_i \quad . \tag{16}$$

For simplicity the equation is here only written for the U component of the velcity vector $\mathbf{U} = (U,V,W)^T$ belonging to the spacial coordinate x ($\mathbf{x} = (x,y,z)^T$). The equations of the other components V, W look accordingly. In eqn. 16 $\frac{\partial\bar{p}}{\partial x_i}$ denotes the mean pressure gradient, $\langle U_i\rangle$ the mean particle velocity, t the time, $\mathrm{d}W_i$ a differential Wiener increment, C_0 a model constant,

k and τ the turbulent kinetic energy and the turbulent time scale, respectively. Finally the evolution of the composition vector can be calculated as

$$\frac{\mathrm{d}\boldsymbol{\Psi}}{\mathrm{d}t} = S + M \tag{17}$$

in which S is the chemical source term (appearing in closed form) and M denotes the effect of molecular mixing. As previously mentioned this term is unclosed und needs further modeling assumptions. For this a modified Curl model is used [32].

2.3 Chemical Kinetics

The source term appearing in eqn. 17 is calculated from a lookup table which is created using automatically reduced chemical mechanisms. The deployed technique to create these tables is the ILDM method ("Intrinsic Low-Dimensional Manifold") by Maas and Pope [17, 18].
The basic idea of this method is the identification and separation of fast and slow time scales. In typical turbulent flames the time scales governing the chemical kinetics range from 10^{-9}s to 10^{2}s. This is a much larger spectrum than that of the physical processes (e.g. molecular transport) which vary only from 10^{-1}s to 10^{-5}s. Reactions that occur in the very fast chemical time scales are in partial equilibrium and the species are in steady state. These are usually responsible for equilibrium processes. Making use of this fact it is possible to decouple the fast time scales. The main advantage of decoupling the fast time scales is that the chemical system can be described with a much smaller number of variables (degrees of freedom).
In our test case the chemical kinetics are described with only three parameters namely the mixure fraction, the mole fraction of CO_2 and the mole fraction of H_2O instead of the 34 species (degrees of freedom) appearing in the detailed methane reaction mechanism. Further details of the method and its implementation can be found in [17, 18].

3 Results and Discussion

As a test case for the presented model simulations of a premixed, swirling, confined flame are performed. A sketch of the whole test rig is shown in Fig. 2. Details of the test rig and the experimental data can be found in [20, 21, 22].

The test rig consists of a plenum containing a premixed methane-air mixture, a swirl generator, a premixing duct and the combustion chamber itself. In general three different modes exist to stabilize flames. Flames can be stabilized by a small stable burning pilot, by bluff-bodies inserted into the main

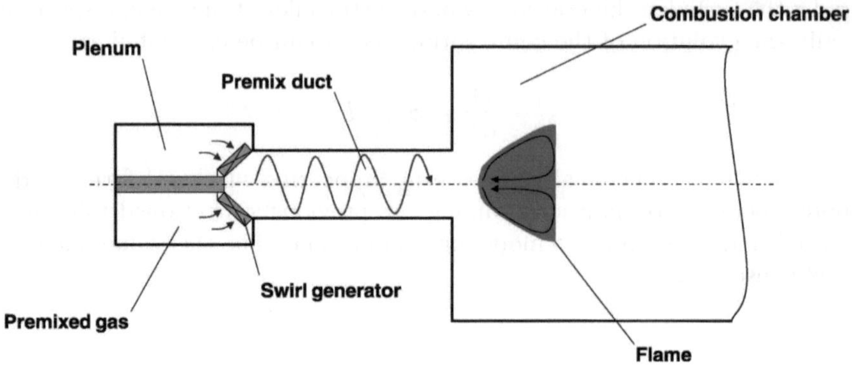

Fig. 2. Sketch of the investigated combustion chamber

flow or by aerodynamic arrangements creating a recirculation zone above the burner exit. The last possibility has been increasingly employed for flame stabilization in the gas turbine industry. The recirculation zone (often also abreviated IRZ [4]) is a region of negative axial velocity close to the symmetry line (see Fig. 2). Heat and radicals are transported upstream towards the flame tip causing a stable operation of the flame. The occurrence and stability of the IRZ depend crucially on the swirl number, the geometry, and the profiles of the axial and tangential velocity.
Simulations were performed using a 2D axi-symmetric grid with approximately 15000 cells. The PDF is discretized with 50 particles per cell. The position of the simulated domain is shown in Fig. 3. Only every forth grid line is shown for clarity. In this case the mapping of the real geometry (3D) on the 2D axi-symmetric solution domain is possible since the experiments show that all essential features of the flow field exhibit the two dimensional axi-symmetric behaviour [36]. The mapping approach has shown to be valid for the modeling also in [19]. In order to consider the influence of the velocity profiles created by the swirl gernerator radial profiles of all flow quantities

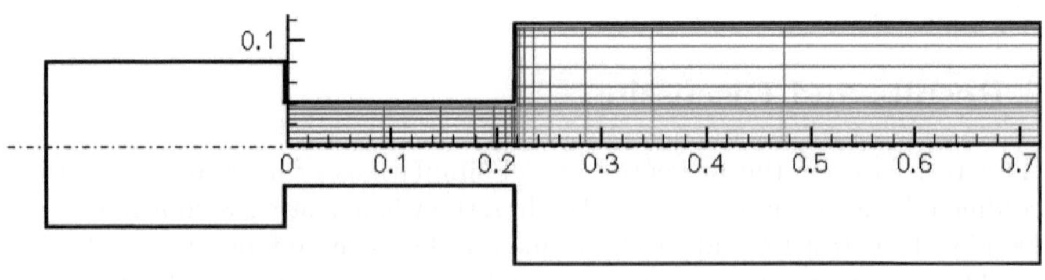

Fig. 3. Position of the mesh in the combustion chamber

[4] Internal Recirculation Zone

served as inlet boundary conditions. These profiles stem from detailed 3D simulations of the whole test rig using a Reynolds stress turbulence closure and have been taken from the literature [22].

The global operation parameters are an equivalence ratio of $\phi = 1$, an inlet mass flow of $70\frac{\text{g}}{\text{s}}$, a preheated temperature of 373K and a swirl number of $S = 0.5$.

First of all simulations of the non-reacting case were done to validate the CFD model and the used boundary conditions which are mapped from the detailed 3D simulations. Fig. 4 shows an example of the achieved results. From the steamtraces one can see two areas with negative axial velocity. One in the upper left corner of the combustion chamber is caused by the step in the geometry and one close to the symmetry line which is caused aerodynamically by the swirl. This area is the internal recirculation zone described above which is in the reactive case used to stabilize the flame. These simulations are validated with experimental results from [20, 21]. The comparison of the experimental data and the results of the simulations for one case are exemplarily shown in Fig. 5 and Fig. 6. In all figures the radial coordinate is plotted over the velocity. Both upper figures show the axial velocity, both lower show the tangential velocity. The lines denote the results of the simulations the scatters denote the results of the measurements. The two axial positions are arbitrarily chosen from the available experimental data. The (relative) x coordinates refer to the beginning of the premixing duct (Fig. 2). In both cases the profiles of the simulations seem to match reasonably well with the measured data. So the presented model gives a sound description of the flow field of the investigated test case.

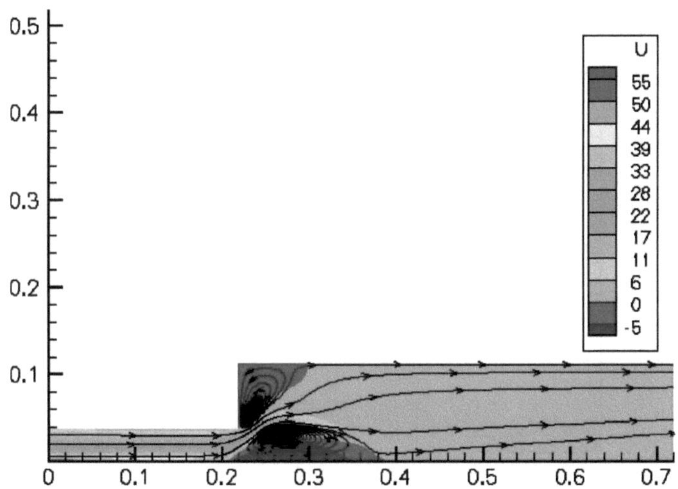

Fig. 4. Contourplot of the axial velocity component (with steamtraces)

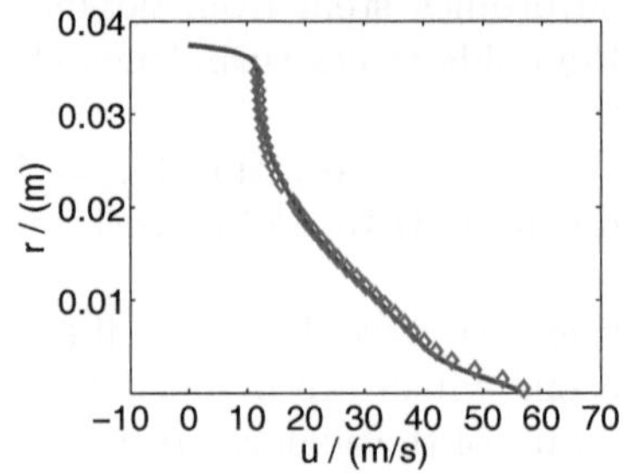

(a) Axial velocity

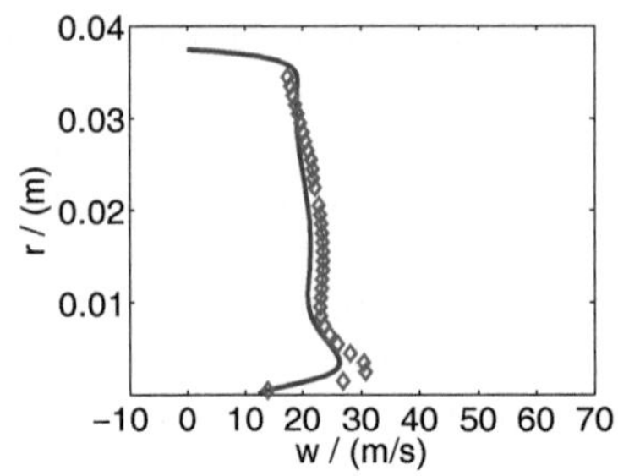

(b) Tangential velocity

Fig. 5. $x = 29\text{mm}$

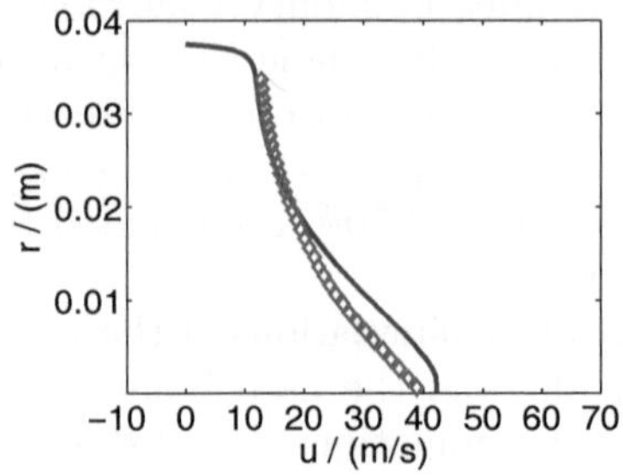

(a) Axial velocity

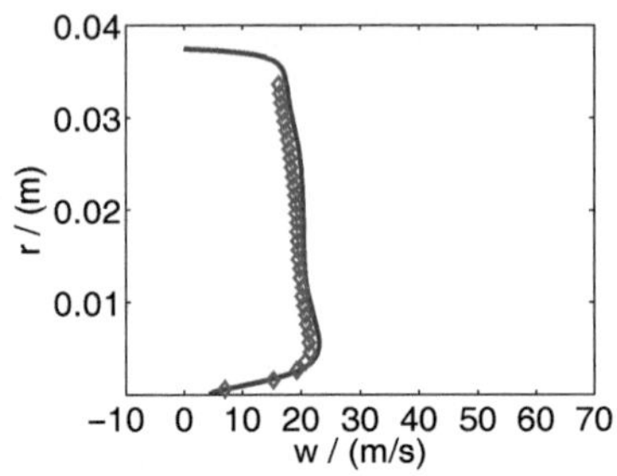

(b) Tangential velocity

Fig. 6. $x = 170\text{mm}$

As an example for the reacting case the calculated temperature field of the flame is shown in Fig. 7 which can not be compared to quantitative experiments due to the lack of data. But the qualitative behaviour of the flame is predicted correctly. As one can see the tip of the flame is located at the start of the inner recirulation zone. It shows a turbulent flame brush in which the

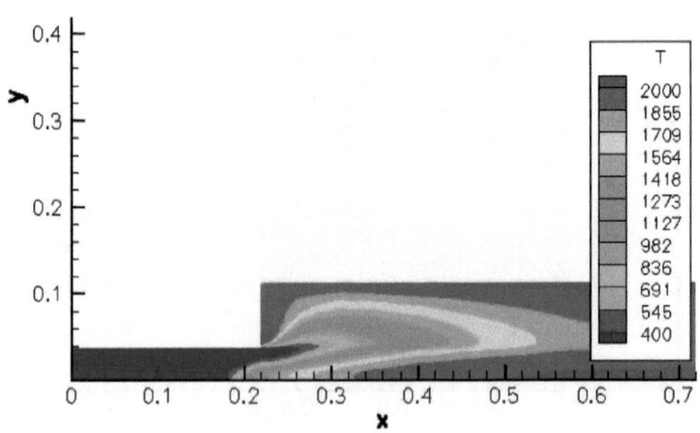

Fig. 7. Temperature field

reaction occurs which can be seen in the figure by the rise of temperature. It can not be assessed whether the thickness of the reaction zone is predicted well because no measurements of the temperature field are available.

4 Conclusion

Simulations of a premixed swirling methane-air flame are presented. To account for the strong turbulence chemistry interaction occuring in these flames a hybrid finite-volume/transported PDF model is used. This model consists of two parts: a finte volume solver for the mean velocities and the mean pressure gradient and a Monte Carlo solver for the transport equation of the joint PDF of velocity and compostion vector. Chemical kinetics are described by automatically reduced mechanisms created with the ILDM method.
The presented results show the validity of the model. The simulated velocity profiles match well with the experimental results. The calculations of the reacting case also show a qualitatively correct behaviour of the flame. A quantitative analysis is subject of future research work.

References

1. S.B. Pope. Pdf methods for turbulent reactive flows. *Progress in Energy Combustion Science*, 11:119–192, 1985.
2. S.B Pope. Lagrangian pdf methods for turbulent flows. *Annual Review of Fluid Mechanics*, 26:23–63, 1994.
3. Z. Ren and S.B. Pope. An investigation of the performence of turbulent mixing models. *Combustion and Flame*, 136:208–216, 2004.
4. P.R. Van Slooten and S.B Pope. Application of pdf modeling to swirling and nonswirling turbulent jets. *Flow Turbulence and Combustion*, 62(4):295–334, 1999.
5. V. Saxena and S.B Pope. Pdf simulations of turbulent combustion incorporating detailed chemistry. *Combustion and Flame*, 117(1-2):340–350, 1999.
6. S. Repp, A. Sadiki, C. Schneider, A. Hinz, T. Landenfeld, and J. Janicka. Prediction of swirling confined diffusion flame with a monte carlo and a presumed-pdf-model. *International Journal of Heat and Mass Transfer*, 45:1271–1285, 2002.
7. K. Liu, S.B. Pope, and D.A. Caughey. Calculations of bluff-body stabilized flames using a joint probability density function model with detailed chemistry. *Combustion and Flame*, 141:89–117, 2005.
8. J.H. Ferziger and M. Peric. *Computational Methods for Fluid Dynamics*. Springer Verlag, 2 edition, 1997.
9. S.B Pope. A monte carlo method for pdf equations of turbulent reactive flow. *Combustion, Science and Technology*, 25:159–174, 1981.
10. P. Jenny, M. Muradoglu, K. Liu, S.B. Pope, and D.A. Caughey. Pdf simulations of a bluff-body stabilized flow. *Journal of Computational Physics*, 169:1–23, 2000.

11. A.K. Tolpadi, I.Z. Hu, S.M. Correa, and D.L. Burrus. Coupled lagrangian monte carlo pdf-cfd computation of gas turbine combustor flowfields with finite-rate chemistry. *Journal of Engineering for Gas Turbines and Power*, 119:519–526, 1997.
12. M. Muradoglu, P. Jenny, S.B Pope, and D.A. Caughey. A consistent hybrid finite-volume/particle method for the pdf equations of turbulent reactive flows. *Journal of Computational Physics*, 154:342–370, 1999.
13. M. Muradoglu, S.B. Pope, and D.A. Caughey. The hybid method for the pdf equations of turbulent reactive flows: Consistency conditions and correction algorithms. *Journal of Computational Physics*, 172:841–878, 2001.
14. G. Li and M.F. Modest. An effective particle tracing scheme on structured/unstructured grids in hybrid finite volume/pdf monte carlo methods. *Journal of Computational Physics*, 173:187–207, 2001.
15. V. Raman, R.O. Fox, and A.D. Harvey. Hybrid finite-volume/transported pdf simulations of a partially premixed methane-air flame. *Combustion and Flame*, 136:327–350, 2004.
16. H.S. Zhang, R.M.C. So, C.G. Speziale, and Y.G. Lai. A near-wall two-equation model for compressible turbulent flows. In *Aerospace Siences Meeting and Exhibit, 30th, Reno, NV*, page 23. AIAA, 1992.
17. U. Maas and S. B. Pope. Simplifying chemical kinetics: Intrinsic low-dimensional manifolds in composition space. *Combustion and Flame*, 88:239–264, 1992.
18. U. Maas and S.B. Pope. Implementation of simplified chemical kinetics based on intrinsic low-dimensional manifolds. In *Twenty-Fourth Symposium (International) on Combustion*, pages 103–112. The Combustion Institute, 1992.
19. F. Kiesewetter, C. Hirsch, J. Fritz, M. Kröner, and T. Sattelmayer. Two-dimensional flashback simulation in strongly swirling flows. In *Proceedings of ASME Turbo Expo 2003*.
20. M. Kröner. *Einfluss lokaler Löschvorgänge auf den Flammenrückschlag durch verbrennungsinduziertes Wirbelaufplatzen.* PhD thesis, Technische Universität München, Fakultät für Maschinenwesen, 2003.
21. J. Fritz. *Flammenrückschlag durch verbrennungsinduziertes Wirbelaufplatzen.* PhD thesis, Technische Universität München, Fakultät für Maschinenwesen, 2003.
22. F. Kiesewetter. *Modellierung des verbrennungsinduzierten Wirbelaufplatzens in Vormischbrennern.* PhD thesis, Technische Universität München, Fakultät für Maschinenwesen, 2005.
23. D.C. Haworth and S.H. El Tahry. Propbability density function approach for multidimensional turbulent flow calculations with application to in-cylinder flows in reciproating engines. *AIAA Journal*, 29:208, 1991.
24. S.M. Correa and S.B. Pope. Comparison of a monte carlo pdf finite-volume mean flow model with bluff-body raman data. In *Twenty-Fourth Symposium (International) on Combustion*, page 279. The Combustion Institute, 1992.
25. W.C. Welton and S.B. Pope. Pdf model calculations of compressible turbulent flows using smoothed particle hydrodynamics. *Journal of Computational Physics*, 134:150, 1997.
26. D.C. Haworth and S.B. Pope. A generalized langevin model for turbulent flows. *Physics of Fluids*, 29:387–405, 1986.
27. H.A. Wouters, T.W. Peeters, and D. Roekaerts. On the existence of a generalized langevin model representation for second-moment closures. *Physics of Fluids*, 8, 1996.

28. T.D. Dreeben and S.B. Pope. Pdf/monte carlo simulation of near-wall turbulent flows. *Journal of Fluid Mechanics*, 357:141–166, 1997.
29. C. Dopazo. Relaxation of initial probability density functions in the turbulent convection of scalar flieds. *Physics of Fluids*, 22:20–30, 1979.
30. P.A. Libby and F.A. Williams. *Turbulent Reacting Flows.* Academic Press, 1994.
31. R.L. Curl. Dispersed phase mixing: 1. theory and effects in simple reactors. *A.I.Ch.E. Journal*, 9:175,181, 1963.
32. J. Janicka, W. Kolbe, and W. Kollmann. Closure of the transport equation of the probability density function of turbulent scalar flieds. *Journal of Non-Equilibrium Thermodynamics*, 4:47–66, 1979.
33. S.B Pope. An improved turbulent mixing model. *Combustion, Science and Technology*, 28:131–135, 1982.
34. S. Subramaniam and S.B Pope. A mixing model for turbulent reactive flows based on euclidean minimum spanning trees. *Combustion and Flame*, 115(4):487–514, 1999.
35. S. Subramaniam and S.B Pope. Comparison of mixing model performance for nonpremixed turbulent reactive flow. *Combustion and Flame*, 117(4):732–754, 1999.
36. J. Fritz, M. Kroñer, and T. Sattelmayer. Flashback in a swirl burner with cylindrical premixing zone. In *Proceedings of ASME Turbo Expo 2001*.

Supernova Simulations with the Radiation Hydrodynamics Code PROMETHEUS/VERTEX

B. Müller[1], A. Marek[1], K. Benkert[2], K. Kifonidis[1], and H.-Th. Janka[1]

[1] Max-Planck-Institut für Astrophysik, Karl-Schwarzschild-Strasse 1, Postfach 1317, D-85741 Garching bei München, Germany
bjmuellr@mpa-garching.mpg.de
[2] High Performance Computing Center Stuttgart (HLRS), Nobelstrasse 19, D-70569 Stuttgart, Germany

Summary. We give an overview of the problems and the current status of our two-dimensional (core collapse) supernova modelling, and discuss the system of equations and the algorithm for its solution that are employed in our code. In particular we report our recent progress, and focus on the ongoing calculations that are performed on the NEC SX-8 at the HLRS Stuttgart. We also discuss recent optimizations carried out within the framework of the Teraflop Workbench, and comment on the parallel performance of the code, stressing the importance of developing a MPI version of the employed hydrodynamics module.

1 Introduction

A star more massive than about 8 solar masses ends its live in a cataclysmic explosion, a supernova. Its quiescent evolution comes to an end, when the pressure in its inner layers is no longer able to balance the inward pull of gravity. Throughout its life, the star sustained this balance by generating energy through a sequence of nuclear fusion reactions, forming increasingly heavier elements in its core. However, when the core consists mainly of iron-group nuclei, central energy generation ceases. The fusion reactions producing iron-group nuclei relocate to the core's surface, and their "ashes" continuously increase the core's mass. Similar to a white dwarf, such a core is stabilised against gravity by the pressure of its degenerate gas of electrons. However, to remain stable, its mass must stay smaller than the Chandrasekhar limit. When the core grows larger than this limit, it collapses to a neutron star, and a huge amount ($\sim 10^{53}$ erg) of gravitational binding energy is set free. Most ($\sim 99\%$) of this energy is radiated away in neutrinos, but a small fraction is transferred to the outer stellar layers and drives the violent mass ejection which disrupts the star in a supernova.

Despite 40 years of research, the details of how this energy transfer happens and how the explosion is initiated are still not well understood. Observational evidence about the physical processes deep inside the collapsing star is sparse and almost exclusively indirect. The only direct observational access is via measurements of neutrinos or gravitational waves. To obtain insight into the events in the core, one must therefore heavily rely on sophisticated numerical simulations. The enormous amount of computer power required for this purpose has led to the use of several, often questionable, approximations and numerous ambiguous results in the past. Fortunately, however, the development of numerical tools and computational resources has meanwhile advanced to a point, where it is becoming possible to perform multi-dimensional simulations with unprecedented accuracy. Therefore there is hope that the physical processes which are essential for the explosion can finally be unravelled.

An understanding of the explosion mechanism is required to answer many important questions of nuclear, gravitational, and astro-physics like the following:

- How do the explosion energy, the explosion timescale, and the mass of the compact remnant depend on the progenitor's mass? Is the explosion mechanism the same for all progenitors? For which stars are black holes left behind as compact remnants instead of neutron stars?
- What is the role of the – poorly known – equation of state (EoS) for the proto neutron star? Do softer or stiffer EoSs favour the explosion of a core collapse supernova?
- What is the role of rotation during the explosion? How rapidly do newly formed neutron stars rotate?
- How do neutron stars receive their natal kicks? Are they accelerated by asymmetric mass ejection and/or anisotropic neutrino emission?
- What are the generic properties of the neutrino emission and of the gravitational wave signal that are produced during stellar core collapse and explosion? Up to which distances could these signals be measured with operating or planned detectors on earth and in space? And what can one learn about supernova dynamics from a future measurement of such signals in case of a Galactic supernova?

2 Numerical models

2.1 History and constraints

According to theory, a shock wave is launched at the moment of "core bounce" when the neutron star begins to emerge from the collapsing stellar iron core. There is general agreement, supported by all "modern" numerical simulations, that this shock is unable to propagate directly into the stellar mantle and envelope, because it looses too much energy in dissociating iron into free nucleons while it moves through the outer core. The "prompt" shock ultimately stalls.

Thus the currently favoured theoretical paradigm needs to exploit the fact that a huge energy reservoir is present in the form of neutrinos, which are abundantly emitted from the hot, nascent neutron star. The absorption of electron neutrinos and antineutrinos by free nucleons in the post shock layer is thought to reenergize the shock, and lead to the supernova explosion.

Detailed *spherically symmetric* hydrodynamic models, which recently include a very accurate treatment of the time-dependent, multi-flavour, multi-frequency neutrino transport based on a numerical solution of the Boltzmann transport equation [1, 2, 3], reveal that this "delayed, neutrino-driven mechanism" does not work as simply as originally envisioned. Although in principle able to trigger the explosion (e.g., [4], [5], [6]), neutrino energy transfer to the postshock matter turned out to be too weak. For inverting the infall of the stellar core and initiating powerful mass ejection, an increase of the efficiency of neutrino energy deposition is needed.

A number of physical phenomena have been pointed out that can enhance neutrino energy deposition behind the stalled supernova shock. They are all linked to the fact that the real world is multi-dimensional instead of spherically symmetric (or one-dimensional; 1D) as assumed in the work cited above:

(1) Convective instabilities in the neutrino-heated layer between the neutron star and the supernova shock develop to violent convective overturn [7]. This convective overturn is helpful for the explosion, mainly because (a) neutrino-heated matter rises and increases the pressure behind the shock, thus pushing the shock further out, and (b) cool matter is able to penetrate closer to the neutron star where it can absorb neutrino energy more efficiently. Both effects allow multi-dimensional models to explode easier than spherically symmetric ones [8, 9, 10].
(2) Recent work [11, 12, 13, 14] has demonstrated that the stalled supernova shock is also subject to a second non-radial low-mode instability, called SASI, which can grow to a dipolar, global deformation of the shock [14, 15].
(3) Convective energy transport inside the nascent neutron star [16, 17, 18, 19] might enhance the energy transport to the neutrinosphere and could thus boost the neutrino luminosities. This would in turn increase the neutrino-heating behind the shock.

This list of multi-dimensional phenomena awaits more detailed exploration in multi-dimensional simulations. Until recently, such simulations have been performed with only a grossly simplified treatment of the involved microphysics, in particular of the neutrino transport and neutrino-matter interactions. At best, grey (i.e., single energy) flux-limited diffusion schemes were employed. All published successful simulations of supernova explosions by the convectively aided neutrino-heating mechanism in two [8, 9, 20] and three dimensions [21, 22] used such a radical approximation of the neutrino transport.

Since, however, the role of the neutrinos is crucial for the problem, and because previous experience shows that the outcome of simulations is indeed very sensitive to the employed transport approximations, studies of the explosion mechanism require the best available description of the neutrino physics. This implies that one has to solve the Boltzmann transport equation for neutrinos.

2.2 Recent calculations and the need for TFLOP simulations

We have recently advanced to a new level of accuracy for supernova simulations by generalising the VERTEX code, a Boltzmann solver for neutrino transport, from spherical symmetry [23] to multi-dimensional applications [24, 25]. The corresponding mathematical model, and in particular our method for tackling the integro-differential transport problem in multi-dimensions, will be summarised in Sect. 3.

Results of a set of simulations with our code in 1D and 2D for progenitor stars with different masses have recently been published by [25, 26], and with respect to the expected gravitational-wave signals from rotating and convective supernova cores by [27]. The recent progress in supernova modelling was summarised and set in perspective in a conference article by [24].

Our collection of simulations has helped us to identify a number of effects which have brought our two-dimensional models close to the threshold of explosion. This makes us optimistic that the solution of the long-standing problem of how massive stars explode may be in reach. In particular, we have recognised the following aspects as advantageous:

- The details of the stellar progenitor (i.e. the mass of the iron core and its radius–density relation) have substantial influence on the supernova evolution. Especially, we found explosions of stellar models with low-mass (i.e. small) iron cores [28, 26], whereas more massive stars resist the explosion more persistent [25]. Thus detailed studies with different progenitor models are necessary.
- Stellar rotation, even at a moderate level, supports the expansion of the stalled shock by centrifugal forces and instigates overturn motion in the neutrino-heated postshock matter by meridional circulation flows in addition to convective instabilities.

All these effects are potentially important, and some (or even all of them) may represent crucial ingredients for a successful supernova simulation. So far no multi-dimensional calculations have been performed, in which two or more of these items have been taken into account simultaneously, and thus their mutual interaction awaits to be investigated. It should also be kept in mind that our knowledge of supernova microphysics, and especially the EoS of neutron star matter, is still incomplete, which implies major uncertainties for supernova modelling. Unfortunately, the impact of different descriptions for this input physics has so far not been satisfactorily explored with respect to the neutrino-heating mechanism and the long-time behaviour of the supernova shock, in particular in multi-dimensional models. However, first multi-dimensional simulations of core collapse supernovae with different nuclear EoSs [29, 19] show a strong dependence of the supernova evolution on the EoS.

In recent simulations – partly performed on the SX-8 at HLRS, typically on 8 processors with 22000 MFLOP per second – we have found a developing explosion for a rotating 15 $M_\odot$ progenitor star at a time of roughly 500 ms after shock formation (see Fig. 1). The reason for pushing this simulation to such late times is that rotation and angular momentum become more and more important at later times as matter has fallen from larger radii to the shock position. However, it is not yet clear whether the presence of rotation is crucial for the explosion of this 15 $M_\odot$ model, or whether this model would also explode without rotation. Since the comparison of the rotating and a corresponding non-rotating model reveals qualitatively the same behaviour, see e.g. Fig. 2, it is absolutely necessary to evolve both models to a time of more than 500 ms after the shock formation in order to answer this question. In any case, our results suggest that the neutrino-driven mechanism may work at rather late times, at least as long as the simulations remain limited to axisymmetry.

From this it is clear that rather extensive parameter studies carrying multidimensional simulations until late times are required to identify the physical processes which are essential for the explosion. Since on a dedicated machine performing at a sustained speed of about 30 GFLOPS already a single 2D simulation has a turn-around time of more than a year, these parameter studies are hardly feasible without TFLOP capability of the code.

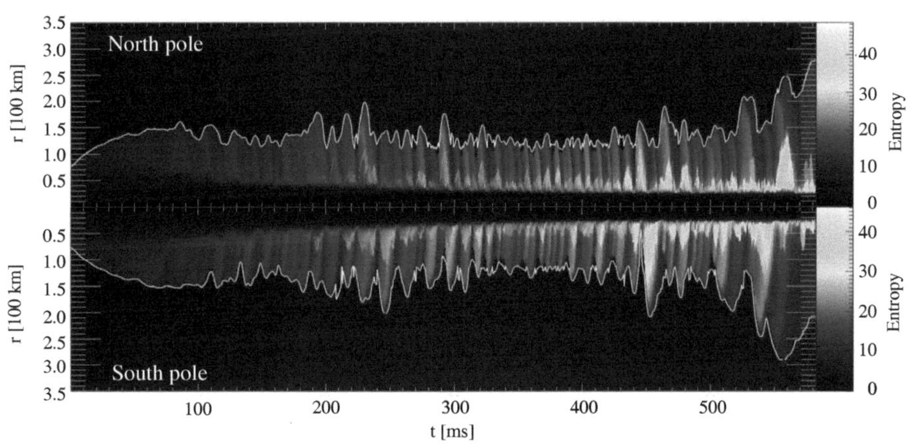

Fig. 1. The shock position (solid white line) at the north pole (upper panel) and south pole (lower panel) of the rotating 15 $M_\odot$ model as function of postbounce time. Colour coded is the entropy of the stellar fluid.

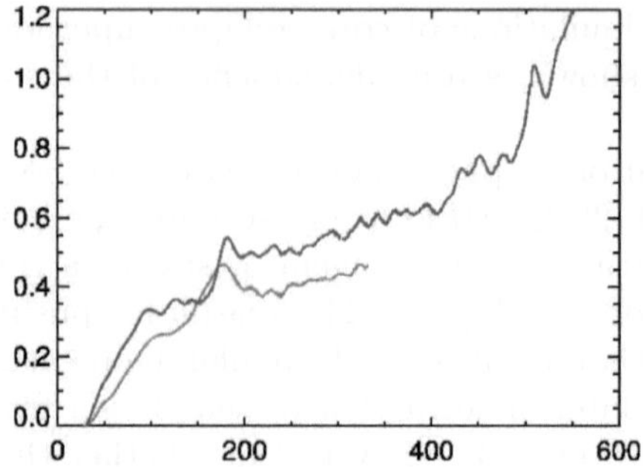

Fig. 2. The ratio of the advection timescale to the heating timescale for the rotating model L&S-rot and the non-rotating model L&S-2D. Also shown is model L&S-rot-90 which is identical to model L&S-rot except for the computational domain that does not extend from pole to pole but from the north pole to the equator. The advection timescale is the characteristic timescale that matter stays inside the heating region before it is advected to the proto-neutron star. The heating timescale is the typical timescale that matter needs to be exposed to neutrino heating for observing enough energy to become gravitationally unbound.

3 The mathematical model

The non-linear system of partial differential equations which is solved in our code consists of the following components:

- The Euler equations of hydrodynamics, supplemented by advection equations for the electron fraction and the chemical composition of the fluid, and formulated in spherical coordinates;
- the Poisson equation for calculating the gravitational source terms which enter the Euler equations, including corrections for general relativistic effects;
- the Boltzmann transport equation which determines the (non-equilibrium) distribution function of the neutrinos;
- the emission, absorption, and scattering rates of neutrinos, which are required for the solution of the Boltzmann equation;
- the equation of state of the stellar fluid, which provides the closure relation between the variables entering the Euler equations, i.e. density, momentum, energy, electron fraction, composition, and pressure.

For the integration of the Euler equations, we employ the time-explicit finite-volume code PROMETHEUS, which is an implementation of the third-order Piecewise Parabolic Method (PPM) of Colella and Woodward [30], and is described elsewhere in more detail [31].

In what follows we will briefly summarise the neutrino transport algorithms. For a more complete description of the entire code we refer the reader to [25], and the references therein.

3.1 "Ray-by-ray plus" variable Eddington factor solution of the neutrino transport problem

The crucial quantity required to determine the source terms for the energy, momentum, and electron fraction of the fluid owing to its interaction with the neutrinos is the neutrino distribution function in phase space, $f(r, \vartheta, \phi, \epsilon, \Theta, \Phi, t)$. Equivalently, the neutrino intensity $I = c/(2\pi\hbar c)^3 \cdot \epsilon^3 f$ may be used. Both are seven-dimensional functions, as they describe, at every point in space (r, ϑ, ϕ), the distribution of neutrinos propagating with energy ϵ into the direction (Θ, Φ) at time t (Fig. 3).

The evolution of I (or f) in time is governed by the Boltzmann equation, and solving this equation is, in general, a six-dimensional problem (as time is usually not counted as a separate dimension). A solution of this equation by direct discretisation (using an S_N scheme) would require computational resources in the Petaflop range. Although there are attempts by at least one group in the United States to follow such an approach, we feel that, with the currently available computational resources, it is mandatory to reduce the dimensionality of the problem.

Actually this should be possible, since the source terms entering the hydrodynamic equations are *integrals* of I over momentum space (i.e. over ϵ, Θ, and Φ), and thus only a fraction of the information contained in I is truly required to compute the dynamics of the flow. It makes therefore sense to consider angular moments of I, and to solve evolution equations for these moments, instead of dealing with the Boltzmann equation directly. The 0th to 3rd order moments are defined as

$$J, \boldsymbol{H}, \mathbf{K}, \mathbf{L}, \ldots (r, \vartheta, \phi, \epsilon, t) = \frac{1}{4\pi} \int I(r, \vartheta, \phi, \epsilon, \Theta, \Phi, t)\, \boldsymbol{n}^{0,1,2,3,\ldots}\, \mathrm{d}\Omega \qquad (1)$$

where $\mathrm{d}\Omega = \sin\Theta\, \mathrm{d}\Theta\, \mathrm{d}\Phi$, $\boldsymbol{n} = (\cos\Theta, \sin\Theta\cos\Phi, \sin\Theta\sin\Phi)$, and exponentiation represents repeated application of the dyadic product. Note that the moments are *tensors* of the required rank.

This leaves us with a four-dimensional problem. So far no approximations have been made. In order to reduce the size of the problem even further,

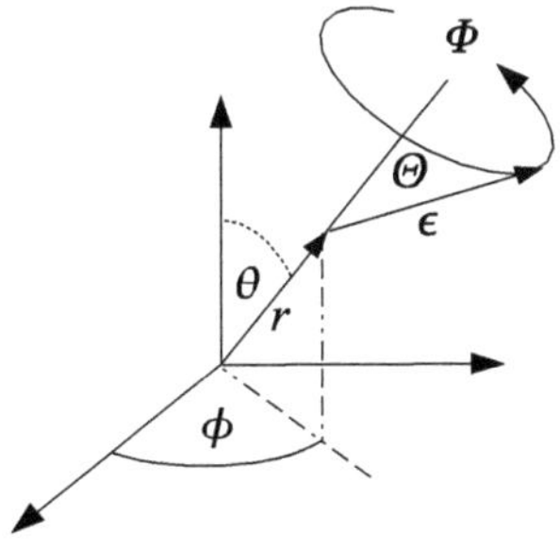

Fig. 3. Illustration of the phase space coordinates (see the main text).

one needs to resort to assumptions on its symmetry. At this point, one usually employs azimuthal symmetry for the stellar matter distribution, i.e. any dependence on the azimuth angle ϕ is ignored, which implies that the hydrodynamics of the problem can be treated in two dimensions. It also implies $I(r,\vartheta,\epsilon,\Theta,\Phi) = I(r,\vartheta,\epsilon,\Theta,-\Phi)$. If, in addition, it is assumed that I is even independent of Φ, then each of the angular moments of I becomes a *scalar*, which depends on two spatial dimensions, and one dimension in momentum space: $J,H,K,L = J,H,K,L(r,\vartheta,\epsilon,t)$. Thus we have reduced the problem to three dimensions in total.

The system of equations

With the aforementioned assumptions it can be shown [25], that in order to compute the source terms for the energy and electron fraction of the fluid, the following two transport equations need to be solved:

$$\begin{aligned}
&\left(\frac{1}{c}\frac{\partial}{\partial t} + \beta_r\frac{\partial}{\partial r} + \frac{\beta_\vartheta}{r}\frac{\partial}{\partial \vartheta}\right) J + J\left(\frac{1}{r^2}\frac{\partial(r^2\beta_r)}{\partial r} + \frac{1}{r\sin\vartheta}\frac{\partial(\sin\vartheta\beta_\vartheta)}{\partial\vartheta}\right)\\
&+\frac{1}{r^2}\frac{\partial(r^2 H)}{\partial r} + \frac{\beta_r}{c}\frac{\partial H}{\partial t} - \frac{\partial}{\partial\epsilon}\left\{\frac{\epsilon}{c}\frac{\partial\beta_r}{\partial t}H\right\} - \frac{\partial}{\partial\epsilon}\left\{\epsilon J\left(\frac{\beta_r}{r} + \frac{1}{2r\sin\vartheta}\frac{\partial(\sin\vartheta\beta_\vartheta)}{\partial\vartheta}\right)\right\}\\
&-\frac{\partial}{\partial\epsilon}\left\{\epsilon K\left(\frac{\partial\beta_r}{\partial r} - \frac{\beta_r}{r} - \frac{1}{2r\sin\vartheta}\frac{\partial(\sin\vartheta\beta_\vartheta)}{\partial\vartheta}\right)\right\} + J\left(\frac{\beta_r}{r} + \frac{1}{2r\sin\vartheta}\frac{\partial(\sin\vartheta\beta_\vartheta)}{\partial\vartheta}\right)\\
&\qquad + K\left(\frac{\partial\beta_r}{\partial r} - \frac{\beta_r}{r} - \frac{1}{2r\sin\vartheta}\frac{\partial(\sin\vartheta\beta_\vartheta)}{\partial\vartheta}\right) + \frac{2}{c}\frac{\partial\beta_r}{\partial t}H = C^{(0)}, \quad (2)
\end{aligned}$$

$$\begin{aligned}
&\left(\frac{1}{c}\frac{\partial}{\partial t} + \beta_r\frac{\partial}{\partial r} + \frac{\beta_\vartheta}{r}\frac{\partial}{\partial \vartheta}\right) H + H\left(\frac{1}{r^2}\frac{\partial(r^2\beta_r)}{\partial r} + \frac{1}{r\sin\vartheta}\frac{\partial(\sin\vartheta\beta_\vartheta)}{\partial\vartheta}\right)\\
&\qquad + \frac{\partial K}{\partial r} + \frac{3K-J}{r} + H\left(\frac{\partial\beta_r}{\partial r}\right) + \frac{\beta_r}{c}\frac{\partial K}{\partial t} - \frac{\partial}{\partial\epsilon}\left\{\frac{\epsilon}{c}\frac{\partial\beta_r}{\partial t}K\right\}\\
&\qquad - \frac{\partial}{\partial\epsilon}\left\{\epsilon L\left(\frac{\partial\beta_r}{\partial r} - \frac{\beta_r}{r} - \frac{1}{2r\sin\vartheta}\frac{\partial(\sin\vartheta\beta_\vartheta)}{\partial\vartheta}\right)\right\}\\
&-\frac{\partial}{\partial\epsilon}\left\{\epsilon H\left(\frac{\beta_r}{r} + \frac{1}{2r\sin\vartheta}\frac{\partial(\sin\vartheta\beta_\vartheta)}{\partial\vartheta}\right)\right\} + \frac{1}{c}\frac{\partial\beta_r}{\partial t}(J+K) = C^{(1)}. \quad (3)
\end{aligned}$$

These are evolution equations for the neutrino energy density, J, and the neutrino flux, H, and follow from the zeroth and first moment equations of the comoving frame (Boltzmann) transport equation in the Newtonian, $\mathcal{O}(v/c)$ approximation. The quantities $C^{(0)}$ and $C^{(1)}$ are source terms that result from the collision term of the Boltzmann equation, while $\beta_r = v_r/c$ and $\beta_\vartheta = v_\vartheta/c$, where v_r and v_ϑ are the components of the hydrodynamic velocity, and c is the speed of light. The functional dependences $\beta_r = \beta_r(r,\vartheta,t)$, $J = J(r,\vartheta,\epsilon,t)$, etc. are suppressed in the notation. This system includes four unknown moments (J,H,K,L) but only two equations, and thus needs to be supplemented by two more relations. This is done by substituting $K = f_K \cdot J$ and $L = f_L \cdot J$, where f_K and f_L are the variable Eddington factors, which for the moment may be regarded as being known, but in our case are indeed determined from the formal solution of a simplified ("model") Boltzmann equation. For the adopted coordinates, this amounts to the solution of *independent* one-dimensional PDEs (typically more than 200 for each ray), hence very efficient vectorization is possible [23].

A finite volume discretisation of Eqs. (2–3) is sufficient to guarantee exact conservation of the total neutrino energy. However, and as described in detail in [23], it is not sufficient to guarantee also exact conservation of the neutrino number. To achieve this, we discretise and solve a set of two additional equations. With $\mathcal{J} = J/\epsilon$, $\mathcal{H} = H/\epsilon$, $\mathcal{K} = K/\epsilon$, and $\mathcal{L} = L/\epsilon$, this set of equations reads

$$\begin{aligned}
&\left(\frac{1}{c}\frac{\partial}{\partial t} + \beta_r\frac{\partial}{\partial r} + \boldsymbol{\frac{\beta_\vartheta}{r}\frac{\partial}{\partial\vartheta}}\right)\mathcal{J} + \mathcal{J}\left(\frac{1}{r^2}\frac{\partial(r^2\beta_r)}{\partial r} + \boldsymbol{\frac{1}{r\sin\vartheta}\frac{\partial(\sin\vartheta\beta_\vartheta)}{\partial\vartheta}}\right) \\
&+ \frac{1}{r^2}\frac{\partial(r^2\mathcal{H})}{\partial r} + \frac{\beta_r}{c}\frac{\partial\mathcal{H}}{\partial t} - \frac{\partial}{\partial\epsilon}\left\{\frac{\epsilon}{c}\frac{\partial\beta_r}{\partial t}\mathcal{H}\right\} - \frac{\partial}{\partial\epsilon}\left\{\epsilon\mathcal{J}\left(\frac{\beta_r}{r} + \boldsymbol{\frac{1}{2r\sin\vartheta}\frac{\partial(\sin\vartheta\beta_\vartheta)}{\partial\vartheta}}\right)\right\} \\
&\quad - \frac{\partial}{\partial\epsilon}\left\{\epsilon\mathcal{K}\left(\frac{\partial\beta_r}{\partial r} - \frac{\beta_r}{r} - \boldsymbol{\frac{1}{2r\sin\vartheta}\frac{\partial(\sin\vartheta\beta_\vartheta)}{\partial\vartheta}}\right)\right\} + \frac{1}{c}\frac{\partial\beta_r}{\partial t}\mathcal{H} = \mathcal{C}^{(0)},
\end{aligned} \tag{4}$$

$$\begin{aligned}
&\left(\frac{1}{c}\frac{\partial}{\partial t} + \beta_r\frac{\partial}{\partial r} + \boldsymbol{\frac{\beta_\vartheta}{r}\frac{\partial}{\partial\vartheta}}\right)\mathcal{H} + \mathcal{H}\left(\frac{1}{r^2}\frac{\partial(r^2\beta_r)}{\partial r} + \boldsymbol{\frac{1}{r\sin\vartheta}\frac{\partial(\sin\vartheta\beta_\vartheta)}{\partial\vartheta}}\right) \\
&\quad + \frac{\partial\mathcal{K}}{\partial r} + \frac{3\mathcal{K}-\mathcal{J}}{r} + \mathcal{H}\left(\frac{\partial\beta_r}{\partial r}\right) + \frac{\beta_r}{c}\frac{\partial\mathcal{K}}{\partial t} - \frac{\partial}{\partial\epsilon}\left\{\frac{\epsilon}{c}\frac{\partial\beta_r}{\partial t}\mathcal{K}\right\} \\
&\quad - \frac{\partial}{\partial\epsilon}\left\{\epsilon\mathcal{L}\left(\frac{\partial\beta_r}{\partial r} - \frac{\beta_r}{r} - \boldsymbol{\frac{1}{2r\sin\vartheta}\frac{\partial(\sin\vartheta\beta_\vartheta)}{\partial\vartheta}}\right)\right\} \\
&- \frac{\partial}{\partial\epsilon}\left\{\epsilon\mathcal{H}\left(\frac{\beta_r}{r} + \boldsymbol{\frac{1}{2r\sin\vartheta}\frac{\partial(\sin\vartheta\beta_\vartheta)}{\partial\vartheta}}\right)\right\} - \mathcal{L}\left(\frac{\partial\beta_r}{\partial r} - \frac{\beta_r}{r} - \boldsymbol{\frac{1}{2r\sin\vartheta}\frac{\partial(\sin\vartheta\beta_\vartheta)}{\partial\vartheta}}\right) \\
&\quad - \mathcal{H}\left(\frac{\beta_r}{r} + \boldsymbol{\frac{1}{2r\sin\vartheta}\frac{\partial(\sin\vartheta\beta_\vartheta)}{\partial\vartheta}}\right) + \frac{1}{c}\frac{\partial\beta_r}{\partial t}\mathcal{J} = \mathcal{C}^{(1)}.
\end{aligned} \tag{5}$$

The moment equations (2–5) are very similar to the $\mathcal{O}(v/c)$ equations in spherical symmetry which were solved in the 1D simulations of [23] (see Eqs. 7,8,30, and 31 of the latter work). This similarity has allowed us to reuse a good fraction of the one-dimensional version of VERTEX, for coding the multidimensional algorithm. The additional terms necessary for this purpose have been set in boldface above.

Finally, the changes of the energy, e, and electron fraction, Y_e, required for the hydrodynamics are given by the following two equations

$$\frac{\mathrm{d}e}{\mathrm{d}t} = -\frac{4\pi}{\rho}\int_0^\infty \mathrm{d}\epsilon \sum_{\nu\in(\nu_\mathrm{e},\bar\nu_\mathrm{e},\dots)} C_\nu^{(0)}(\epsilon), \tag{6}$$

$$\frac{\mathrm{d}Y_\mathrm{e}}{\mathrm{d}t} = -\frac{4\pi\, m_\mathrm{B}}{\rho}\int_0^\infty \mathrm{d}\epsilon\left(\mathcal{C}_{\nu_\mathrm{e}}^{(0)}(\epsilon) - \mathcal{C}_{\bar\nu_\mathrm{e}}^{(0)}(\epsilon)\right) \tag{7}$$

(for the momentum source terms due to neutrinos see [25]). Here m_B is the baryon mass, and the sum in Eq. (6) runs over all neutrino types. The full system consisting of Eqs. (2–7) is stiff, and thus requires an appropriate discretisation scheme for its stable solution.

Method of solution

In order to discretise Eqs. (2–7), the spatial domain $[0, r_{\text{max}}] \times [\vartheta_{\text{min}}, \vartheta_{\text{max}}]$ is covered by N_r radial, and N_ϑ angular zones, where $\vartheta_{\text{min}} = 0$ and $\vartheta_{\text{max}} = \pi$ correspond to the north and south poles, respectively, of the spherical grid. (In general, we allow for grids with different radial resolutions in the neutrino transport and hydrodynamic parts of the code. The number of radial zones for the hydrodynamics will be denoted by N_r^{hyd}.) The number of bins used in energy space is N_ϵ and the number of neutrino types taken into account is N_ν.

The equations are solved in two operator-split steps corresponding to a lateral and a radial sweep.

In the first step, we treat the boldface terms in the respectively first lines of Eqs. (2–5), which describe the lateral advection of the neutrinos with the stellar fluid, and thus couple the angular moments of the neutrino distribution of neighbouring angular zones. For this purpose we consider the equation

$$\frac{1}{c}\frac{\partial \Xi}{\partial t} + \frac{1}{r \sin\vartheta}\frac{\partial(\sin\vartheta\, \beta_\vartheta\, \Xi)}{\partial\vartheta} = 0\,, \tag{8}$$

where Ξ represents one of the moments J, H, $\mathcal{J}$, or $\mathcal{H}$. Although it has been suppressed in the above notation, an equation of this form has to be solved for each radius, for each energy bin, and for each type of neutrino. An explicit upwind scheme is used for this purpose.

In the second step, the radial sweep is performed. Several points need to be noted here:

- terms in boldface not yet taken into account in the lateral sweep, need to be included into the discretisation scheme of the radial sweep. This can be done in a straightforward way since these remaining terms do not include derivatives of the transport variables (J, H) or $(\mathcal{J}, \mathcal{H})$. They only depend on the hydrodynamic velocity v_ϑ, which is a *constant* scalar field for the transport problem.
- the right hand sides (source terms) of the equations and the coupling in energy space have to be accounted for. The coupling in energy is non-local, since the source terms of Eqs. (2–5) stem from the Boltzmann equation, which is an integro-differential equation and couples all the energy bins
- the discretisation scheme for the radial sweep is *implicit* in time. Explicit schemes would require very small time steps to cope with the stiffness of the source terms in the optically thick regime, and the small CFL time step dictated by neutrino propagation with the speed of light in the optically thin regime. Still, even with an implicit scheme $\gtrsim 10^5$ time steps are required per simulation. This makes the calculations expensive.

Once the equations for the radial sweep have been discretized in radius and energy, the resulting solver is applied ray-by-ray for each angle ϑ and for each type of neutrino, i.e. for constant ϑ, N_ν two-dimensional problems need to be solved.

The discretisation itself is done using a second order accurate scheme with backward differencing in time according to [23]. This leads to a non-linear system of algebraic equations, which is solved by Newton-Raphson iteration with explicit construction and inversion of the corresponding block-pentadiagonal Jacobian matrix. For the construction of the Jacobian, which entails the calculation of neutrino-matter interactions rates, the vector capabilities on the NEC SX-8 are a major asset, and allow FLOP rates of 7-9 GFLOP per second and per CPU for routines that are major bottlenecks on scalar machines. On the other hand, the Block-Thomas algorithm used for the solution of the linear system suffers from rather small block sizes (up to 70×70) and cannot fully exploit the available vector length of the SX-8. Since the bulk of the computational time (around 70% on SX-8) is consumed by the linear solver, an optimal implementation of the solution algorithm is crucial for obtaining a good performance.

4 Optimization of the block-pentadiagonal solver

The Thomas algorithm [32, 33] is an adaption of Gaussian elimination to (block) tri- and (block) pentadiagonal systems reducing computational complexity. The optimizations carried out are twofold: at the algorithmic level, the computational steps of the Thomas algorithm are reordered and at the implementation level, BLAS and LAPACK calls are replaced by self-written, highly optimized code. The reader is referred to [34] for details.

4.1 Thomas algorithm

The block-pentadiagonal (BPD) linear system of equations with solution vector $\mathbf{x}$ and right hand side (RHS) $\mathbf{f}$

$$\begin{array}{lllllll} \mathsf{C}_1\,\mathbf{x}_1 & +\,\mathsf{D}_1\,\mathbf{x}_2 & +\,\mathsf{E}_1\,\mathbf{x}_3 & & & & = \mathbf{f}_1 \\ \mathsf{B}_2\,\mathbf{x}_1 & +\,\mathsf{C}_2\,\mathbf{x}_2 & +\,\mathsf{D}_2\,\mathbf{x}_3 & +\,\mathsf{E}_2\,\mathbf{x}_4 & & & = \mathbf{f}_2 \\ \mathsf{A}_i\,\mathbf{x}_{i-2} & +\,\mathsf{B}_i\,\mathbf{x}_{i-1} & +\,\mathsf{C}_i\,\mathbf{x}_i & +\,\mathsf{D}_i\,\mathbf{x}_{i+1} & +\,\mathsf{E}_i\,\mathbf{x}_{i+2} & & = \mathbf{f}_i \\ & \mathsf{A}_{n-1}\,\mathbf{x}_{n-3} & +\,\mathsf{B}_{n-1}\,\mathbf{x}_{n-2} & +\,\mathsf{C}_{n-1}\,\mathbf{x}_{n-1} & +\,\mathsf{D}_{n-1}\,\mathbf{x}_n & & = \mathbf{f}_{n-1} \\ & & \mathsf{A}_n\,\mathbf{x}_{n-2} & +\,\mathsf{B}_n\,\mathbf{x}_{n-1} & +\,\mathsf{C}_n\,\mathbf{x}_n & & = \mathbf{f}_n, \end{array} \tag{9}$$

where $3 \leq i \leq n-2$, consists of n block rows resp. block columns, each block being of size $k \times k$. To simplify notation and implementation, the system (9) is artifically enlarged entailing a compact form

$$\mathsf{A}_i\mathbf{x}_{i-2} + \mathsf{B}_i\mathbf{x}_{i-1} + \mathsf{C}_i\mathbf{x}_i + \mathsf{D}_i\mathbf{x}_{i+1} + \mathsf{E}_i\mathbf{x}_{i+2} = \mathbf{f}_i, \quad 1 \leq i \leq n. \tag{10}$$

The vectors $\mathbf{x} = (\mathbf{x}_{-1}^T\,\mathbf{x}_0^T \ldots \mathbf{x}_{n+2}^T)^T$ and $\mathbf{f} = (\mathbf{f}_{-1}^T\,\mathbf{f}_0^T \ldots \mathbf{f}_{n+2}^T)^T$ are of size $(n+4)k$ and the BPD matrix is of size $nk \times (n+4)k$, where A_1, B_1, A_2, E_{n-1}, D_n and E_n are set to zero.

If the sub-diagonal matrix blocks A_i and B_i are eliminated and the diagonal matrix blocks C_i are inverted, one would obtain a system of the form

$$\begin{aligned} \mathbf{x}_i + \mathsf{Y}_i\,\mathbf{x}_{i+1} + \mathsf{Z}_i\,\mathbf{x}_{i+2} &= \mathbf{r}_i, \qquad 1 \le i \le n-2, \\ \mathbf{x}_{n-1} + \mathsf{Y}_{n-1}\,\mathbf{x}_n &= \mathbf{r}_{n-1}, \\ \mathbf{x}_n &= \mathbf{r}_n. \end{aligned} \tag{11}$$

Applying the Thomas algorithm signifies that the variables Y_i, Z_i and $\mathbf{r}_i$ are calculated by substituting $\mathbf{x}_{i-2}$ and $\mathbf{x}_{i-1}$ in (10) using the appropriate equations of (11) and comparing coefficients. This results in

$$\begin{aligned} \mathsf{Y}_i &= \mathsf{G}_i^{-1}\,(\mathsf{D}_i - \mathsf{K}_i\mathsf{Z}_{i-1}) \\ \mathsf{Z}_i &= \mathsf{G}_i^{-1}\,\mathsf{E}_i \\ \mathbf{r}_i &= \mathsf{G}_i^{-1}\,(\mathbf{f}_i - \mathsf{A}_i\mathbf{r}_{i-2} - \mathsf{K}_i\mathbf{r}_{i-1}) \end{aligned} \tag{12}$$

for $i = 1, n$, where

$$\begin{aligned} \mathsf{K}_i &= \mathsf{B}_i - \mathsf{A}_i\mathsf{Y}_{i-2} \\ \mathsf{G}_i &= \mathsf{C}_i - \mathsf{K}_i\mathsf{Y}_{i-1} - \mathsf{A}_i\mathsf{Z}_{i-2} \end{aligned} \tag{13}$$

and Y_{-1}, Z_{-1}, Y_0, and Z_0 are set to zero. Backward substitution

$$\begin{aligned} \mathbf{x}_n &= \mathbf{r}_n, \\ \mathbf{x}_{n-1} &= \mathbf{r}_{n-1} - \mathsf{Y}_{n-1}\,\mathbf{x}_n, \\ \mathbf{x}_i &= \mathbf{r}_i - \mathsf{Y}_i\,\mathbf{x}_{i+1} - \mathsf{Z}_i\mathbf{x}_{i+2}, \qquad i = n-2, -1, 1. \end{aligned} \tag{14}$$

yields the solution $\mathbf{x}$.

4.2 Algorithmic improvements

Reordering the computational steps of the Thomas algorithm (12) and (13) reduces memory traffic, which is the limiting factor for performance in case of small matrix blocks. More precisely, computing Y_i, Z_i and $\mathbf{r}_i$ via

$$\begin{aligned} \mathsf{K}_i &= \mathsf{B}_i - \mathsf{A}_i\,\mathsf{Y}_{i-2} & \mathsf{G}_i &= \mathsf{G}'_i - \mathsf{K}_i\,\mathsf{Y}_{i-1} \\ \mathsf{G}'_i &= \mathsf{C}_i - \mathsf{A}_i\,\mathsf{Z}_{i-2} & \mathsf{H}_i &= \mathsf{D}_i - \mathsf{K}_i\,\mathsf{Z}_{i-1} \\ \mathbf{r}'_i &= \mathbf{f}_i - \mathsf{A}_i\,\mathbf{r}_{i-2} & \mathbf{r}''_i &= \mathbf{r}'_i - \mathsf{K}_i\,\mathbf{r}_{i-1} \end{aligned} \tag{15}$$

and

$$\begin{aligned} \mathsf{Y}_i &= \mathsf{G}_i^{-1} \cdot \mathsf{H}_i \\ \mathsf{Z}_i &= \mathsf{G}_i^{-1} \cdot \mathsf{E}_i \\ \mathbf{r}_i &= \mathsf{G}_i^{-1} \cdot \mathbf{r}''_i, \end{aligned} \tag{16}$$

has the following advantages:

- during elimination of the subdiagonal matrix blocks (15), the matrices A_i and K_i are loaded only k times from main memory instead of $2k+1$ times for a straight forward implementation
- H_i, E_i and $\mathbf{r}''_i$ can be stored contiguously in memory allowing the inverse of G to be applied simultaneously to a combined RHS $(\mathsf{H}_i\,\mathsf{E}_i\,\mathbf{r}''_i)$ of size $k \times (2k+1)$ during blockrow-wise Gaussian elimination and backward substitution (16).

4.3 Implementation-based improvements

To compute (16), LAPACK's xGETRF (factorization) and parts of xGETRS (forward and backward substitution) routines [35] are replaced by self-written code. This offers the possiblity to combine factorization $\mathsf{G}_i = \mathsf{L}_i\mathsf{U}_i$ and forward substitution, so that L_i is applied to the combined RHS during the elimination process and not afterwards. Furthermore, the code can be tuned specifically for the case of small block sizes. An efficient implementation for (15) was also introduced in [34].

4.4 Solution of a sample BPD system

The compact multiplication scheme and the improved solver for (16) are integrated into a new BPD solver. Its execution times are compared to a traditional BLAS/LAPACK solver in table 1 for 100 systems with block sizes $k = 20, 55$ and 85 and $n = 500$ block rows resp. columns. The diagonals of the BPD matrix are stored as five vectors of matrix blocks.

5 Parallelization

The ray-by-ray approximation readily lends itself to parallelization over the different angular zones. For the radial transport sweep, we presently use an OpenMP/MPI hybrid approach, while the hydrodynamics module PROMETHEUS can only exploit shared-memory parallelism as yet. For a small number of MPI processes, this does not severely affect parallel scaling, as the neutrino transport takes 90% to 99% (heavily model-dependent) of the total serial time. This is a reasonable strategy for systems with a large number of processors per shared-memory node on which our code has been used in the past, such as the IBM Regatta at the Rechenzentrum Garching of the Max-Panck-Gesellschaft (32 processors per node) or the Altix 3700 Bx2 (MPA, ccNUMA architecture with 112 processors). However, this approach does not allow us to fully exploit the capabilities of the NEC SX-8 with its 8 CPUs per node. While the neutrino transport algorithm can be expected to exhibit good scaling for up to N_ϑ processors (128-256 for a typical setup), the lack of MPI parallelism in PROMETHEUS prohibits the use of more than

Table 1. Execution times for the BPD solver for 100 systems with $n = 500$

$k =$	20	55	85
BLAS + LAPACK [s]	6.43	33.80	76.51
comp. mult. + new solver [s]	3.79	23.10	55.10
decrease in runtime [%]	54.5	42.6	35.4

four nodes. Full MPI functionality is clearly desirable, as it could reduce the turnaround time by another factor of 3–4 on the SX-8. As the code already profits from the vector capabilities of NEC machines, this amounts to a runtime of several weeks as compared to more than a year required on the scalar machines mentioned before, i. e. the overall reduction is even larger. For this reason, a MPI version of the hydrodynamics part is currently being developed within the framework of the Teraflop Workbench.

6 Conclusions

After reporting on recent developments in supernova modelling and briefly describing the numerics of the ray-by-ray method employed in our code PROMETHEUS/VERTEX, we addressed the issue of serial optimization. We presented benchmarks for the improved implementation of the Block-Thomas algorithm, finding reductions in runtime of about 1/3 or more for the relevant block sizes. Finally, we discussed the limitations of the current parallelization approach and emphasized the potential and importance of a fully MPI-capable version of the code.

Acknowledgements

Support from the SFB 375 "Astroparticle Physics", SFB/TR7 "Gravitationswellenastronomie", and SFB/TR27 "Neutrinos and Beyond" of the Deutsche Forschungsgemeinschaft, and computer time at the HLRS and the Rechenzentrum Garching are acknowledged.

References

1. Rampp, M., Janka, H.T.: Spherically Symmetric Simulation with Boltzmann Neutrino Transport of Core Collapse and Postbounce Evolution of a 15 $M_{\odot}$ Star. Astrophys. J. **539** (2000) L33–L36
2. Mezzacappa, A., Liebendörfer, M., Messer, O.E., Hix, W.R., Thielemann, F., Bruenn, S.W.: Simulation of the Spherically Symmetric Stellar Core Collapse, Bounce, and Postbounce Evolution of a Star of 13 Solar Masses with Boltzmann Neutrino Transport, and Its Implications for the Supernova Mechanism. Phys. Rev. Letters **86** (2001) 1935–1938
3. Liebendörfer, M., Mezzacappa, A., Thielemann, F., Messer, O.E., Hix, W.R., Bruenn, S.W.: Probing the gravitational well: No supernova explosion in spherical symmetry with general relativistic Boltzmann neutrino transport. Phys. Rev. D **63** (2001) 103004–+
4. Bethe, H.A.: Supernova mechanisms. Reviews of Modern Physics **62** (1990) 801–866

5. Burrows, A., Goshy, J.: A Theory of Supernova Explosions. Astrophys. J. **416** (1993) L75
6. Janka, H.T.: Conditions for shock revival by neutrino heating in core-collapse supernovae. Astron. Astrophys. **368** (2001) 527–560
7. Herant, M., Benz, W., Colgate, S.: Postcollapse hydrodynamics of SN 1987A - Two-dimensional simulations of the early evolution. Astrophys. J. **395** (1992) 642–653
8. Herant, M., Benz, W., Hix, W.R., Fryer, C.L., Colgate, S.A.: Inside the supernova: A powerful convective engine. Astrophys. J. **435** (1994) 339
9. Burrows, A., Hayes, J., Fryxell, B.A.: On the nature of core-collapse supernova explosions. Astrophys. J. **450** (1995) 830
10. Janka, H.T., Müller, E.: Neutrino heating, convection, and the mechanism of Type-II supernova explosions. Astron. Astrophys. **306** (1996) 167–+
11. Thompson, C.: Accretional Heating of Asymmetric Supernova Cores. Astrophys. J. **534** (2000) 915–933
12. Foglizzo, T.: Non-radial instabilities of isothermal Bondi accretion with a shock: Vortical-acoustic cycle vs. post-shock acceleration. Astron. Astrophys. **392** (2002) 353–368
13. Blondin, J.M., Mezzacappa, A., DeMarino, C.: Stability of Standing Accretion Shocks, with an Eye toward Core-Collapse Supernovae. Astrophys. J. **584** (2003) 971–980
14. Scheck, L., Plewa, T., Janka, H.T., Kifonidis, K., Müller, E.: Pulsar Recoil by Large-Scale Anisotropies in Supernova Explosions. Phys. Rev. Letters **92** (2004) 011103–+
15. Scheck, L.: Multidimensional simulations of core collapse supernovae. PhD thesis, Technische Universität München (2006)
16. Keil, W., Janka, H.T., Müller, E.: Ledoux Convection in Protoneutron Stars—A Clue to Supernova Nucleosynthesis? Astrophys. J. **473** (1996) L111
17. Burrows, A., Lattimer, J.M.: The birth of neutron stars. Astrophys. J. **307** (1986) 178–196
18. Pons, J.A., Reddy, S., Prakash, M., Lattimer, J.M., Miralles, J.A.: Evolution of Proto-Neutron Stars. Astrophys. J. **513** (1999) 780–804
19. Marek, A.: Multi-dimensional simulations of core collapse supernovae with different equations of state for hot proto-neutron stars. PhD thesis, Technische Universität München (2007)
20. Fryer, C.L., Heger, A.: Core-Collapse Simulations of Rotating Stars. Astrophys. J. **541** (2000) 1033–1050
21. Fryer, C.L., Warren, M.S.: Modeling Core-Collapse Supernovae in Three Dimensions. Astrophys. J. **574** (2002) L65–L68
22. Fryer, C.L., Warren, M.S.: The Collapse of Rotating Massive Stars in Three Dimensions. Astrophys. J. **601** (2004) 391–404
23. Rampp, M., Janka, H.T.: Radiation hydrodynamics with neutrinos. Variable Eddington factor method for core-collapse supernova simulations. Astron. Astrophys. **396** (2002) 361–392
24. Janka, H.T., Buras, R., Kifonidis, K., Marek, A., Rampp, M.: Core-Collapse Supernovae at the Threshold. In Marcaide, J.M., Weiler, K.W., eds.: Supernovae, Procs. of the IAU Coll. 192, Berlin, Springer (2004)
25. Buras, R., Rampp, M., Janka, H.T., Kifonidis, K.: Two-dimensional hydrodynamic core-collapse supernova simulations with spectral neutrino transport. I.

Numerical method and results for a 15 Mȯ star. Astron. Astrophys. **447** (2006) 1049–1092
26. Buras, R., Janka, H.T., Rampp, M., Kifonidis, K.: Two–dimensional hydrodynamic core–collapse supernova simulations with spectral neutrino transport. II. Models for different progenitor stars. Astron. Astrophys. **457** (2006) 281–308
27. Müller, E., Rampp, M., Buras, R., Janka, H.T., Shoemaker, D.H.: Toward Gravitational Wave Signals from Realistic Core-Collapse Supernova Models. Astrophys. J. **603** (2004) 221–230
28. Kitaura, F.S., Janka, H.T., Hillebrandt, W.: Explosions of O–Ne–Mg cores, the Crab supernova, and subluminous type II–P supernovae. Astron. Astrophys. **450** (2006) 345–350
29. Marek, A., Kifonidis, K., Janka, H.T., Müller, B.: The supern-project: Understanding core collapse supernovae. In Nagel, W.E., Jäger, W., Resch, M., eds.: High Performance Computing in Science and Engineering 06, Berlin, Springer (2006)
30. Colella, P., Woodward, P.R.: The piecewise parabolic method for gas-dynamical simulations. Jour. of Comp. Phys. **54** (1984) 174
31. Fryxell, B.A., Müller, E., Arnett, W.D.: Hydrodynamics and nuclear burning. Max-Planck-Institut für Astrophysik, Preprint **449** (1989)
32. Thomas, L.H.: Elliptic problems in linear difference equations over a network. Watson Sci. Comput. Lab. Rept., Columbia University, New York (1949)
33. Bruce, G.H., Peaceman, D.W., Jr. Rachford, H.H., Rice, J.D.: Calculations of unsteady-state gas flow through porous media. Petrol. Trans. AIME **198** (1953) 79–92
34. Benkert, K., Fischer, R.: An efficient implementation of the Thomas-algorithm for block penta-diagonal systems on vector computers. In Shi, Y., van Albada, G.D., Dongarra, J., Sloot, P.M., eds.: Computational Science – ICCS 2007. Volume 4487 of LNCS., Springer (2007) 144–151
35. Anderson, E., Blackford, L.S., Sorensen, D., eds.: Lapack User's Guide. Society for Industrial & Applied Mathematics (2000)

Applications V

Material Science

Green Chemistry from Supercomputers: Car–Parrinello Simulations of Emim-Chloroaluminates Ionic Liquids

Barbara Kirchner[1] and Ari P Seitsonen[2]

[1] Lehrstuhl für Theoretische Chemie, Universität Leipzig, Linnestr. 2 , D-04103 Leipzig bkirchner@uni-leipzig.de
[2] CNRS & Université Pierre at Marie Curie, 4 place Jussieu, case 115, F-75252 Paris Ari.P.Seitsonen@iki.fi

1 Introduction

Ionic liquids (IL) or room temperature molten salts are alternatives to "more toxic" liquids. [1] Their solvent properties can be adjusted to the particular problem by combining the right cation with the right anion, which makes them designer liquids. Usually an ionic liquid is formed by an organic cation combined with an inorganic anion. [2, 3] For a more detailed discussion on the definition we refer to the following review articles. [4, 5, 6]

Despite of this continuing interest in ionic liquids their fundamental properties and microscopic behavior are still only poorly understood. unresolved questions regarding those liquids are still controversially discussed. A large contribution to the understanding of the microscopic aspects can come from the investigation of these liquids by means of theoretical methods. [7, 8, 9, 10, 11, 12, 13, 14, 15, 16, 17, 18, 19, 20, 21, 22]

In our project **AIMD-IL** at HLRS/NEC SX-8 we have investigated a prototypical ionic liquid using *ab initio* molecular dynamics methods, where the interaction between the ions is solved by explicitly treating the electronic structure during the simulation. The huge investigation in terms of computing time is more than justified due to the increased accuracy and reliability compared to simulations employing parameterized, classical potentials.

In this summary we will describe the results obtained within our project of a Car–Parrinello simulation of 1-ethyl-3-methylimidazolium ($[C_2C_1im]^+$, see Fig. 1) chloroaluminates ionic liquids; for a snapshot of the liquid see Fig. 2. Depending on the mole fraction of the $AlCl_3$ to $[C_2C_1im]Cl$ these liquids can behave from acidic to basic. Welton describes the nomenclature of these fascinating liquids in his review article as follows: [5] "Since Cl^- is a Lewis base and $[Al_2Cl_7]^-$ and $[Al_3Cl_{10}]^-$ are both Lewis acids, the Lewis acidity/basicity of the ionic liquid may be manipulated by altering its composition. This leads

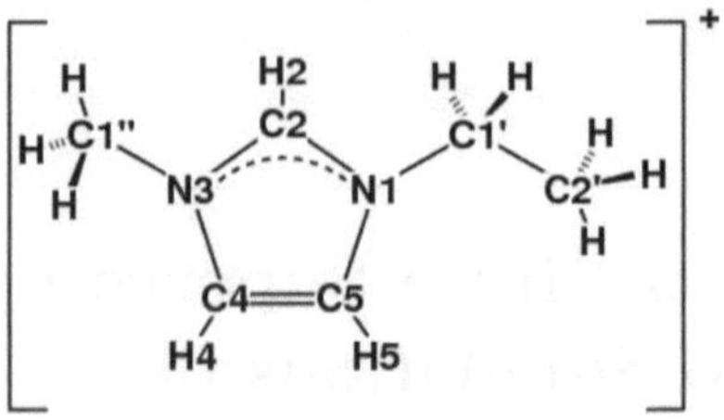

Fig. 1. Lewis structure of 1-ethyl-3-methylimidazolium, or $[C_2C_1im]^+$. Blue spheres: nitrogen; cyan: carbon; white: hydrogen

to a nomenclature of the liquids in which compositions with an excess of Cl^- (i. e. $x(AlCl_3) < 0.5$) are called basic, those with an excess of $[Al_2Cl_7]^-$ (i. e. $x(AlCl_3) > 0.5$) are called acidic, and those at the compound formation point ($x(AlCl_3) = 0.5$) are called neutral.“ In this report we concentrate on the neutral liquid. In a previous analysis we determined the $Al_4Cl_{13}^-$ to be the most abundant species in the acidic mixture as a result of the electron deficiency property. [23]

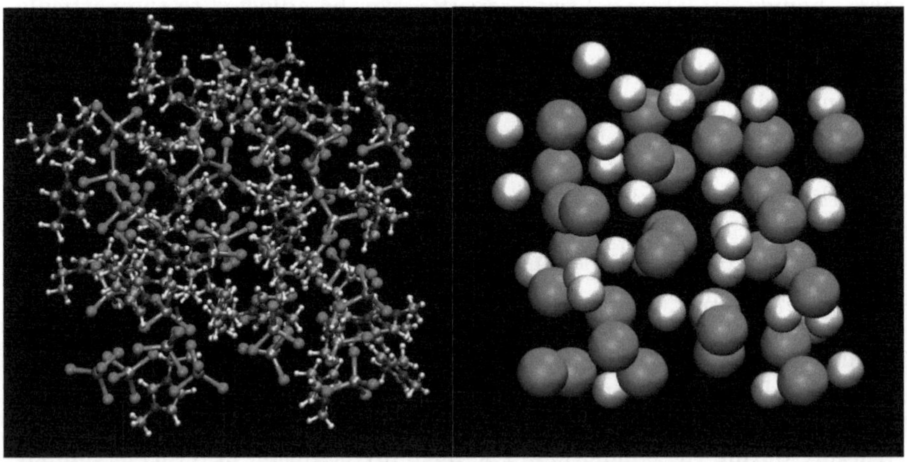

Fig. 2. A snapshot from the Car–Parrinello simulation of the "neutral" ionic liquid $[C_2C_1im]AlCl_4$. Left panel: The system in atomistic resolution. Blue spheres: nitrogen; cyan: carbon; white: hydrogen; silver: aluminium; green: chlorine. Right panel: Center of mass of $[C_2C_1im]^+$, white spheres, and $AlCl_4^-$, green spheres

2 Method

In order to model our liquid we use Car–Parrinello molecular dynamics (CPMD) simulations. The atoms are propagated along the Newtonian trajectories, with forces acting on the ions. These are obtained using density functional theory solved "on the fly". We shall shortly describe the two main ingredients of this method in the following. [24, 25].

2.1 Density functional theory

Density functional theory (DFT) [26, 27] is nowadays the most-widely used electronic-structure method. DFT combines reasonable accuracy in several different chemical environments with minimal computational effort.

The most frequently applied form of DFT is the Kohn–Sham method. There one solves the set of equations

$$\left\{-\frac{1}{2}\nabla^2 + V_{\mathrm{KS}}[n](\mathbf{r})\right\}\psi_i(\mathbf{r}) = \varepsilon_i \psi_i(\mathbf{r})$$
$$n(\mathbf{r}) = \sum_i |\psi_i(\mathbf{r})|^2$$
$$V_{\mathrm{KS}}[n](\mathbf{r}) = V_{\mathrm{ext}}(\{\mathbf{R}_I\}) + V_{\mathrm{H}}(\mathbf{r}) + V_{\mathrm{xc}}[n](\mathbf{r})$$

Here $\psi_i(\mathbf{r})$ are the Kohn–Sham orbitals, or the wave functions of the electrons; ε_i are the Kohn–Sham eigenvalues, $n(\mathbf{r})$ the electron density (can be interpreted also as the probability of finding an electron at position $\mathbf{r}$) and $V_{\mathrm{KS}}[n](\mathbf{r})$ is the Kohn–Sham potential, consisting of the attractive interaction with the ions in $V_{\mathrm{ext}}(\{\mathbf{R}_I\})$, the electron-electron repulsion $V_{\mathrm{H}}(\mathbf{r})$ and the so-called exchange-correlation potential $V_{\mathrm{xc}}[n](\mathbf{r})$.

The Kohn–Sham equations are in principle *exact*. However, whereas the analytic expression for the exchange term is known, it is not the case for the correlation, and even the exact expression for the exchange is too involved to be evaluated in practical calculations for large systems. Thus one is forced to rely on approximations. The mostly used one is the generalized gradient approximation, GGA, where one at a given point includes not only the magnitude of the density – like in the local density approximation, LDA – but also its first gradient as an input variable for the approximate exchange correlation functional. even very good,

In order to solve the Kohn–Sham equations with the aid of computers they have to be discretiszed using a basis set. A straight-forward choice is to sample the wave functions on a real-space grid at points $\{\mathbf{r}\}$. Another approach, widely used in condensed phase systems, is the expansion in the plane wave basis set,

$$\psi_i(\mathbf{r}) = \sum_{\mathbf{G}} c_i(\mathbf{G})\, e^{\mathbf{i}\mathbf{G}\cdot\mathbf{r}}$$

Here $\mathbf{G}$ are the wave vectors, whose possible values are given by the unit cell of the simulation.

One of the advantages of the plane wave basis set is that there is *only one* parameter controlling the quality of the basis set. This is the so-called cut-off energy E_{cut}: All the plane waves within a given radius from the origin,

$$\frac{1}{2}|\mathbf{G}|^2 < E_{\text{cut}} ,$$

are included in the basis set. Typical number of plane wave coefficients in practice is of the order of 10^5 per electronic orbital.

The use of plane waves necessitates a reconsideration of the spiked external potential due to the ions, $-Z/r$. The standard solution is to use pseudo potentials instead of these hard, very strongly changing functions around the nuclei [28]. This is a well controlled approximation, and reliable pseudo potentials are available for most of the elements in the periodic table.

When the plane wave expansion of the wave functions is inserted into the Kohn–Sham equations it becomes obvious that some of the terms are most efficiently evaluated in the reciprocal space, whereas other terms are better executed in real space. Thus it is advantageous to use fast Fourier transforms (FFT) to exchange between the two spaces. Because one usually wants to study realistic, three-dimensional models, the FFT in the DFT codes is also three dimensional. This can, however, be considered as three subsequent one-dimensional FFT's with two transpositions between the application of the FFT in the different directions.

The numerical effort of applying a DFT plane wave code mainly consists of basic linear algebra subprograms (BLAS) and fast Fourier transform (FFT) operations. The previous one generally require quite little communication. However the latter one requires more complicated communication patterns since in larger systems the data on which the FFT is performed needs to be distributed on the processors. Yet the parallellisation is quite straightforward and can yield an efficient implementation, as recently demonstrated in IBM Blue Gene machines [29]; combined with a suitable grouping of the FFT's one can achieve good scaling up to tens of thousands of processors with the computer code `CPMD`. [30]

Car–Parrinello method

The Car–Parrinello Lagrangean reads as

$$\mathcal{L}_{\text{CP}} = \sum_I \frac{1}{2} M_I \dot{\mathbf{R}}_I^2 + \sum_i \frac{1}{2}\mu \left\langle \dot{\psi}_i \middle| \dot{\psi}_i \right\rangle - E_{\text{KS}} + \textit{constraints} \tag{1}$$

where $\mathbf{R}_I$ is the coordinate of ion I, μ is the fictitious electron mass, the dots denote time derivatives, E_{KS} is the Kohn–Sham total energy of the system and the holonomic constraints keep the Kohn–Sham orbitals orthonormal as required by the Pauli exclusion principle. From the Lagrangean the equations of motions can be derived via Euler-Lagrange equations:

$$\begin{aligned} M_I \ddot{\mathbf{R}}_I(t) &= -\frac{\partial E_{\mathrm{KS}}}{\partial \mathbf{R}_I} \\ \mu \ddot{\psi}_i &= -\frac{\delta E_{\mathrm{KS}}}{\delta \langle \psi_i |} + \frac{\delta}{\delta \langle \psi_i |} \{constraints\} \end{aligned} \tag{2}$$

The velocity Verlet is an example of an efficient and accurate algorithm widely used to propagate these equations in time.

The electrons can be seen to follow fictitious dynamics in the Car–Parrinello method, *i. e.* they are not propagated in time physically. However, this is generally not needed, since the electronic structure varies much faster than the ionic one, and the ions see only "an average" of the electronic structure. In the Car–Parrinello method the electrons remain close to the Born-Oppenheimer surface, thus providing accurate forces on the ions but simultaneously abolishing the need to solve the electronic structure exactly at the Born–Oppenheimer surface.studies have demonstrated the high

For Born–Oppenheimer simulations there always exists a residual deviation from the minimum due to insufficient convergence in the self-consistency, and thus the ionic forces calculated contain some error. This leads to a drift in the total conserved energy. On the other hand in the Car–Parrinello method one has to make sure that the electrons and ions do not exchange energy, *i. e.* that they are adiabatically decoupled. Also the time step used to integrate the equations of motion in the Car–Parrinello molecular dynamics has to be 6-10 times shorter than in the Born-Oppenheimer dynamics due to the rapidly oscillating electronic degrees of freedom. In practice the two methods are approximately as fast, and the Car–Parrinello method has a smaller drift in the conserved quantities, but the ionic forces are weakly affected by the small deviation from the Born-Oppenheimer surface.

2.2 Technical details

For the simulations we used density functional theory with the generalized gradient approximation of Perdew, Burke and Ernzerhof, PBE [31] as the exchange-correlation term in the Kohn–Sham equations, and we replaced the action of the core electrons on the valence orbitals with norm-conserving pseudo potentials of Troullier-Martins type [32]; they are the same ones as in [33] for Al and Cl. We expanded the wave functions with plane waves up to the cut-off energy of 70 Ry. We sampled the Brillouin zone at the Γ point, employing periodic boundary conditions.

We performed the simulations in the NVT ensemble, employing a Nosé-Hoover thermostat at a target temperature of 300 K and a characteristic frequency of 595 cm^{-1}, a stretching mode of the $AlCl_3$ molecules. We propagated the velocity Verlet equations of motion with a time step of 5 a.t.u. = 0.121 fs, and the fictitious mass in the Car–Parrinello dynamics for the electrons was 700 a.u. A cubic simulation cell with a edge length of 22.577 Å containing 32 molecules of cations and anions each, equaling to the experimental density of 1.293 g/cm^3. We ran our trajectory employing the Car–Parrinello molecular dynamics for 20 ps.

3 Results: Ionic structure

3.1 Radial pair distribution functions

In order to characterise the ionic structure in our simulation we first consider the radial pair distribution functions. Fig. 3 depicts the radial distribution function of the $AlCl_4^-$ anion in our ionic liquid and of $AlCl_3$ in the pure $AlCl_3$ liquid from Ref. [33]. It should be noted that both simulations were carried out at different temperature, which results in different structured functions. In the case of the neutral $[C_2C_1im]AlCl_4$ ionic liquid there will be hardly a possibility for larger anions to be formed. In contrast to this the pure $AlCl_3$ liquid shows mostly the dimer (45%), but also larger clusters such as trimers (30%), tetramers (10%) and pentamers as well as even larger units (<10%). [33]

It can be recognised from Fig. 3 that the first Al-Al peak is missing when comparing the pure $AlCl_3$ simulations to the one from the ionic liquid. This

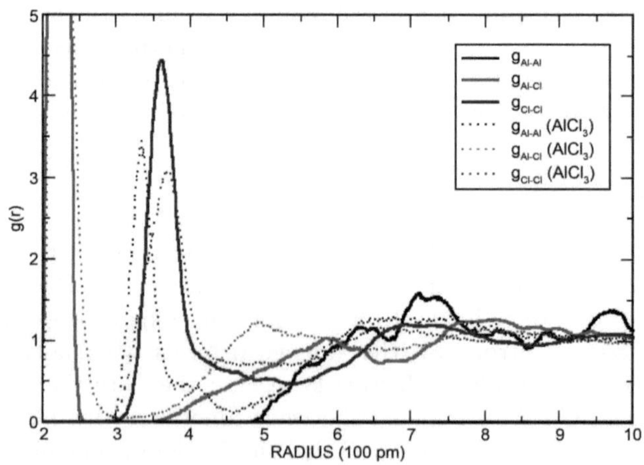

Fig. 3. The radial distribution function of the $AlCl_4^-$ anion (bold lines) together with the corresponding function from the pure $AlCl_3$ simulations (dotted lines) of Ref. [33]. Distances are in Å. Black: Al-Al; red: Al-Cl; blue: Cl-Cl

is because only monomer units ($(AlCl_3)_nCl^-$ with n=1) exists and these monomers are separated from each other by the cations. The more structured functions of the ionic liquid can be attributed to the lower temperature at which it was simulated. The first Al-Cl peak (black solid line) appears at 222.4 pm while the corresponding peak in the pure $AlCl_3$ simulations occurs already at 214.0 pm. There is no shoulder in the Al-Cl function at the first peak and the second peak occurs at larger distances. The Cl-Cl function presents its first peak at 361.1 pm which is approximately 10 pm earlier than what was observed for the pure $AlCl_3$ liquid.

In Fig. 4 we concentrate on the radial distribution functions of the imidazolium protons to the chlorine atoms. For each of the three ring protons we show an individual function in the left panel of Fig. 4. Because the H2-Cl function shows the most pronounced peak and the first peak appears at shorter distances than for the functions of H4-Cl and H5-Cl it is clear that this proton is the most popular coordination site for the chlorine atoms. However, the other two protons from the rear show also peaks at slightly larger distances which indicates an involved network instead of individual pairs. It should be noted that from this structural behavior it can not be deduced how long lived the coordination partners are. Considering the protons of the ethyl and the methyl group it is striking that here also small pronounced peaks can be observed. While the ethyl group hydrogen atoms-Cl functions like the H4-Cl functions are least pronounced, the methyl-group-Cl function show a maximum height larger than that of the H4-Cl function. Obviously this functional group is also in touch with the chlorine atoms of the anion.

3.2 Intramolecular tetrahedral structure of $AlCl_4^-$

In Table 1 we list some of the characteristic distances of $AlCl_4^-$ from isolated molecule calculation as well as from simulations.

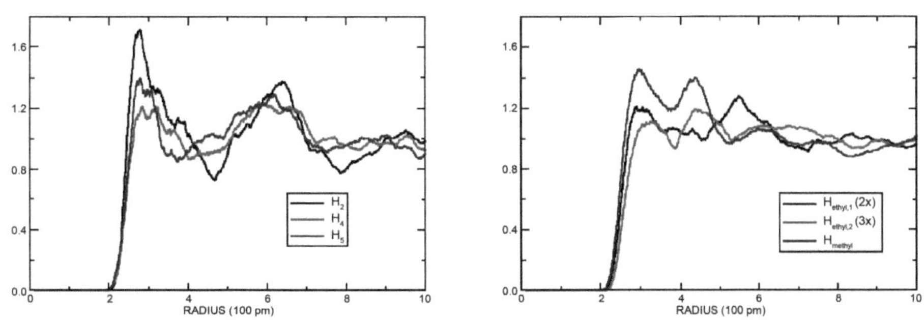

Fig. 4. The radial distribution function of chlorine atoms from $AlCl_4^-$ anion to the protons from $[C_2C_1im]$. Distances are in Å. Left: H2-Cl (black), H4-Cl (red), H5-Cl (blue); Right: Terminal ethyl-H-Cl (black), α-ethyl-H-Cl (red), methyl-H-Cl (blue)

Table 1. Geometrical parameters of the isolated and the average $AlCl_4^-$ in the ionic liquid. Distances r are in pm. r_{min} indicates the shortest while r_{max} indicates the longest distance. r is the average over all configurations. The abbreviation "iso/dyn" indicates a dynamic calculation of the isolated anion. "liq" denotes the average values from the simulations of the neutral liquid

	AlCl			ClCl		
	r_{min}	r_{max}	$r_{min/max}$	r_{min}	r_{max}	$r_{min/max}$
iso/dyn	218.1	230.2	1.06	354.4	376.9	1.06
liq	216.6	229.1	1.06	344.2	381.8	1.11

For the $AlCl_4^-$ anion we observe a perturbation from the ideal tetrahedral symmetry both in the isolated system and in the anions in the liquid. Whereas the shortest and longest Al-Cl distances vary only by 10 pm, the Cl-Cl distances show larger deviations of 30 pm (iso/dyn) to 40 pm (liq). This means that the perturbation is already induced by temperature; in the liquid the perturbation from the optimal geometry is somewhat more enhanced.

3.3 Intermolecular structure: Proton-Cl coordination

We now turn to the intermolecular structure of the hydrogen atom-chlorine atom distances in order to shed more light onto the interesting coordination behaviour of the imidazolium protons as observed before in the radial distribution function. Therefore we collected the shortest chlorine distance from a particular proton into a histogram which is shown in Fig. 5. On the average the shortest H-Cl distances is 279 pm for H2 and 290 pm for both H4 and H5. A broad range of distances from 200 pm to over 400 pm can be seen. This variety is typical for a weak to a medium ranged hydrogen bond distance. It is obvious from Fig. 5 that the Cl atoms can approach the acidic H2 atom closer than the other two protons H4 and H5. The distribution of H4 and H5 are almost identical, as expected from their similar local geometry in the molecule.

4 Results: Electronic structure

One of the advantages of the Car–Parrinello simulations over the traditional molecular dynamics simulations is that the electronic structure is available on the fly in each step of the simulations. This allows for several ways of analysis of the electronic structure.

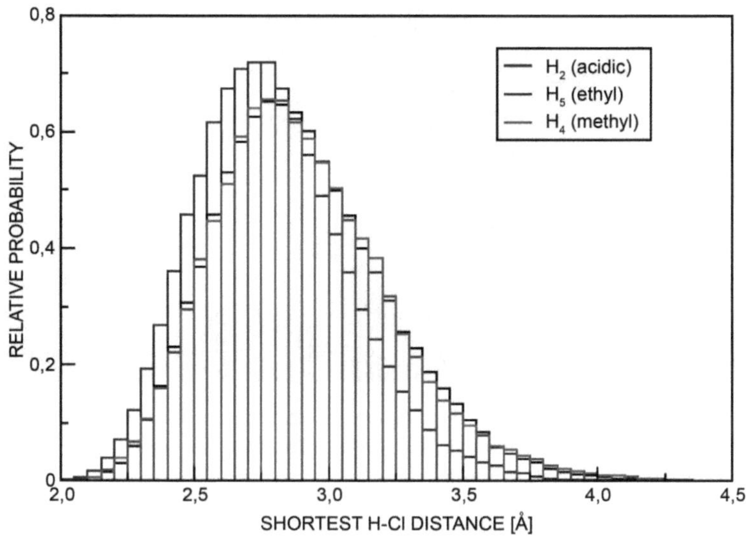

Fig. 5. The distribution of the shortest proton-Cl distance from a particular proton (H2, H4, H5) to any Cl atom

Electrostatic potential

We begin by considering the electrostatic potential mapped onto the isosurface of the electron density of the two individual ions. From Fig. 6 we obtain insight into the charge distribution according to the particular part of each ion.

For the $AlCl_4^-$ we recognize in the blue color (low electro-static potential) the consequence of the negative charge that is associated with this ion. The negative charge is distributed all over the chlorine atoms. Towards the center, consisting of the aluminium atom, we find a decreased negative charge showing as the green color in the left panel of Fig. 6. The opposite is the case for the $[C_2C_1im]^+$. Here the positive charge leads to the red color (high electro-static potential) around this ion. Upon closer inspection we find that the ring protons all hold the same red color as most of the molecule. A slight decrease of the charges can be found in the methyl group protons and a stronger decrease (yellow color) can be found at the terminal ethyl protons. This is in accordance with the observation from the radial pair distribution functions and with chemical intuition that these protons are less acidic than the other protons of the cation.

Wannier centers

We used the maximally localized Wannier functions and their geometric centers, also called Wannier centers, to characterize the distribution of electrons

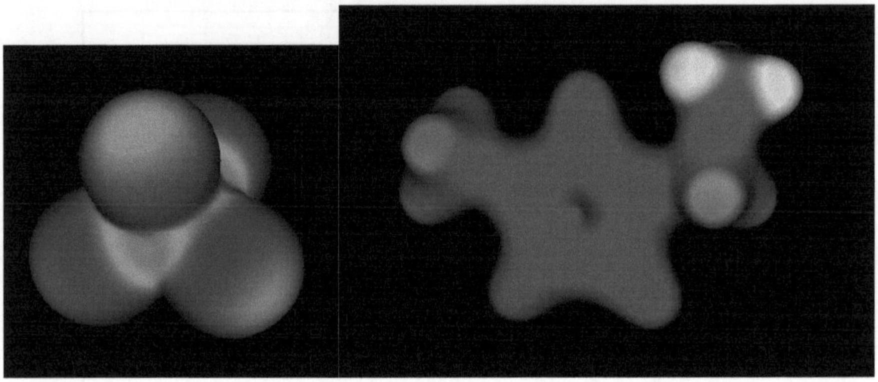

Fig. 6. The electrostatic potential mapped onto the electron density with a surface values of 0.067 $e^-/\text{Å}^3$. The colour scale ranges from −0.1 (blue) to +0.1 (red) atomic units. Left: $AlCl_4^-$, right: $[C_2C_1im]^+$

in the condensed phase. An example is shown in Fig. 7, which demonstrates how the Wannier centers can be used to interpret the chemical nature of the bonds, for example the polarity or the alternating single-double bonds (please compare to the scheme in Fig. 1). By observing the average distance of the Wannier centare between the carbon atom in the imidazolium ring and the

Fig. 7. The Wannier centers, denoted as red spheres. Top: $[C_2C_1im]^+$, bottom: $AlCl_4^-$.

corresponding proton we can see that in the C2-H2 pair the electrons are closer to the carbon than in the C4-H4 and C5-H5 pairs, pointing towards a larger polarity of the C2-H2 pair. Thus the H2 is more positive than the H4 and H5, and can electro-statically attract the negative Cl atoms from the anion molecules towards itself, as was seen in the Section 3.3.

5 Computational performance

For the simulation of our system, i. e. 32 $[C_2C_1im]AlCl_4$ pairs, we have to treat 768 atoms and 1216 electronic states in each time step. The amount of atoms is by far larger than in a usual single-molecule static calculation. Therefore the use of GGA-density functional theory is necessary in order to make the simulation computationally tractable. It should be noted that 32 molecules is more or less the lowest limit of a calculation employing periodic boundary conditions, as smaller sized systems would result in artificial finite-size effects due to interactions with the mirror images. Regarding these circumstances our simulation provides the first real *ab initio* molecular dynamics simulations of an ionic liquid. Due to the computational constraints previous simulations treated a smaller amount of molecular pairs or only employed simplified models of ionic liquids (for example $[C_1C_1im]Cl$). For the obvious reasons it was necessary for our calculations to be carried out on a large amount of efficient processors. We used 128 processors on the NEC SX-8. Therefore we were able to carry out our simulations within just two months. The size of our system leads to restart files of 14 GB in size.

Before starting the real production runs we measured the scaling of the computing time and computational efficiency when changing the system size and/or the number of processors incorporated in a job. The results of these tests are shown in Table 2 and Fig. 8. The smallest system contains 32 (IL-32) pairs. The next system contains 48 (IL-48) and the largest system 64 (IL-64) pairs.

We see very good scaling in the computing time still when going from 64 to 128 processors. We did not go beyond 128 processors still, but we estimate

Table 2. Scaling of the wall-clock time in seconds per iteration and performance in GFLOPs versus number of processors. IL-x indicates the system size of the ionic liquid

	Time per iteration (s)			Performance (GFLOPs)		
	processors			processors		
system	32	64	128	32	64	128
IL-32	46.0	23.5	13.2	381	729	1307
IL-48	–	72.6	37.9	–	784	1496
IL-64	–	159.4	83.3	–	817	1561

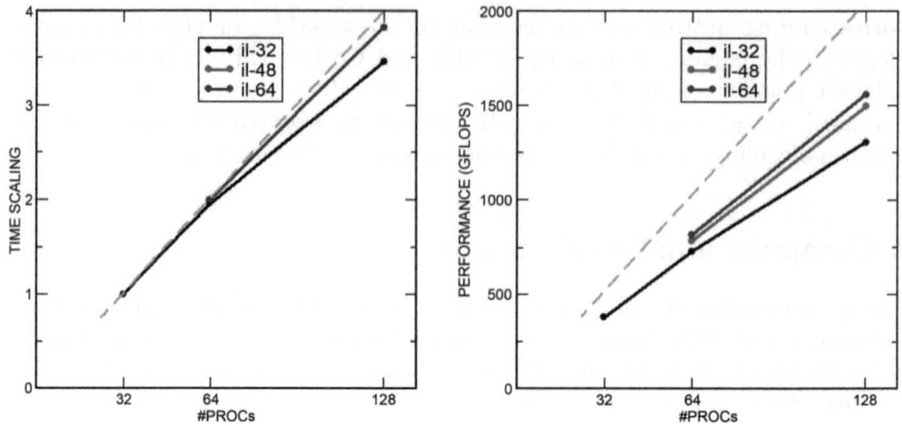

Fig. 8. The scaling of the wall clock time per iteration – left – and numerical performance in GFLOPs – right – plotted against the number of processors. The green, dashed lines denote the ideal scaling and theoretical peak performance, respectively

a decent or a good scaling in the IL-32 system, or very good scaling in the IL-48 and IL-64 cases. We note that at very large processor counts a different parallellisation using OpenMP built in the CPMD code could be tried if the scaling otherwise is no longer satisfactory. A concrete limitation is met if the number of processors is larger than the length of the FFT grid in the first direction; however, further scaling is achieved by applying *task groups*, yet-another efficient method inside CPMD; thereby the FFT's over different electronic states are grouped to set of processors, thus overcoming the limitation on the maximum number of processors due to the length of the FFT grid. The task groups can be incorporated particularly well on the NEC SX-8 computer at HLRS due to the large amount of memory at each node, because the task groups increase the memory requirement per node somewhat. We report here always the best performance over the different number of task groups; typically its optimum is at four or eight groups.

The numerical performance exceeds one 10^{12} floating point operations per second (tera-FLOPs or TFLOPs) in all the systems studied at 128 processors. Furthermore, the performance still scales very favorably when going from 64 to 128 processors. Thus from the efficiency point of view processor counts exceeding 128 could also be used. However, due to the limited number of processors available, and because we already hit the "magic target" of one TFLOPs we restricted our production to 128 processors.

Overall we were more than satisfied with the performance and with the prospect of performing the production calculations for the IL-48 or even IL-64 systems. However, due to the total time of the simulation is a multiple of the number of molecular dynamics steps, we were forced to choose the IL-32 system for the production job, as otherwise we would not have been able to simulate a trajectory of $\approx$ 20 ps like we managed to do now.

We also want to note that our calculations profit from the computer architecture of the NEC SX-8 at HLRS not only due to the high degree of vectorization and very good single-processor computing power in as evidenced in the high numerical efficiency (over 10 GFLOPs/processors; this number also includes the I/O), but also due to the large memory as we could stored some large intermediate results in the memory, thus avoiding the need to recalculate part of the results; this would be unavoidable in a machine with smaller amount of memory per processor. This way almost one third of the FFT's, and thus of the most demanding all-to-all parallel operations can be avoided, improving the parallel scaling still somewhat over a normal calculation where this option could not be used.

6 Conclusions

We have simulated a prototypical ionic liquid $[C_2C_1im]AlCl_4$ using Car–Parrinello molecular dynamics methods. 768 atoms were included in the simulation cell. The computational efficiency on the NEC SX-8 at HLRS allowed us to simulate the system for about 20 ps at realistic conditions. We achieved a sustained performance of over 1 TFLOPs on 128 processor, clearly exceeding an efficiency of 50 %.

The high throughput in the NEC SX-8 allowed us to execute the simulation in a short project time. This is of great advantage, when we are not forced to wait for extended periods of time in order to execute a simulation.

Our simulations indicate a distorted tetrahedral structure for the $AlCl_4^-$ anion. The most favorable coordination site is the acidic C-H group between the two nitrogen atoms. However coordination to the other protons is also possible. Thus we are dealing most likely with an extended network.

Acknowledgements

We thank the HLRS for the allocation of computing time; without this our project would not have been feasible!

We are grateful to Prof. Jürg Hutter for several discussions, and to Stefan Haberhauer (NEC) for executing the benchmarks on the NEC SX-8 and optimising `CPMD` on the vector machines. BK would like to thank T. Welton, A. East, K. E. Johnson and J. S. Wilkes for helpful discussion. BK acknowledges the financial support of the DFG priority program SPP 1191 "Ionic Liquids", the ERA program and the financial support from the collaborative research center SFB 624 "Templates" at the University of Bonn.

References

1. Ed, P. Wasserscheid and T. Welton. *Ionic Liquids in Synthesis.* VCH-Wiley, Weinheim, 2003.
2. J. H. Davis. Task-specific ionic liquids . *Chem. Lett.*, 33:1072–1077, 2004.
3. A. E. Visser, R. P. Swatloski, W. M. Reichert, R. Mayton, S. Sheff, A. Wierzbicki, J. H. Davis, and R. D. Rogers. Task-specific ionic liquids for the extraction of metal ions from aqueous solutions. *Chem. Commun.*, 01:135–136, 2001.
4. V. A. Cocalia, K. E. Gutowski, and R. D. Rogers. The coordination chemistry of actinides in ionic liquids: A review of experiment and simulation. *Coord. Chem. Rev.*, 150:755–764, 2006.
5. T. Welton. Room-Temperature Ionic Liquids. Solvents for Synthesis and Catalysis. *Chem. Rev.*, 99:2071–2083, 1999.
6. T. Welton. Ionic Liquids in catalysis. *Coord. Chem. Rev.*, 248:2459–2477, 2004.
7. P.A. Hunt and I. Gould. *J. Phys. Chem. A*, 110:2269, 2006.
8. S. Kossmann, J. Thar, B. Kirchner, P. A. Hunt, and T. Welton. Cooperativity in ionic liquids. *J. Chem. Phys.*, 124:174506, 2006.
9. Z. Liu, S. Haung, and W. Wang. A refined force field for molecular simulation of imidazolium-based ionic liquids. *J. Phys. Chem. B*, 108:12978, 2004.
10. J.K. Shah and E.J. Maginn. *Fluid Phase Equlib*, 222-223:195, 2004.
11. T.I. Morrow and E.J. Maginn. Molecular dynamics study of the ionic liquid 1-n-butyl-3-methylimidazolium hexafluorophosphate. *J. Phys. Chem. B*, 106:12807, 2002.
12. C.J. Margulis, H.A. Stern, and B.J. Berne. Computer simulation of a "green chemistry" room-temperature ionic solvent. *J. Phys. Chem. B*, 106:12017, 2002.
13. J. Lopes, J. Deschamps, and A. Padua. Modeling ionic liquids using a systematic all-atom force field. *J. Chem. Phys. B*, 108:2038, 2004.
14. S. Urahata and M. Ribeiro. Structure of ionic liquids of 1-alkyl-3-methylimidazolium cations: A systematic computer simulation study. *J. Chem. Phys.*, 120(4):1855, 2004.
15. T. Yan, C.J. Burnham, M.G. Del Popolo, and G.A. Voth. Molecular dynamics simulation of ionic liquids: The effect of electronic polarizability. *J. Phys. Chem. B*, 108:11877, 2004.
16. S. Takahashi, K. Suzuya, S. Kohara, N. Koura, L.A. Curtiss, and M. Saboungi. Structure of 1-ethyl-3-methylimidazolium chloroaluminates: Neutron diffraction measurements and ab initio calculations. *Z. fur Phys. Chem.*, 209:209, 1999.
17. Z. Meng, A. Dölle, and W.R. Carper. *J. Mol. Struct.*, 585:119, 2002.
18. A. Chaumont and G. Wipff. Solvation of uranyl(ii) and europium(iii) cations and their chloro complexes in a room-temperature ionic liquid. a theoretical study of the effect of solvent "humidity". *Inorg. Chem.*, 43:5891, 2004.
19. F.C. Gozzo, L.S. Santos, R. Augusti, C.S. Consorti, J. Dupont, and M.N. Eberlin. *Chem. Eur. J.*, 10:6187, 2004.
20. E.R. Talaty, S. Raja, V.J. Storhaug, A. Dölle, and W.R. Carper. *J. Phys. Chem. B*, 108:13177, 2004.
21. Y.U. Paulechka, G.J. Kabo, A.V. Blokhin, A.O. Vydrov, J.W. Magee, and M. Frenkel. *J. Chem. Eng. Data*, 48:457, 2003.
22. J. de Andrade, E.S. Böes, and H. Stassen. Computational study of room temperature molten salts composed by 1-alkyl-3-methylimidazolium cations-force-field proposal and validation. *J. Phys. Chem. B*, 106:13344, 2002.

23. B. Kirchner and A. P. Seitsonen. Ionic liquids from car-parrinello simulations, part ii: Structural diffusion leading to large anions in chloraluminate ionic liquids. *Inorg. Chem.*, 47:2751–2754, 2007. DOI 10.1021/ic0624874.
24. J. Hutter and D. Marx. Proceeding of the february conference in jülich. In J. Grotendorst, editor, *Modern Methods and algorithms of Quantum chemistry*, page 301, Jülich, 2000. John von Neumann Institute for Computing. `http://www.fz-juelich.de/nic-series/Volume1/`.
25. J. Thar, W. Reckien, and B. Kirchner. Car–parrinello molecular dynamics simulations and biological systems. In M. Reiher, editor, *Atomistic Approaches in Modern Biology*, volume 268, pages 133–171, Top. Curr. Chem., 2007. Springer.
26. P. Hohenberg and W. Kohn. Inhomogeneous electron gas. *Phys. Rev.*, 136:B864 – B871, 1964.
27. W. Kohn and L. J. Sham. Self-consistent equations including exchange and correlation effects. *Phys. Rev.*, 140:A1133 – A1139, 1965.
28. W. E. Pickett. Pseudo potential methods in condensed matter applications. *Comput. Phys. Rep.*, 115, 1989.
29. Jürg Hutter and Alessandro Curioni. Car-parrinello molecular dynamics on massively parallel computers. *ChemPhysChem*, 6:1788–1793, 2005.
30. CPMD V3.8 Copyright IBM Corp 1990-2003, Copyright MPI für Festkörperforschung Stuttgart 1997-2001. see also `www.cmpd.org`.
31. J. P. Perdew, K. Burke, and M. Ernzerhof. Generalized gradient approximation made simple. *Physical Review Letters*, 77:3865–3868, 1996.
32. N. Troullier and J. L. Martins. Efficient pseudopotentials for plane-wave calculations. *Physical Review B*, 43:1993–2006, 1991.
33. B. Kirchner, A. P. Seitsonen, and J. Hutter. Ionic Liquids from Car–Parrinello Simulations, Part I: Liquid $AlCl_3$. *J. Phys. Chem. B*, 110:11475–11480, 2006.

Micromagnetic Simulations of Magnetic Recording Media

Simon Greaves[1]

Research Institute of Electrical Communication, Tohoku University, Katahira 2-1-1, Sendai, 980-8577, Japan. simon@riec.tohoku.ac.jp

In recent years the capacity of magnetic hard disk drives used for data storage has increased at annual rates of between 30% - 60%. Behind this rapid increase in areal density lies a constant process of innovation and technological improvement.

Micromagnetic models can be used to simulate the recording process in magnetic data storage media. Such models are useful because they allow new head and media designs to be evaluated and optimised prior to the fabrication of prototypes, speeding up the development cycle.

This paper describes the components of a micromagnetic model running on the NEC SX-7 supercomputer installed at Information Synergy Centre of Tohoku University. The model is mainly used for the simulation of magnetic recording.

1 Fundamentals of micromagnetics

1.1 Discretisation

Micromagnetics is often used in studies of ferromagnetic materials to investigate phenomena such as magnetisation reversal processes, interactions between magnetic bodies and magnetic recording. In an ideal model of a magnetic material, each individual magnetic moment in a body would be modelled, i.e. the model would take into account the location and magnetic moment of every atom in the body. Such models already exist [EB05], but are currently restricted to modelling relatively small volumes of material due to the huge numbers of atoms involved.

An alternative, micromagnetic, approach is to sub-divide (discretise) the body into small units (cells) and assume that the magnetisation within each cell is uniform and represented by a single moment $\mathbf{M}$ of magnitude M_sV, where M_s is the saturation magnetisation and V the cell volume. Depending on the problem under consideration the cell volumes might lie in a range from 1 nm^3 to 1000 nm^3.

The data storage layer of a typical magnetic recording medium consists of polycrystalline grains with an average diameter of 6 nm - 8 nm and a thickness of 10 nm - 20 nm. A transmission electron microscope (TEM) image showing a plan view of a magnetic recording medium is shown in Fig. 1(a). The grains are irregular shapes and are separated by non-magnetic material. Early micromagnetic models were restricted to modelling grains of uniform size with square or hexagonal cross sections. Increased memory and computing power have enabled the modelling of irregular grains, often represented by Voronoi cells. To create a set of Voronoi cells a set of seed points is first distributed over a plane. A Voronoi cell is defined as the region of the plane which is nearest to a particular seed point. The boundaries of the Voronoi cells are the loci of points equidistant between seed points. The average size and size distribution of the Voronoi cells can be controlled through the density and location of the seed points. Non-magnetic boundary regions can be created by moving the vertices of each Voronoi cell some distance towards the seed point. An example of Voronoi cells used in the micromagnetic model is shown in Fig. 1(b). The Voronoi cell microstructure is a good approximation of the real medium. If the grains are sufficiently small each grain can be modelled by a single cell. If more detail is required the Voronoi cells can be subdivided into smaller cells both in the plane and along the axis normal to the plane. An algorithm for the generation of Voronoi cells can be found in [SF01].

Each of the grains in a recording medium can be considered as a small, permanent magnet. The important magnetic properties of the grains can be obtained from a hysteresis loop, such as the one shown in Fig. 2. The magnetisation of the grain along an axis is measured as an external magnetic field

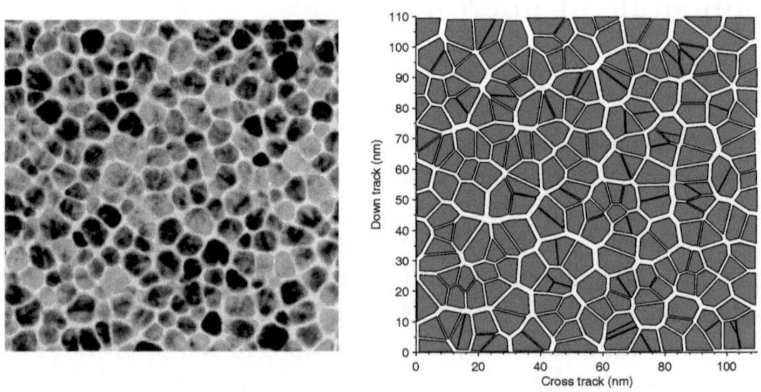

(a) TEM image (b) Voronoi model

Fig. 1. TEM image of grains in a recording medium (plan view) (left) and a Voronoi cell representation of grains (right). The side of each image is 110 nm and the average grain size is about 8 nm.

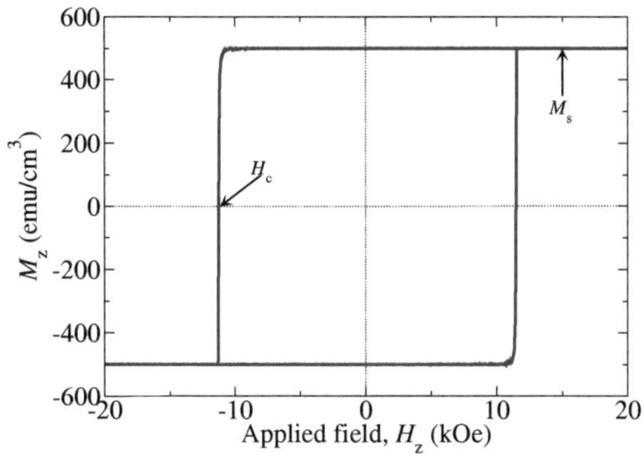

Fig. 2. A hysteresis loop of a single grain. $M_s = 500$ emu/cm^3, $H_c = 11.5$ kOe.

is applied along the same axis. When the magnetic field is sufficiently large the magnetisation of the grain will reverse direction. The field at which this reversal occurs is called the coercive field, or coercivity H_c. In order to provide a stable storage medium H_c should be large enough to resist the effect of stray fields and temperature fluctuations. Another important parameter is the saturation magnetisation M_s which is a measure of the maximum strength of the grain magnetisation. A ferromagnetic element such as iron has M_s of 1700 emu/cm^3 at room temperature. The saturation magnetisation of grains in a recording medium is typically in the range 400 - 750 emu/cm^3. The poles of magnetic write heads are made of alloy materials with M_s as high as 2100 emu/cm^3, allowing them to produce a large magnetic field.

1.2 The LLG equation

Having discretised the body into suitably sized cells, the time variation of the magnetisation of each cell under the influence of internal and external magnetic fields is calculated using the Landau-Lifshitz-Gilbert (LLG) equation [TG55].

$$\frac{d\mathbf{M}}{dt} = -\gamma \left(\mathbf{M} \times \left(\mathbf{H} - \frac{\alpha}{\gamma M_s} \frac{d\mathbf{M}}{dt} \right) \right) \tag{1}$$

In Eq. 1 $\mathbf{M}$ is the magnetic moment in each cell and $\mathbf{H}$ is the magnetic field acting on $\mathbf{M}$. γ is the gyromagnetic constant ($1.76 \times 10^7 s^{-1} Oe^{-1}$) and α is the damping constant; typical values lie between 0.01 and 1. The time variation of $\mathbf{M}$ is obtained by computing $\mathbf{H}$ and calculating $d\mathbf{M}$ using the LLG equation.

Eq. 1 consists of two terms: a precession term and a damping term. To see the effect of the precession term we set $\alpha = 0$ and the LLG equation becomes

$$\frac{d\mathbf{M}}{dt} = -\gamma \left(\mathbf{M} \times \mathbf{H} \right). \tag{2}$$

The magnetic field **H** exerts a torque on **M** which is perpendicular to the $\mathbf{M} - \mathbf{H}$ plane, causing **M** to precess about **H**. However, if $\alpha = 0$ **M** will precess about **H** indefinitely and will never align with **H**. The damping term in the LLG equation containing the constant α damps the precessional motion, causing **M** to align with **H**. The higher the value of α the sooner **M** aligns with **H**. Fig. 3 shows an example of damped precessional motion for a single magnetic vector **M** which initially lies in the $x - y$ plane, as indicated by the arrow. A magnetic field **H** of magnitude 2 kOe is then applied along the z axis and **M** begins to precess, tracing the path indicated by the spiral. The magnitude of **M** is constant, meaning that as the in-plane component of **M** decreases the out of plane component increases, and eventually **M** ends up pointing along the z axis i.e. perpendicular to the page.

1.3 Magnetic field sources

The **H** term in the LLG equation is a summation of multiple magnetic field sources, many of which are intrinsic to the magnetic material. Other field sources may be generated externally, e.g. in magnetic recording the field used to record data is generated by a write head. Field sources can be confined to each cell, a result of nearest-neighbour interactions, or due to long range interactions. A micromagnetic model needs to calculate all of these fields accurately. The main field sources, together with examples of how they may be calculated are outlined below.

Magnetostatic field

The magnetostatic field $\mathbf{H_d}$ (also called the demagnetising, or demag. field) is a long range interaction between cells. It is usually the most computationally

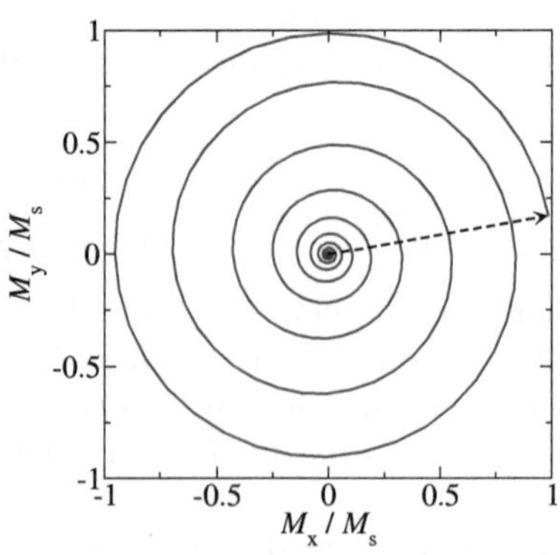

Fig. 3. Projection of a magnetic vector in the $x - y$ plane when a magnetic field is applied along the z axis. $\alpha = 0.1$, $M_s = 500$ emu/cm^3.

intensive part of a micromagnetic model since the calculation of the magnetostatic field in a target cell involves a summation of the magnetostatic field components from every other cell in the model. For a model containing N cells, N^2 calculations must be performed to calculate the magnetostatic field in all cells every time the LLG equation is evaluated. For regularly-shaped cells the calculation time can be reduced using Fourier transforms, but such techniques cannot be applied to irregularly-shaped cells. Alternatively, cells which are far from the target cell can be merged into groups, reducing the total number of cells for the purposes of the magnetostatic field calculation.

To calculate the magnetostatic field, first consider a single magnetic dipole moment $\mathbf{m}$. The field from the dipole at a point $\mathbf{r}$ is given by

$$\mathbf{H}(\mathbf{r}) = \frac{1}{r^3}\left(3\left(\frac{\mathbf{m}\cdot\mathbf{r}}{r}\right)\frac{\mathbf{r}}{r} - \mathbf{m}\right). \tag{3}$$

The magnetostatic field decays rapidly with distance from the dipole, but the amount of magnetic material contained within an annulus of radius r and width dr increases in proportion to r^2, so the calculation cannot simply be terminated at some limiting radius without sacrificing accuracy.

Eq. 3 can be integrated over the cell volume to obtain the total magnetostatic field from each cell. $\mathbf{H_d}$ can be split into orthogonal components and rewritten in tensor form to simplify the calculation. The total magnetostatic field in cell i is then given by

$$\begin{pmatrix} H_{dx}(i) \\ H_{dy}(i) \\ H_{dz}(i) \end{pmatrix} = \sum_{j=0}^{N} \begin{pmatrix} K_{xx} & K_{xy} & K_{xz} \\ K_{yx} & K_{yy} & K_{yz} \\ K_{zx} & K_{zy} & K_{zz} \end{pmatrix} \begin{pmatrix} M_x(j) \\ M_y(j) \\ M_z(j) \end{pmatrix}. \tag{4}$$

Expressions for the tensors K_{xx} etc. for cuboid cells are given in [YN89] and [HF98]. For irregular shapes the cells can be segmented into cuboids of various sizes and the tensors for the individual cuboids summed. This technique only works using the expressions in [YN89]. The result is a tensor which describes the boundary of the irregular cell. Fig. 4 shows an example of segmenting a hexagonal cell into cuboids. The smallest cuboids have an edge length of 0.1 nm, equivalent to atomic resolution. Generally, the tensors K_{xx} etc. are calculated at the start of the program and stored as the tensor calculation can be time consuming, particularly if the cells are irregularly shaped. An additional issue to consider is the fact that for irregular cells the tensors are unique, being a function of cell shape and distance from the target cell. For a model containing N cells it is necessary to store $6N^2$ tensors (off-diagonal tensors are equivalent, i.e. $K_{xy} = K_{yx}$ etc.). The amount of memory required to store the tensors increases rapidly with N, unless distant cells are merged into groups.

Anisotropy field

The anisotropy field $\mathbf{H_k}$ is a local field specific to each cell which reflects a preference for the magnetisation vector to lie in a particular direction (the

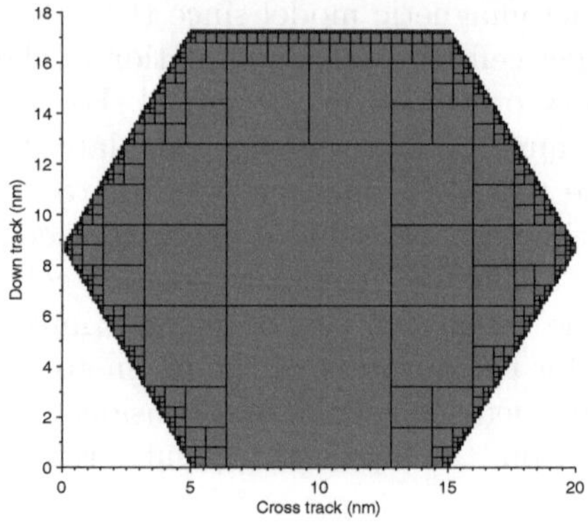

Fig. 4. Segmentation of a hexagonal cell into cuboids for the purposes of calculating the magnetostatic tensors.

easy axis). The magnitude of $\mathbf{H_k}$ is proportional to the strength of the uniaxial anisotropy, which is a material-dependent property. For example, current hard disk media have perpendicular anisotropy, meaning that $\mathbf{M}$ prefers to lie perpendicular to the disk plane. $\mathbf{H_k}$ is given by

$$\mathbf{H_k} = \frac{2\mathbf{K_u} \cdot \mathbf{m}}{M_s} \tag{5}$$

where $\mathbf{K_u}$ is a vector representing both the direction of the easy axis and magnitude of the uniaxial anisotropy, $\mathbf{m}$ is a unit vector indicating the direction of the magnetic moment in the cell and M_s is the saturation magnetisation of the cell. In the absence of other magnetic fields, the total energy will be minimised when $\mathbf{m}$ lies along $\mathbf{K}$. In the model, both the magnitude and direction of $\mathbf{K}$ can be varied from cell to cell, reflecting compositional and microstructural variations inherent to experimental recording media.

Exchange coupling field

Exchange coupling is a nearest-neighbour interaction between a cell i and its neighbours j. The exchange coupling field is given by

$$\mathbf{H_{ex}}(i) = \frac{2}{M_s(i)} \sum_j A_{ij} \nabla^2 \mathbf{m}(j). \tag{6}$$

Here, A_{ij} is the exchange coupling constant between the two cells i and j. For a regular microstructure with cell centres separated by a distance d_{ij} we can expand the $\nabla^2 \mathbf{m}$ term in Eq. 6 and approximate $\mathbf{H_{ex}}$ using

$$\mathbf{H_{ex}}(i) = \frac{2}{M_s(i)} \sum_j A_{ij} \left(\frac{\mathbf{m}(j) - \mathbf{m}(i)}{d_{ij}^2} \right). \tag{7}$$

However, if the cells are irregular, we must correct for variations in the length of the common boundary L_{ij} between each pair of cells. The exchange coupling strength between cells depends upon the ratio of L_{ij} to the total boundary length of the cell L_{total}. This is shown schematically in Fig. 5(a) by the thicknesses of the lines connecting cell i to neighbouring cells. A distribution of common boundary lengths gives rise to a distribution of $\mathbf{H_{ex}}$ characterised by σ_L. Eq. 8 takes account of the boundary length dependence of $\mathbf{H_{ex}}$ with the addition of an extra term $4L_{ij}/L_{total}(i)$. For cuboid cells $L_{ij} = L_{total}(i)/4$, the last term becomes unity and Eq. 8 is identical to Eq. 7. Differences in intergranular spacing will also give rise to a distribution of $\mathbf{H_{ex}}$ characterised by a parameter σ_A. Thus, $\mathbf{H_{ex}}$ distributions have two origins: edge length and intergranular spacing variations, as shown in Fig. 5(d).

$$\mathbf{H_{ex}}(i) = \frac{2}{M_s(i)} \sum_j A_{ij} \left(\frac{\mathbf{m}(j) - \mathbf{m}(i)}{d_{ij}^2} \right) \times \left(\frac{4L_{ij}}{L_{total}(i)} \right) \tag{8}$$

Thermal field

The thermal field is a stochastic term, localised to each cell, which represents the effect of heat on the magnetic moment $\mathbf{M}$. In the same way that a small particle suspended in a liquid undergoes random position fluctuations due to Brownian motion, the magnetic moment of a cell also fluctuates randomly under the influence of the thermal field. New values of the thermal field are chosen each time the LLG equation is evaluated. Orthogonal components of the thermal field form a Gaussian distribution with a standard deviation given by

$$\sigma = \sqrt{\frac{2k_b T \alpha}{V M_s \gamma dt}} \tag{9}$$

where T is the temperature, V the cell volume, k_b is Boltzmann's constant and dt the time step used in the LLG equation. There is no correlation between successive values of the thermal field and the time-averaged value is zero.

There are several advantages to including the thermal field in a micromagnetic model which are founded on the avoidance of anomalous magnetic states.

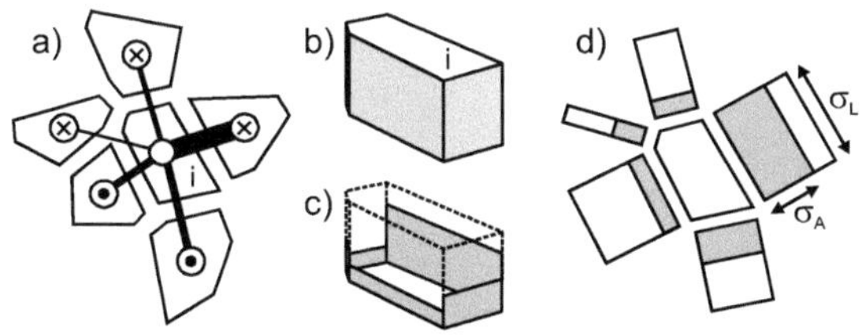

Fig. 5. Irregular cells result in a distribution of exchange coupling strengths.

For example, consider a situation in which **H** is anti-parallel to **M**. According to the LLG equation, the torque acting on **M** would be zero ($\mathbf{M}\times\mathbf{H}=0$) and as a result **M** would never reverse direction to align with **H**. A thermal field term prevents the occurrence of anti-parallel states by constantly perturbing **M**. For the same reason, **M** is less likely to end up in a local energy minimum if thermal fluctuations are constantly probing the stability of such minima. Inclusion of the thermal field also allows experimental phenomena, such as time dependence, to be reproduced. The effect of the thermal field is shown in Fig. 6, which shows the motion of a magnetic moment in a magnetic field at a temperature of 77 K. All parameters, apart from the temperature, are the same as in Fig. 3. Comparing the two figs. it can be seen that the thermal field causes **M** to follow an erratic path as it precesses about the applied field axis. In fact, the thermal field guarantees that **M** never completely aligns with the applied field axis. Fig. 7 shows the probability distribution of the angle between **M** and **H** for the data in Fig. 6, excluding the first few precessions of **M**. A Boltzmann probability distribution is obtained, with the angle at which $P(\theta)$ is a maximum increasing with temperature.

Generation of the random field terms is typically the second most computationally intensive part of a micromagnetic simulation. Uniformly distributed random numbers are generated using the Mersenne twister algorithm [MT01] and converted into a Gaussian distribution of random numbers using a Box-Muller transform. Using different initial seeds on each processor allows the random numbers to be generated in parallel. Vectorisation is improved by generating two large arrays of uniformly distributed random numbers and applying the Box-Muller transform to the two arrays, instead of to individual pairs of random numbers.

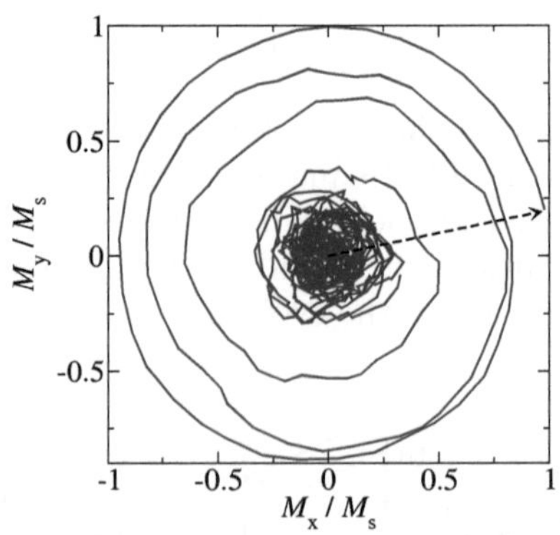

Fig. 6. Precession of **M** in a 8 nm cube. $T=77$ K, $\alpha=0.1$, $M_s=500$ emu/cm^3, $H_z=2$ kOe.

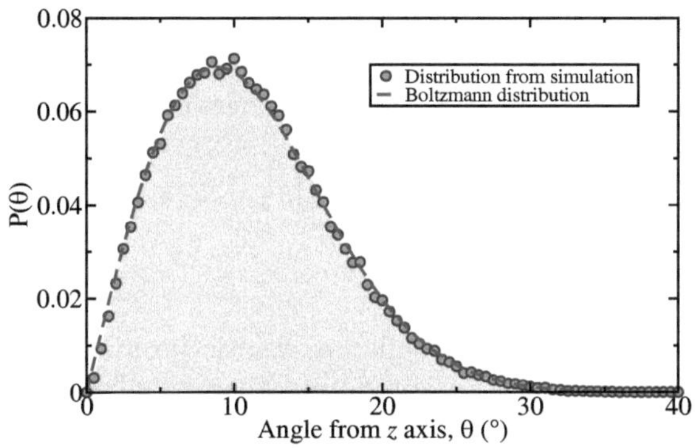

Fig. 7. Angle between **M** and **H** for an 8 nm cube. The distribution was obtained from the data in Fig. 6.

Spin torque

If an electric current flows through a uniformly magnetised material the current will become spin-polarised as the electron spins align with the magnetisation vector of the material. Similarly, if the magnetisation vector changes along the direction of current flow the electron spins will exert a torque on **M**. The magnitude of this spin torque is given by

$$\frac{d\mathbf{M}}{dt} = -(\mathbf{u} \cdot \boldsymbol{\nabla})\,\mathbf{m} \tag{10}$$

where **u** is the direction of current flow with a magnitude proportional to the current density and spin polarisation rate and **m** is a unit vector indicating the direction in which the material is magnetised. The spin torque effect can be important in magnetic sensors, such as the read heads used in hard disk drives as the read current flows through magnetic layers used to sense the stray field from the recording medium.

Applied field

Applied fields could be generated externally by a current flowing through a coil or wire, or by another magnetic object. In magnetic recording the field used to write data onto the hard disk is generated by a write head, typically a small, single pole magnet. The stray field from the pole writes data bits with the same polarity as the pole magnetisation. The pole magnetisation is reversed by passing a current through a coil wrapped around the pole. Ideally, all magnetostatic interactions between the head and medium would be included in a micromagnetic model. However, it is often sufficient to calculate the head field distribution in a separate model and use the distribution in the recording model. When interactions between the head and medium are included the computation time increases significantly. This is partly because of the increased number of calculations required to evaluate the magnetostatic field, but mainly because the write head moves relative to the medium during the simulation, requiring time-consuming updates of the magnetostatic tensors linking the head and medium.

2 A micromagnetic model

There are several freely available micromagnetic programs. Two of the most commonly used are the object oriented micromagnetic framework (*OOMMF*) program developed at NIST [OO01] and the finite element micromagnetics package *magpar*, available at [WS01]. There are also many commercial, closed source programs. A micromagnetic model has also been developed by the author, based on the description in Sect. 1, mainly for the purpose of carrying out simulations of magnetic recording. The code is written in C and runs on SX-7, TX-7 and Linux/UNIX clusters, usually on 8 - 32 processors depending on the problem size. On the SX-7 machine vectorisation exceeds 99% and memory usage for a 19000 cell model without cell grouping for the magnetostatic calculation is about 10 GB. With cell grouping the memory requirement is reduced to around 4 GB for a 54000 cell model. All arrays are single dimensional to eliminate nested loops and improve vectorisation. Other than the standard C libraries, no external libraries are required unless MPI routines are used, making the program highly portable. A diagram depicting program flow is shown in Fig. 8. First of all the model geometry is generated. Voronoi cells are generated using an implementation of the Fortune algorithm [SF01].

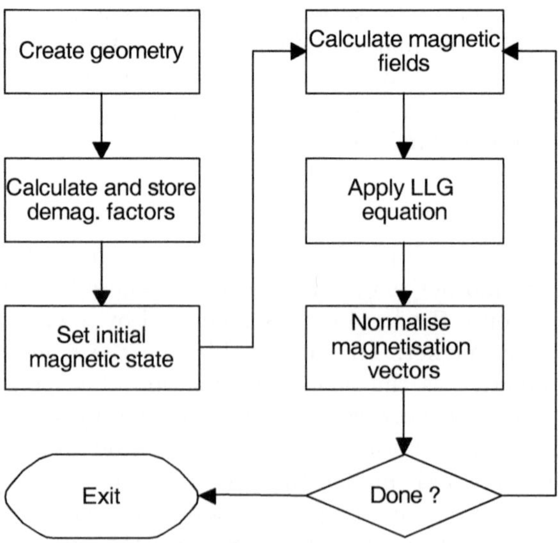

Fig. 8. A flowchart of the micromagnetic model execution.

Next, the magnetostatic tensors are calculated and stored and the initial magnetic state is applied. The head field distribution used to write data tracks can be generated internally or loaded from a pre-existing file. The program then enters a loop in which the magnetic field in each cell is calculated and the magnetisation vectors updated by applying the LLG equation. The LLG equation does not preserve the magnitude of $\mathbf{M}$, so the magnetisation vectors must be re-normalised each time the LLG equation has been applied. The write head moves along the recording medium as time elapses and the polarity and magnitude of the write field is varied to record data bits onto the recording medium.

The program information from a typical micromagnetic simulation run on the SX-7 supercomputer is shown in Fig. 9. The simulation involved the writing of data tracks on media with various parameters; in total, ten tracks were written. Each medium consisted of eight layers and 28400 cells. The vectorisation level exceeded 99.6% and the load was well distributed across the eight processors used. Further optimisations to reduce the Lock Wait and Bank times are required.

```
   ******  Program Information  ******
Real Time (sec)            :        169943.531490
User Time (sec)            :       1274368.077014
Sys  Time (sec)            :         13307.835458
Vector Time (sec)          :       1016207.386925
Inst. Count                :      100143742924520
V. Inst. Count             :       54861150503006
V. Element Count           :    13787293186921640
FLOP Count                 :     5514501004250929
MOPS                       :         10854.458793
MFLOPS                     :          4327.243521
MOPS   (concurrent)        :         81923.446046
MFLOPS (concurrent)        :         32659.638573
A.V. Length                :           251.312505
V. Op. Ratio (%)           :            99.672638
Memory Size (MB)           :          3648.000000
Max Concurrent Proc.       :                    8
 Conc. Time(>= 1)(sec):             168847.582065
 Conc. Time(>= 2)(sec):             161167.838704
 Conc. Time(>= 3)(sec):             160839.803197
 Conc. Time(>= 4)(sec):             160408.811717
 Conc. Time(>= 5)(sec):             159692.126165
 Conc. Time(>= 6)(sec):             158504.944232
 Conc. Time(>= 7)(sec):             156742.837009
 Conc. Time(>= 8)(sec):             148181.017262
Event Busy Count           :                    0
Event Wait (sec)           :             0.000000
Lock Busy Count            :              6880572
Lock Wait (sec)            :          5232.423251
Barrier Busy Count         :                    0
Barrier Wait (sec)         :             0.000000
MIPS                       :            78.583060
MIPS (concurrent)          :           593.101433
I-Cache (sec)              :          1316.809521
O-Cache (sec)              :          9392.596166
Bank (sec)                 :          3035.494823
```

Fig. 9. Program information for a typical simulation.

Solving the LLG equation

The easiest way to solve the LLG equation is to use the Euler method. However, a small time step dt is required in order to maintain stability of $\mathbf{M}$ and this increases the overall computation time. The Heun (or improved Euler) method allows larger time steps to be used and also has the advantage of being compatible with the expressions for the thermal field, given earlier. Higher order solvers are redundant due to the stochastic nature of the model which imposes a minimum noise base on solutions to the LLG equation. An example of the effectiveness of the Heun method is shown in Fig. 10 which depicts small oscillations of the magnetisation in a soft magnetic material. When the time step dt is increased from $0.6\times10^{-14}s$ to $1.2\times10^{-14}s$ the solution obtained using the Euler method diverges from the correct solution, but significant deviations occur only after the simulation has ran for 6000 steps. The Heun method allows a time step at least five times larger than the Euler method at the cost of a less than twofold increase in per-step execution time (the thermal field only needs to be calculated once, the other fields are calculated twice).

Simulations of multiple magnetic objects

If it is necessary to consider the interactions between the write head and recording medium the magnetostatic interactions between the two objects must be calculated. In the micromagnetic model each magnetic object runs in a separate thread as a self-contained model. Interactions between objects are calculated based on the component of $\mathbf{M}$ normal to the surfaces of each object using the approach described in [DL84]. A flow chart for the multiple object program is shown in Fig. 11. The interactions between objects are calculated and included as an extra magnetic field term, $\mathbf{H_{int}}$. The execution of the threads is synchronised via a barrier wait routine. As noted above, if the

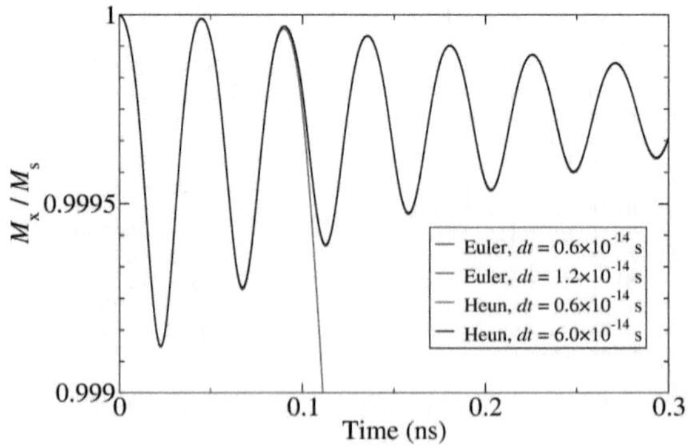

Fig. 10. Effect of LLG time step and integration method when calculating small oscillations in a soft magnetic material.

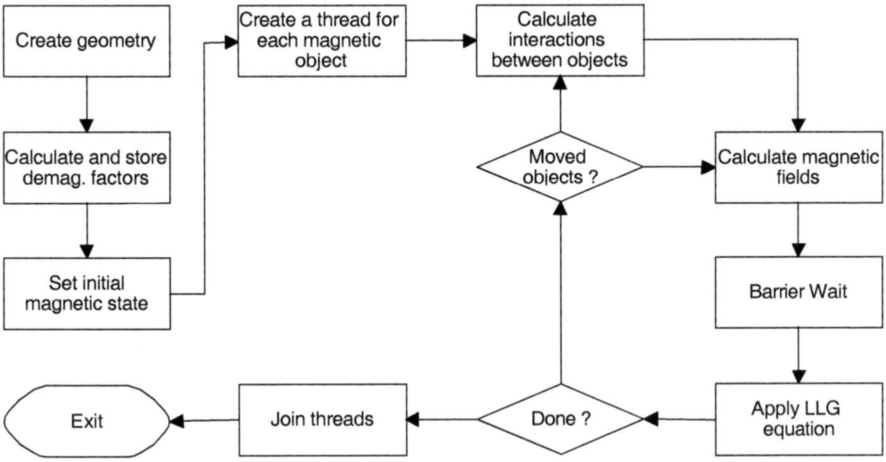

Fig. 11. A flowchart for a model containing multiple, interacting magnetic objects.

relative position of the objects changes during the simulation the interaction tensors must be recalculated and this can take a considerable amount of time. Further optimisation of these routines is required; e.g. interpolating the interaction tensors for intermediate head-medium positions, load balancing to make the thread execution times similar for different sized objects etc.

Heat flow model

Future data storage products may make use of heat assisted recording. The motivation is to improve the thermal stability of recorded data bits by increasing the strength of the uniaxial anisotropy. At room temperature the recording medium will have a very high coercivity and a head would be unable to record on it, the head field being less than the coercivity. During the recording process the medium is heated by a laser, reducing the coercivity and allowing the head to write data bits. To enable the simulation of heat assisted recording a heat flow model has been added to the micromagnetic model. The heat flow between two cells i and j is written as

$$\frac{dQ_i}{dt} = K_j S_{ij} \left(\frac{T_j - T_i}{d_{ij}} \right) \tag{11}$$

where Q = heat, K = thermal conductivity, S = common surface area of the two cells, T = temperature and d = centre-centre distance between the two cells. The heat flow equation and LLG equation are synchronised and solved in parallel. The constant parts of Eq. 11 are encapsulated in an array of values H_{ij} for pairs of neighbouring cells, i and j.

$$H_{ij} = K_j S_{ij} / d_{ij} \tag{12}$$

The H_{ij} values can be varied for each pair of neighbouring cells. This is useful for simulating patterned media in which the heat flow is much larger within patterned elements than between patterned elements separated by air or some other material. Fig. 12 shows the result of a heat flow calculation for a 5 nm thick CoCrPt recording medium on a 10 nm Ru seed layer on glass. The mesh used for the heat flow calculation is independent of the geometry used for the micromagnetic calculation and the cells of the two models do not need to be the same size. The micromagnetic geometry can occupy all or part of the space used for the heat flow calculation. The heat flow calculation usually requires a much larger region than the micromagnetic model in order to avoid boundary issues. An accurate heat flow calculation also requires the inclusion of non-magnetic conduction layers, in addition to the recording layer itself. Temperatures are calculated using the heat conduction model and transferred to the equivalent location in the micromagnetic model. The temperature dependence of magnetic properties such as M_s and K_u is described by a function or data table.

3 Applications of micromagnetics

Magnetic recording simulations

The micromagnetic model is most often used for simulations of perpendicular magnetic recording. The objective is to find ways to increase the areal recording density, either through improvements to the media design, or changes to the structure of the write head. Micromagnetic modelling allows theoretical media designs and material properties to be investigated without worrying about experimental variables. Fig. 13 shows the result of a simulation in which a track was recorded along the centre of a medium. Important parameters to

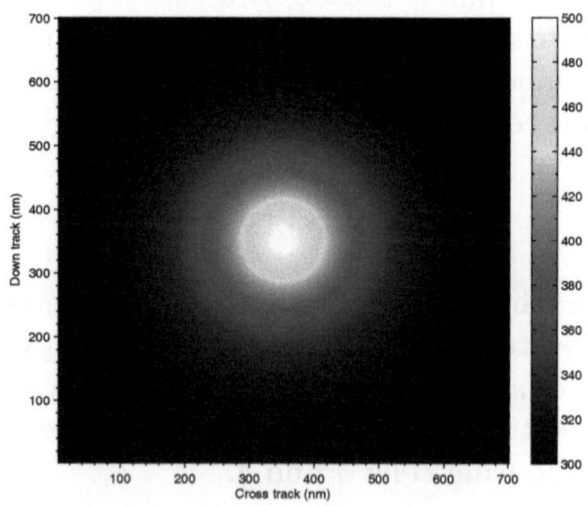

Fig. 12. Temperature of a recording medium (in K) after irradiation by a 6.5 mW laser for 1 ns

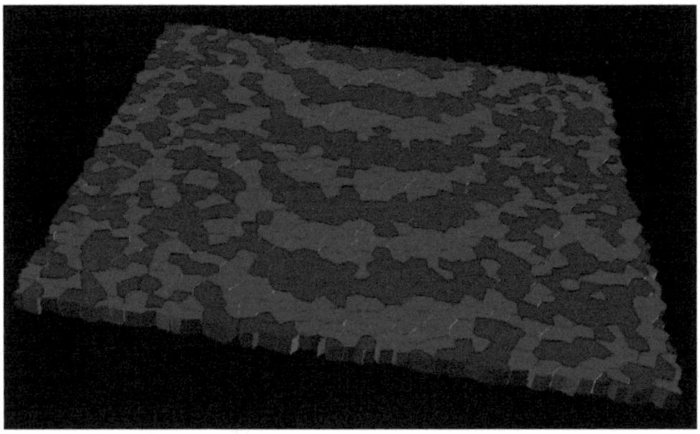

Fig. 13. A written track, the result of a recording simulation.

be considered when evaluating such recorded tracks include: signal to noise ratio (SNR), recorded track width, correlation lengths and transition noise etc.

Thermally assisted recording simulations

Adding a heat source to the write head could allow an increase in the recording density. Combined LLG and heat flow models were used to simulate thermally assisted recording. Fig. 14 shows a snapshot of the recording layer during a thermally assisted recording simulation. On the left is the temperature distribution in the recording layer; the maximum temperature is 623 K and the minimum temperature is 300 K. The magnetic region is shown on the right; it is smaller than the heat flow region to reduce computing time. The portion of the magnetic region directly under the laser spot is above the Curie temperature and the magnetisation is disordered. As the laser moves along the

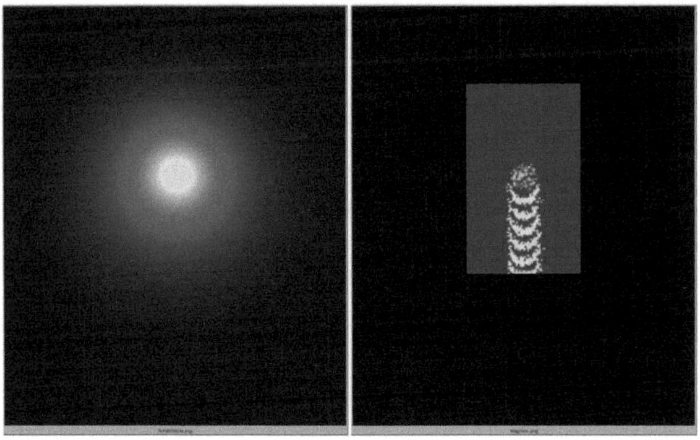

Fig. 14. Simulation of thermally assisted recording, left : temperature, right : magnetisation. Image size : 900 nm × 1100 nm.

medium the recording layer cools below the Curie temperature. M_s and K_u increase as the medium cools and the polarity of the written bits is determined by a magnetic field generated by a write head. The rate of cooling, which can be controlled by the thermal properties of the seed layers, influences the SNR, written track width and transition noise.

4 Conclusions

A high-performance micromagnetic model has been successfully developed for the SX-7 supercomputer. The program allows the simulation of magnetic recording and magnetisation processes in magnetic materials and has been used in collaborative research with academic and industrial partners. In the future, as performance and memory capacity increase, it is expected that the minimum cell size used in the model can be reduced, ultimately allowing the modelling of large systems on an atomic scale.

References

[DL84] Lindholm, D.A.: Three-dimensional magnetostatic fields from point-matched integral equations with linearly varying scalar sources. IEEE Trans. Magn., **20**, 2025–2032 (1984)

[EB05] Boerner, E.D., Chubykalo-Fesenko, O., Mryasov, O.N., Chantrell, R.W., Heinonen, O.: Moving toward an atomistic reader model. IEEE Trans. Magn., **41**, 936–940 (2005)

[HF98] Fukushima., H., Nakatani, Y., Hayashi, N.: Volume average demagnetising tensor of rectangular prisms. IEEE Trans. Magn., **34**, 193–198 (1998)

[MT01] Matsumomto, M.: An algorithm for random number generation. http://www.math.sci.hiroshima-u.ac.jp/ m-mat/MT/emt.html

[OO01] Donahue, M.: The object oriented micromagnetic framework (OOMMF) project at ITL/NIST. http://math.nist.gov/oommf/

[SF01] Fortune, S.J.: An algorithm for Voronoi cell generation. http://netlib.bell-labs.com/cm/cs/who/sjf/index.html

[TG55] Gilbert, T.L.: A Lagrangian formulation of the gyromagnetic equation of the magnetic field. Phys. Rev., **100**, 1243 (1955).

[WS01] Scholz, W.: magpar - Parallel finite element micromagnetics package. http://magnet.atp.tuwien.ac.at/scholz/magpar/

[YN89] Nakatani, Y., Uesaka, Y., Hayashi, N.: Direct solution of the Landau–Lifshitz–Gilbert equation for micromagnetics. Jap. J. Appl. Phys., **28**, 2485–2507 (1989)

Future High Performance Systems

The Potential of On-Chip Memory Systems for Future Vector Architectures

Hiroaki Kobayashi[1,2], Akihiko Musa[2,3], Yoshiei Sato[2], Hiroyuki Takizawa[2,1], and Koki Okabe[1]

[1] Information Synergy Center, Tohoku University, Sendai 980-8578, Japan
[2] Graduate School of Information Sciences, Tohoku University, Sendai 980-8578, Japan
[3] NEC Corporation, Tokyo, 108-8001, Japan
{koba@, musa@sc, yoshiei@sc, tacky@, okabe@}isc.tohoku.ac.jp

1 Introduction

The most advantageous feature of modern vector systems is their outstanding memory performance compared to scalar systems. This feature brings them to their high-sustained system performance when executing real application codes, which are extensively used in the fields of advanced sciences and engineering [9],[10],[1]. However, recent trends in semiconductor technology generate a strong head wind for vector systems. Thanks to the historical growth rate in on-chip silicon budget, named Moore's law, processor performance regarding flop/s rates increases remarkably, but memory performance cannot follow it [2]. Regarding vector systems, their bytes/flop rates that show the balance between flop/s performance and memory bandwidth go down from 8 B/flop in 1998, to 4 in 2003, and to 2 in 2007. We have pointed out that reducing the memory bandwidth seriously affects the sustained system performance even in case of vector systems [3], although their absolute performance increases to a certain degree. Memory performance definitely becomes one of key points for design of future highly-efficient vector architectures to survive in an era of multi-core processors.

To cover the lack of memory performance of future vector systems, this paper discusses the potential of on-chip memory systems such as cache and local (scratchpad) memory, in particular, the their effects on the sustained system performance. Some vector supercomputers have already employed caches [13], however, the quantitative discussion on the effects of on-chip caching for vector architectures by using practical application codes has not been discussed so far. This paper presents some early experimental results of vector on-chip caches in execution of real application codes, and tries to figure out how much capacity of the on-chip cache would be equivalent to B/flop rates.

The rest of the paper is organized as follows. Section 2 reviews the trends in performance of the modern vector systems, and presents that the memory performance seriously affects their sustained system performance in execution of real application codes. Section 3 presents a baseline vector architecture with an on-chip memory subsystem to be discussed in this paper. The architecture is designed based on the NEC SX-7 architecture [6]. We also discuss the pros and cons of on-chip cache and on-chip local memory. In Section 4, we present the experimental results on the vector cache system by using a simulator. For the performance evaluation, five real application codes, which are actually developed in the leading fields of computational science and engineering, are used. We discuss the effects of on-chip caching, an assisting mechanism for vector load/store units, on the sustained performance, and show that on-chip caching is promising even for vector architectures. Finally, Section 5 summarizes the paper.

2 Vector Architecture: Its Light and Shadow

The most important feature that distinguishes vector systems from other high-end computing systems is their outstanding memory performance. Table 1 summarizes the comparison in performance of modern vector and scalar systems. As the table shows, the SX systems have 5 to 10 times higher bandwidth than Itanium-based systems even though they have almost the same flop/s performance.

This higher memory performance can also be confirmed by many memory-related benchmark tests. For example, our SX-7 [6] system has top ranking on the STREAM benchmark tests in 2005, regarding per-processor memory bandwidth test as well as system memory bandwidth tests even with a small number of processors [4]. The memory bandwidth of scalar systems that rely on data caches for memory performance is a factor of 7-25 slower than the NEC SX-7 [2].

The balanced flop/s-memory bandwidth performance of the vector systems definitely contributes to the higher sustain performance when executing real application codes. Figure 1 shows the ratio of the sustained system per-

Table 1. Characteristics of modern vector and scalar systems

System	CPUs /Node	Clock Freq (GHz)	Per CPU			Mem /Node (GB)
			Peak Perf. (Gflop/s)	Memory BW (GB/s)	L3 Cache (MB)	
SX-7	32	1.1	8.83	35.3(dedicated)	-	256
SX-7C/8	8	2.0	16.0	64.0(dedicated)	-	128
TX7/i9510	32	1.6	6.4	6.4(shared)	9	128
Altix3700	64	1.6	6.4	6.4(shared)	6	128
TX7/i9610	64	1.6	6.4	8.5(shared)	12	512

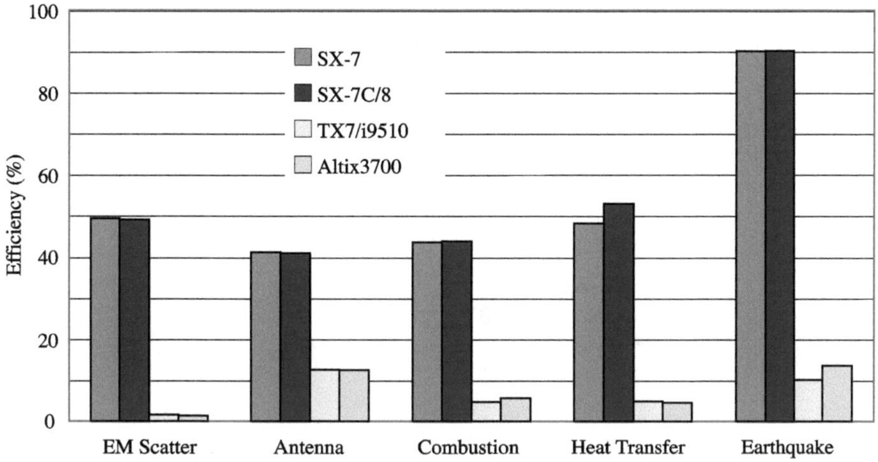

Fig. 1. Efficiency of the four systems for the five simulation codes

formance to the peak performance of vector systems and Itanium-based scalar systems, when executing five application codes used in the fields of electromagnetic, combustion, heat transfer and earthquake analyses. The vector systems achieved 40 to 90% efficiencies of their peak performance. On the other hand, scalar systems stay around 10% or less. This is, of course, due to the difference in memory performance of the vector and scalar systems. Figure 2 presents the ratios of memory processing time exposed in the total processing time. On the scalar systems, most of processing time is spent for data transfer between CPU and memory, and its ratios reach 70% to 90% in the cases that the efficiency is less than 10%. Therefore, memory bandwidth matching with flop/s rates is a key to realize the highly efficient supercomputing of real application codes.

Vector systems have higher memory bandwidth, however, it is getting harder to keep the higher memory bandwidth as processor performance improves at the historical growth rate of 50% per year. Figure 3 shows the trend

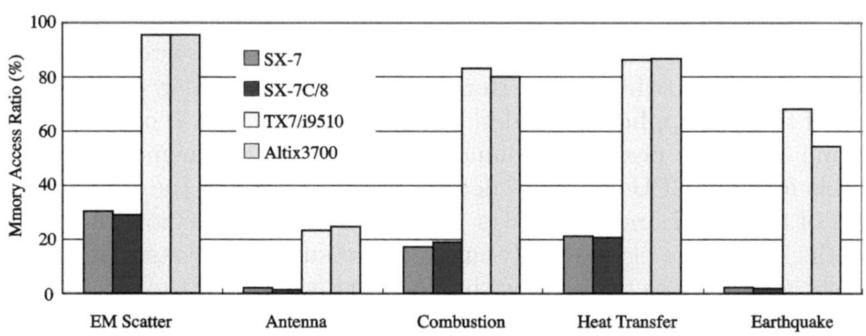

Fig. 2. Memory access ratios to the total execution time

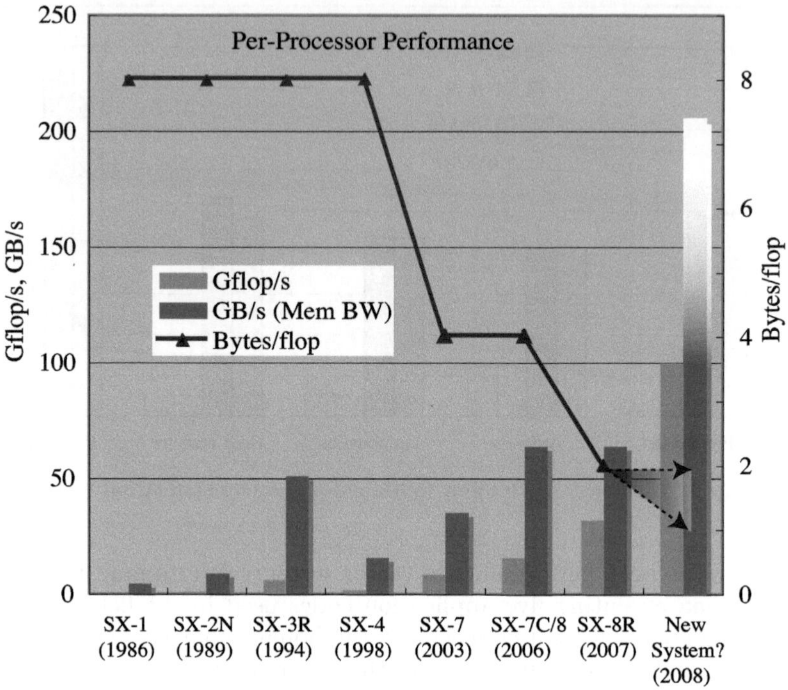

Fig. 3. SX performance trend

in the processor and memory performance of the SX systems, most of which have been installed at Tohoku University. Until SX-4, the systems had 8 bytes per flop rates, but SX-7 and 7C (SX-8) systems have 4 bytes per flop, even though the high performance, relatively low-power single-chip vector processors have boosted their total flop/s performance. The recently released new SX system named SX-8R improves per-processor performance twice compared to the SX-8, but now its bytes per flop rate goes down to only 2. So we are seriously concerning about flop/s-memory bandwidth balance of the next and future SX systems, in particular, the memory bandwidth of of a 100+gflop/s vector processor that is expected to come out next year [12].

We have examined how the memory bandwidths of the vector systems contribute to their overwhelming performance against the scalar systems in execution of the real application codes. Figure 4 shows the ratio of the sustained performance to the peak performance when limiting the memory bandwidth available for each CPU of a SX-7 node, i.e., in the cases of the memory bandwidths of 8.83, 17.7 and 35.3 GB/s per CPU (4B/flop, 2B/flop and 1B/flop, respectively). As the figure clearly suggests, the sustained system performance seriously goes down as the memory bandwidth decreases. If the memory bandwidth is reduced to the level of typical scalar systems, i.e., 1B/flop or lower, the sustained performance is also reduced to the same level as the scalar systems,

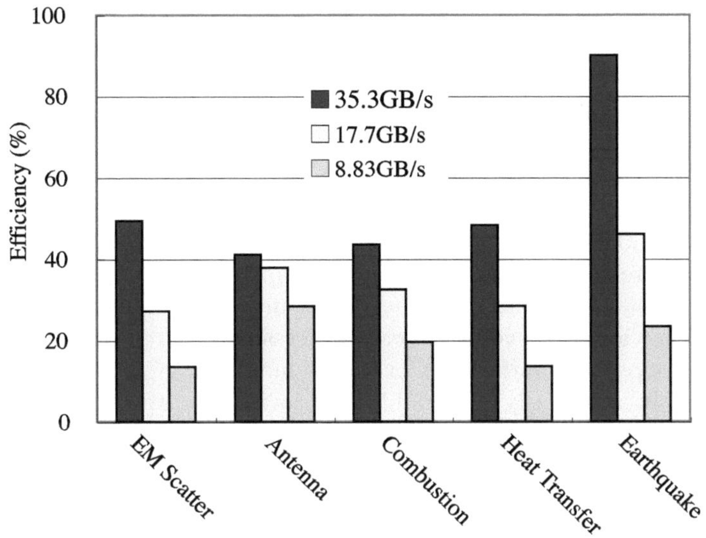

Fig. 4. Effect of memory bandwidths on SX-7 performance

around 10%. These results mean that the outstanding sustained performance of the vector system are strongly supported by the excellent memory performance, and therefore, to keep the sustained performance higher, sufficient memory bandwidth over peak-flop/s is essential for future vector systems to survive in the HPC community.

3 On-Chip Memory Systems for Vector Architectures

As the number of pins on a single chip is limited, spending plenty of silicon budget for temporal storage units on a chip makes sense even for vector architectures. However, a large number of register files is not a good idea because they need a larger area to implement and power to drive. So we are focusing on the potential of on-chip memory systems for future vector architectures as an assisting mechanism for vector load/store units. There are two choices regarding on-chip memory systems: on-chip local memory (scratchpad memory) and on-chip cache. Of course, they have advantages and drawbacks.

Regarding the on-chip memory, it is used as a dedicated local store within a chip and totally controllable by software. Therefore, explicit and efficient data management can be achieved by selectively storing data with higher locality of reference. This leads to a favorable feature of the efficient use of limited capacity; only useful data are stored and no worry about cache eviction of effective data by non-effective ones, which is a problem of conventional cache-based systems. As the data management is explicitly controlled by software, unlike cache systems, complicated hardware mechanisms are not needed, and the power-efficient implementation is also expected. Even though the efficient use of the local memory forces users to have skillful programming, highly regular and predictable memory references of vector codes, sophisticated vector ISAs such as efficient vector loads/stores for gather/scatter-type memory references, and its power and area efficient implementation suggest its great potential for future vector architectures.

On the other hand, the superiority of the on-chip cache against the on-chip local memory approach is its transparency to the programmers. As all the data always are accessed through cache, no programming efforts are needed. Therefore, the on-chip cache is effective for data accesses with unpredictable memory behavior. However, in vector processing, a large number of data of individual loops are handled, for example, hundreds or more MB in a single kernel, and it is actually impossible to provide the on-chip cache with an enough capacity to capture the entire data set of many kernels of application codes. Therefore, it is essential for on-chip vector caches to have mechanisms for the effective use of its limited space such as no-eviction of effective data and cache bypassing of data without the locality. Pre-fetching is also effective for streaming-type accesses with fixed strides.

As the first step, we are designing a vector architecture with an on-chip cache based on the NEC SX vector processor architecture as shown in Figure 5. We introduce the cache between vector registers and load/store units. In the SX vector processor, as main memory is connected to the processor through multiple on-chip memory ports, we provide a sub-cache for each port. Eventually, the on-chip cache consists of multiple n-way set-associative sub-caches. We think that the bandwidth between cache and register files should be kept

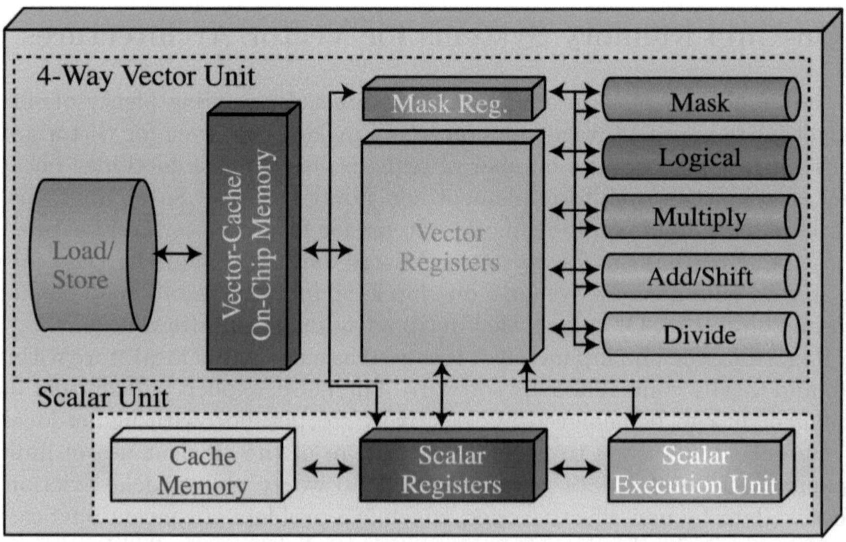

Fig. 5. Vector architecture with on-chip cache/local memory

at least at the 4B/flop rate, (a memory bandwidth of 400GB/s if 100Gflop/s chip is designed), which is the minimum balance point between the flop/s performance and memory bandwidth per vector processor core. We expect the cache to provide data with shorter latency and higher bandwidth compared to the off-chip memory system.

4 Performance Evaluation

4.1 Experimental Environment

To evaluate the effects of the on-chip cache for vector architectures, we are developing a simulator with a cache mechanism based on the NEC SX simulator. The simulator can change its memory bandwidth from 1B/flop to 8B/flop. We discuss the performance of the SX systems with 0.5MB and 2MB vector caches, and assume that the memory latency is 1.5 to 2 times longer than that of vector caches. The line size is 8B, as same as double precision data size. The write-through policy is employed. The number of memory ports are 32, and a 2-way set-associative sub-cache is connected to each port.

Table 2 shows five application codes used for benchmarking. The application codes have been developed by researchers of Tohoku University, and actually used in each research area. Through the performance evaluation, we will discuss how much capacity of vector-cache is equivalent to B/flop metric.

Table 3 summarizes the features of the five simulation codes. In the table, *Arithmetic Intensity* means the ratio of floating-point operations to memory references, and is a metric that suggests computational-intensiveness or

Table 2. Five application codes for benchmarking

Area	Simulation Code Name	Method	Subdivisions
Electro-Magnetic Analysis	*EM Scattering* Analysis of antenna radiation in a bore hole	FDTD	50x750x750
Electro-Magnetic Analysis	*Antenna* Electromagnetic wave analysis	FDTD	612x105x505
CFD/ Heat Analysis	*Combustion* Instability analysis of 2-dimensional premixed reactive flow in combustion	DNS	513x513
CFD/ Heat Analysis	*Heat Transfer* Analysis of separated flow and heat transfer	SMAC	711x91x221
Earth-quake Analysis	*Earthquake* Earthquake analysis of seismic slow slip on the plate boundary	Friction Law	32400x32400

Table 3. Characteristics of the application codes

Code Name	Memory Size	Arithmetic Intensity	Loop Length	Memory Access	VOR	AVL	Remarks
EM Scatter	8.9GB	1.1	500+	576-byte stride	99.7%	AVL	Low Spatial Locality
Antenna	12GB	2.25	255	Sequential	99.5%	255.5	Compute Intensive
Combustion	1.4GB	0.7	513	4K-byte Stride, Sequential	99.3%	179.0	Memory Intensive
Heat Transfer	6.6GB	1.0	349	16-byte stride	99.4%	192.9	Low Spatial Locality
Earthquake	8GB	1.0	32,400	Sequential	99.5%	255.5	Large Data I/O

memory-intensiveness of the application codes. VOR and AVL are the vectorization ratio and average vector length of the codes, respectively. In the following, we quickly review each code.

EM Scatter

This code is for a simulation of an array antenna, named *SAR-GPR (Synthetic Aperture Radar-Ground Penetrating Radar)*, which detects anti personnel mines in shallow subsurface [14]. This simulation method is the three dimensional FDTD (Finite Difference Time Domain) method with Berenger's PML (Perfectly Matched Layer) [7]. The FDTD is a way to describe a finite difference form of Maxwell's equations, and obtains electromagnetic field in a simulation space. The simulation space consists of two regions; air-space and subsurface space with PML of 10 layers. The performance of this code is primarily determined by the performance of electromagnetic field calculation processes. The basic computation structure of the processes consists of triple-nested loops with non-unit stride memory accesses. The ratio of the calculation cost to the total execution is 80%. The innermost loop length is 500+.

Antenna

The antenna code is for studying radiation patterns of an Anti-Podal Fermi Antenna (APFA) to design high gain antennas [15]. The simulation consists of two sections, a calculation of the electromagnetic field around an APFA using the FDTD method with Berenger's PML, and an analysis of the radiation patterns using the Fourier transform. The performance of the simulation is primarily determined by calculations of the radiation patterns. The ratio of its calculation cost to the total execution is 99%. The computation structure of the calculations is triple-nested loops; the innermost loop is a unit-stride loop, and its length is 255. The arithmetic intensity of the innermost loop is 2.25. Therefore, this code is computational-intensive and cache-friendly.

Combustion

This code realizes direct numerical simulations of two-dimensional Premixed Reactive Flow (PRF) in combustion for design of plane engines [8]. The simulation uses the 6th-order compact finite difference scheme and the 3rd-order Runge-Kutta method for time advancement. The performance of this code is primarily determined by calculations of derivations of physical equations. The ratio of its calculation cost to the total execution is 50%, and the rest of the cost has been distributed among various routines. The code has doubly nested loop; the loop of x-derivations induces unit-stride memory accesses, and the loop of y-derivations induces non-unit-stride memory accesses. The length of each innermost loop is 513.

Heat Transfer

This simulation code realizes direct numerical simulations of three-dimensional laminar separated flow and heat transfer on plane surfaces [11]. The finite-difference forms are the 5th-order upwind difference scheme for space derivatives and the Crank-Nicholson method for a time derivative. The resulting finite-difference equations are solved using the Simplified Marker And Cell (SMAC) method. The performance of the code is primarily determined by calculations of the predictor-corrector methods. The ratio of its calculation cost to the total execution is 67%, and the rest of the cost has been distributed among various routines. The code has triple-nested loops; the innermost loop needs unit-stride memory accesses, and its length is 349.

Earthquake

This code uses the three-dimensional numerical plate boundary models to explain an observed variation in propagation speed of post-seismic slip [5]. This is a quasi-static simulation in an elastic half-space including a rate and state-dependent friction. The performance of the simulation is primarily determined by a process of thrust stress with the Green function. The ratio of its calculation cost to the total execution is 99%. The computation structure of the process is a doubly nested loop that calculates a product of matrices. The innermost loop needs unit-stride memory accesses and its length is 32400.

4.2 Experimental Results and Discussion

Before discussion on the effects of vector caching for SX architecture, we evaluated the accuracy of the simulator that we have developed. Figure 6 shows

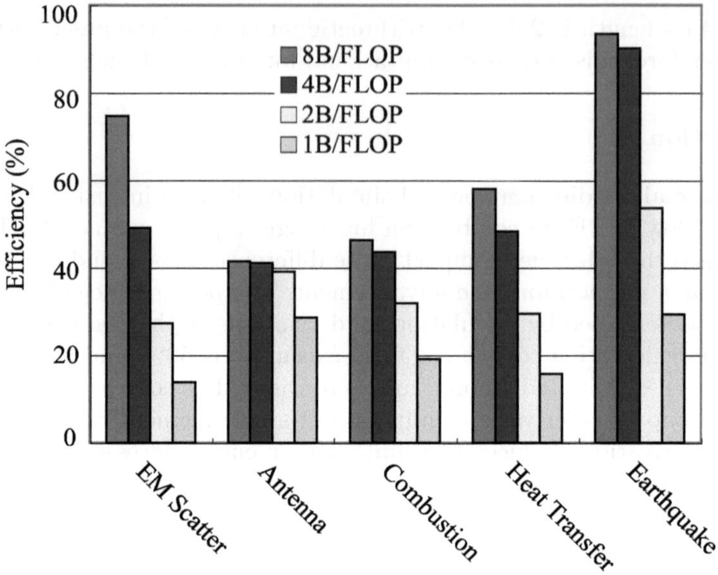

Fig. 6. Experimental results using SX-7 simulator

simulation results obtained by our simulator. By comparing the simulation results with the real results measured by SX-7 in Figure 4, our simulator is quite accurate for all the cases of 4B/flop, 2B/flop and 1B/flop in the five codes. We also examine the performance of the SX-7 in the 8Bytes/flop case, which is actually not available on the real SX-7. Although EM Scatter showed a scaled performance at a memory bandwidth of 8B/flop, most applications suggest the SX-7 has made a reasonable decision regarding its flop-bandwidth rate of 4Bytes/flop as a well-balanced system.

Figure 7 shows the relationship between the cache hit rates of the 2MB cache in the execution of the applications and the rates of the recovered efficiency by vector caching to the 4B/flop performance. The degree of the contribution of the vector caches highly depends on the characteristics of the applications, i.e., the locality of reference. Cache hit rates vary from 1.5% in earthquake at minimum to 96% in Antenna at maximum. In addition, the effect of vector caching to cover the lack of the memory bandwidth becomes remarkable for the 2B/flop system compared to the 1B/flop system. As the cache is an assisting mechanism on the vector system, the performance of load/store units is a key to keep the higher sustained performance, and the cache works as a booster for its base performance.

In the following subsections, we would like to pick up some applications for detailed discussion.

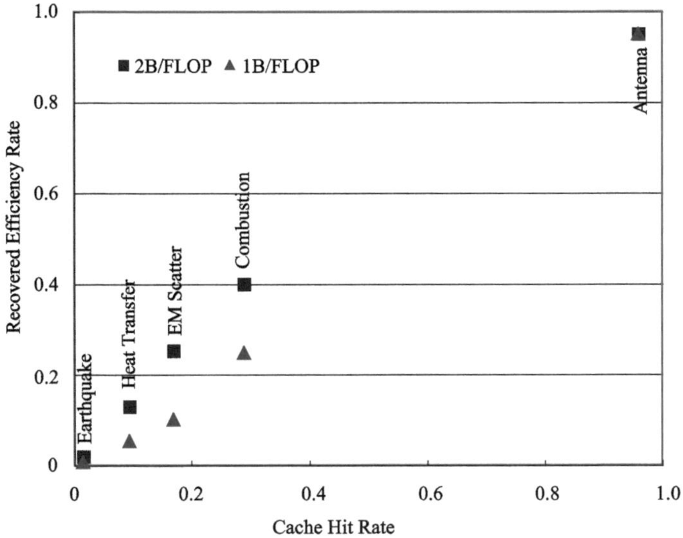

Fig. 7. Relationship between cache hit rates and recovered efficiency rates

EM Scatter

Figure 8 shows the performance of vector on-chip caching for EM Scatter. EM Scatter is a memory-intensive program, so in the execution of EM scatter, if the memory bandwidth is reduced to 2B/flop and 1B/flop, the performance is seriously affected, and the efficiency is reduced half and quarter the performance of the 4B/flop system, respectively, as show in Figure 8-(a). Figure 8-(b) shows the breakdown of processing time into the time for arithmetic operations and the time for load/store operations. Memory operations become dominant when limiting the memory bandwidth; meanwhile, the processing time for processing time stays at the same level, irrespective of the memory bandwidth. Therefore, the memory performance is an important factor to realize the efficient execution of this application. bandwidth.

Figure 9 is the high cost FDTD kernel of EM Scatter, which is the differential equation in space domain. In this kernel, vector loads for colored groups (orange, blue, and purple) of arrays are cache-friendly, and in each group, the preceding vector load fills the cache, and then the following vector load for the same array elements benefits from the cache for the data accesses. For example, in the i-th iteration, the preceding vector load for $H_Z(i,j,k)$ of the orange group brings the elements into the cache, and then cached $H_Z(i,j,k)$ will be reused in successive vector load for $H_z(i-1,j,k)$. Each array of this kernel consume 333MB memory space, but in the split-mined loop with 256 vector elements of each vector register, 2KB space is enough for each array in the innermost do-loop processing.

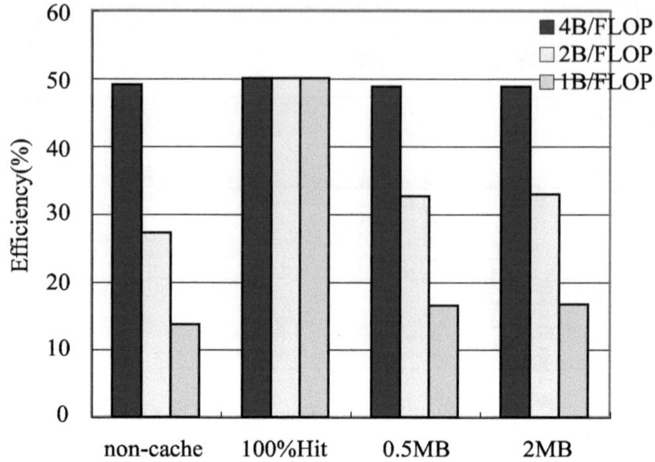

(a) Efficiency

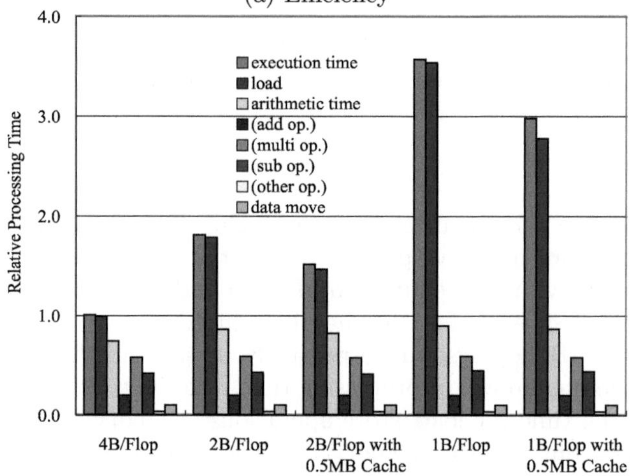

(b) Breakdown of processing time (Normalized)

Fig. 8. EM Scatter

Vector cache hit rates of 13% and 17% are obtained in the EM Scatter case when using 0.5MB and 2MB on-chip caches, respectively. As a result, the efficiency is improved by 5.4% with the 0.5MB cache and 5.7% with the 2MB cache for the 2B/flop case, and 2.8% with the 0.5MB cache and 3.0% with the 2MB cache for 1B/flop case. The experimental results suggest that the 2MB cache covers 24% of the memory bandwidth shortage (2B/flop) when employing the 2B/flop memory bandwidth compared to the 4B/flop system. It also covers 8% of the memory bandwidth shortage (3B/flop) of the 1B/flop system. To obtain the further improvement by vector caching, more capacity is required. The discussion on the effects of using larger caches remains as our future work.

```
DO 10 k=0,Nz
   DO 10 i=0,Nx
      DO 10 j=0,Ny
         E_x(i,j,k) = C_x_a(i,j,k) * E_x(i,j,k)
&              + C_x_b(i,j,k) * ((H_z(i,j,k) - H_z(i,j-1,k))/dy
&              - (H_y(i,j,k) - H_y(i,j,k-1))/dz - E_x_Current(i,j,k))
         E_z(i,j,k) = C_z_a(i,j,k) * E_z(i,j,k)
&              + C_z_b(i,j,k) * ((H_y(i,j,k) - H_y(i-1,j,k))/dx
&              - (H_x(i,j,k)-H_x(i,j-1,k))/dy - E_z_Current(i,j,k))
         E_y(i,j,k) = C_y_a(i,j,k) * E_y(i,j,k)
&              + C_y_b(i,j,k) * ((H_x(i,j,k) - H_x(i,j,k-1))/dz
&              - (H_z(i,j,k) - H_z(i-1,j,k))/dx - E_y_Current(i,j,k))
10 CONTINUE
```

Fig. 9. High cost kernel of EM Scatter

Antenna

Figure 10 show the results of Antenna. As Antenna is a computation intensive code, 2B/flop still works fine, but 1B/flop decreases the efficiency by 12.6%. In this code, there is a large locality of reference, and the cache hit rate reaches 95% even with a 512KB capacity. Figure 11 shows the kernel of Antenna with the highest cost. In the innermost loop, there are many loop independent arrays, colored in red in this kernel, each needs 2KB space, and therefore, effective use of on-chip cache can recover a 13% efficiency loss due to the limited memory bandwidth of 1B/flop, resulting in the 4B/flop equivalent performance obtained.

Effects of Selective Caching in Heat Transfer

As the on-chip size is limited, selective caching might be an effective approach to efficient use of on-chip cache, which is expected to reduce capacity miss. Figure 12 shows the effects of selective caching in Heat Transfer. Here, we discusses the cache sizes from 32KB up to 2MB. Figure 13 shows one of several kernels of Heat Transfer to examine the effects of selective caching. In the case of the 32K cache, no cache hit is obtained if all the arrays in the kernel are cached. However, if caching of data is restricted to selectively store arrays *Phi* on the cache, the capacity misses can be decreased, and some of the following data accesses to these arrays are successively provided from the cache. As a result, a 3% improvement in relative performance is obtained even with the 32KB cache. In addition, as a 1.8MB cache capacity is required to capture all the arrays regarding loop j in Figure 13, the size of 2MB is minimum for the cache with the all-data-caching policy. However, if selective caching

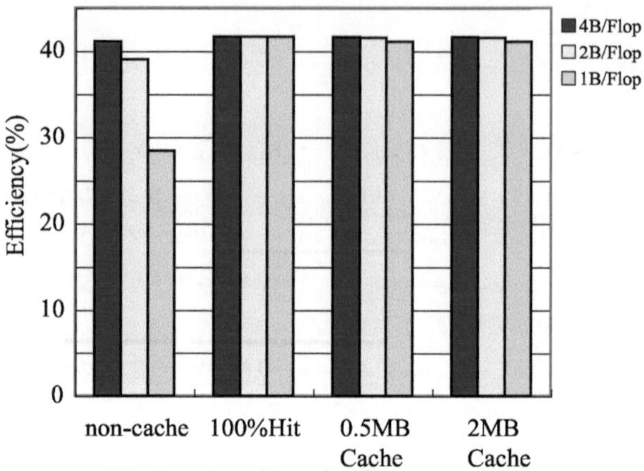

(a) Efficiency

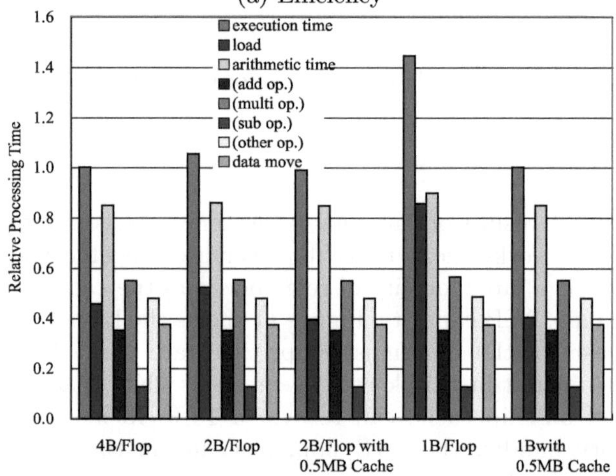

(b) Breakdown of processing time (Normalized)

Fig. 10. Antenna

of the colored arrays *Phi* is performed, vector loads for underlined arrays are effectively provided from the cache, and a 2.8 times higher hit rate is obtained even with the 1MB cache compared to the case where all the arrays are cached, resulting in more than 9% improvement in processing time. Of course more detailed discussion on the results of larger on-chip caches for different kernels is required, but these results suggest a potential of the selective caching for efficient on-chip data management on the limited on-chip capacity.

```
do kio=ki1,ki2,jcbs
do if=1,nf
   do ki=kio,min(kio+jcbsm,ki2)
      m1=NO+mm1(ki,ithixyz)
      m2=NO+mm2(ki,ithixyz)
      x1=x11(ki,ithixyz)
      x2=x12(ki,ithixyz)
      x3=x13(ki,ithixyz)
      y1=y11(ki,ithixyz)
      y2=y12(ki,ithixyz)
      y3=y13(ki,ithixyz)
      CC1=CDEXP(-CI*2.D0*PI*DT*FRE(if)*m1)
&         *(X1+cwexp1(if)*X2+cwexp2(if)*X3)
      CC2=CDEXP(-CI*2.D0*PI*DT*FRE(if)*m2)
&         *(Y1+cwexp1(if)*Y2+cwexp2(if)*Y3)
      ex1=eex1(ki); ex2=eex2(ki)
      ez1=eez1(ki); ez2=eez2(ki)
      CUX(if,ITHIXYZ)=CUX(if,ITHIXYZ)+DXZC*PIC*(EZ1*CC1-EZ2*CC2)
      CUZ(if,ITHIXYZ)=CUZ(if,ITHIXYZ)-DXZC*PIC*(EX1*CC1-EX2*CC2)
      HX1=hhx1(ki); HX2=hhx2(ki)
      HZ1=hhz1(ki); HZ2=hhz2(ki)
      CWX(if,ITHIXYZ)=CWX(if,ITHIXYZ)-DXZC*PIC*(HZ1*CC1-HZ2*CC2)
      CWZ(if,ITHIXYZ)=CWZ(if,ITHIXYZ)+DXZC*PIC*(HX1*CC1-HX2*CC2)
     end do;
   end do;
end do
```

Fig. 11. High cost kernel of Antenna

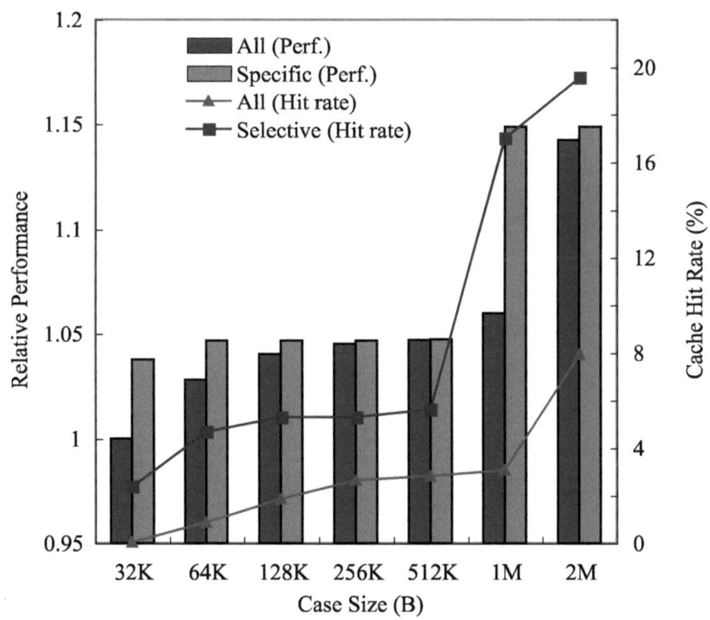

Fig. 12. Effects of selective caching on performance (Heat Transfer)

```
do 110 l=1,8
   do 110 k=KST(l),NK,2
      do 110 j=JST(l),NJ,4
         do 110 i=IST(l),NI,2
            res1=(AN(i,j,k)*Phi(i,j+1,k)+AS(i,j,k)*Phi(i,j-1,k)
&                 +AE(i,j,k)*Phi(i+1,j,k)+AW(i,j,k)*Phi(i-1,j,k)
&                 +AT(i,j,k)*Phi(i,j,k+1)+AB(i,j,k)*Phi(i,j,k-1)
&                 +AP(i,j,k)*Phi(i,j,k)-DIV(i,j,k))/AP(i,j,k)
            res2=(AN(i,j+2,k)*Phi(i,j+3,k)+AS(i,j+2,k)*Phi(i,j+1,k)
&                 +AE(i,j+2,k)*Phi(i+1,j+2,k)+AW(i,j+2,k)*Phi(i-1,j+2,k)
&                 +AT(i,j+2,k)*Phi(i,j+2,k+1)+AB(i,j+2,k)*Phi(i,j+2,k-1)
&                 +AP(i,j+2,k)*Phi(i,j+2,k)-DIV(i,j+2,k))/AP(i,j+2,k)
            res1=res1*RGNc(i,j,k)
            res2=res2*RGNc(i,j+2,k)
            Phi(i,j,k)=Phi(i,j,k)-res1
            Phi(i,j+2,k)=Phi(i,j+2,k)-res2
110   continue
```

Fig. 13. High cost kernel of Heat Transfer for selective caching

Earthquake: Case Causing Conflict between Loop-Unrolling and Caching

As Figure 7 shows, the effect of vector caching for Earthquake is quite low; the hit rates of the 0.5MB and 2MB caches are 0.3% and 1.5%, respectively. The 2MB cache realize only 2% coverage of the bandwidth shortage for the 2B/flop system, and 0.7% coverage for the 1B/flop system, compared with the 4B/flop system. These quite lower hit rates are due to the conflict between loop-unrolling and caching. Loop unrolling is effective for higher utilization of vector pipelines, but it also needs more cache capacity to capture all the arrays of unrolled loops. Therefore, controlling the degree of loop unrolling according to the cache capacity is very important to increase the efficiency when the bandwidth is limited. In addition, selective caching is also effective to have a synergetic effect of loop unrolling and caching. More detailed discussion on the trade-off between loop unrolling and caching when limiting the memory bandwidth remains as our future work.

5 Summary

In this paper, we have discussed the potential of on-chip memory subsystems for future vector architectures. The performance evaluation based on the early experiments suggests that even with moderate-sized on-chip cache with 512KB to 2MB, it covered a lack of the memory bandwidths of vector load/store units with 2B/flop or lower, and boosted the sustained system performance up to the level of the 4B/flop performance. Selective caching, in which only the data with the high locality of reference are cached, is also effective for efficient use of limited on-chip caches.

As there are regular and predicable memory references in many vector codes, and they can be efficiently processed through rich addressing modes of the vector ISA, we think there are incredible opportunities for architectural innovations in design of future vector systems in a multi-core era.

Acknowledgments

This work has been done in collaboration between Tohoku University and NEC, and many colleagues contribute to this project. We would also like to thank Professors Motoyuki Sato, Akira Hasegawa, Goro Masuya, Terukazu Ota, and Kunio Sawaya of Tohoku University for providing the codes for the experiments.

References

1. M.Resh et al. Ed. *High Performance Compting on Vector Systems 2006.* Springer, 2006.
2. J. Hennessy and D.Patterson. *Computer Architecture: A Quantitative Approach:4th Edition.* Morgan Kaufmann Publishers, 2006.
3. H.Kobayashi. Implication of Memory Performance in Vector-Parallel and Scalar-Parallel HEC System. *Proceedings of 4th Teraflop Workshop (High Performance Computing on Vector Systems 2006)*, pages 21–51, 2006.
4. J.D.McCalpin. STREAM: Sustainable Memory Bandwidth in High Performance Computers. *http://www.cs.virginia.edu/stream*, 2005.
5. K.Ariyoshi et al. Spatial variation in propagation speed of postseismic slip on the subducting plate boundary. *Proceedings of 2nd Water Dynamics*, 35, 2004.
6. K.Kitagawa et al. A Hardware Overview of SX-6 and SX-7 Supercomputer. *NEC Research & Development*, 44(1):2–7, 2003.
7. K.S.Kunz and R.J.Luebbers. *The Finite Difference Time Domain Method for Electromagnetics.* CRC Press, 1993.
8. K.Tsuboi and G.Masuya. Direct Numerical Simulations for Instabilities of Remixed Planar Flames. *Proceedings of The Fourth Asia-Pacific Conference on Combustion*, 2003.
9. L.Oliker et al. Scientific Computations on Modern Parallel Vector Systems. *Proceedings of SC2004*, 2004.
10. L.Oliker et al. Leading Computational Methods on Scalar and Vector HEC Platforms. *Proceedings of SC2005*, 2005.
11. M.Nakajima et al. Numerical Simulation of Three-Dimensional Separated Flow and Heat Transfer around Staggerd Suerface-Mounted Rectangular Blocks in a Channel. *Numerical Heat Transfer*, 47(Part A):691–708, 2005.
12. NEC. SX-8R Press Releae. *http://www.hpce.nec.com/*, 2006.
13. T.H.Dunigan Jr. et al. Performance Evaluation of The Cray X1 Distributed Shared-Memory Architecture. *IEEE MICRO*, 25(1):30–40, 2005.
14. T.Kobayashi et al. FDTD simulation on array antenna SAR-GPR for land mine detection. *Proceedings of 1st International Symposium on Systems and Human Science*, pages 279–283, 2003.

15. Y.Takagi et al. Study of High Gain and Broadband Antipodal Fermi Antenna with Corrugation. *Proceedings of 2004 International Symposium on Antennas and Propagation*, 1:69–72, 2004.

The Road to TSUBAME and Beyond

Satoshi Matsuoka

Global Scientific Information and Computing Center,
Tokyo Institute of Technology
matsu@is.titech.ac.jp

TSUBAME (Tokyo-tech Supercomputer and Ubiquitously Accessible Mass-storage Environment) is a new supercomputer installed at Tokyo Institute of Technology in Tokyo, Japan, on April 1st, 2006, facilitating over 85 Teraflops of peak compute power with acceleration, 21 Terabytes of memory, and 1.6 Petabytes (initially 1.1 Petabytes) of online disk storage, "Fat Node" as well as fast parallel interconnect supercomputers. TSUBAME became the fastest and largest supercomputer in Asia in terms of performance, memory and storage capacity etc. It recorded 38.18 Teraflops performance for the June 2006 Top500 [1], making it the 7th fastest supercomputer in the world at the time and fastest in the Asia-Pacific, besting the Earth Simulator. TSUBAME continues to improve performance in three consecutive Top500s, and with lucid use off acceleration, has now marked 48.88 Teraflops. At the same time, being PC architecture-based, TSUBAME, being a large collection of PC servers, allows for offering much broader services than traditional supercomputers resulting in a much wider user base, including incubation of novice students. We term such architectural and operational property of TSUBAME as "Everybody's Supercomputer", as opposed to traditional supercomputers with very limited number of users, thus making their financial justifications increasingly difficult.

TSUBAME was designed, procured, and installed at the end of March, 2006. The contract was awarded to NEC, who jointly with Sun Microsystems built and installed the entire machine, and also jointly provide on-site engineering to operate the machine. Other commercial partners, such as AMD (Opteron CPUs), Voltaire (Infiniband), ClearSpeed (Accelerator), CFS (LUSTRE parallel filesystem), Novell (SUSE Linux), provided their own products and expertise as building blocks. It was installed in just three weeks, and when its operation started on April 3, 2006, it became the largest academic machine in the world to be hosted by a university. The overall architecture of TSUBAME is shown in Figure 1, and the actual photos in Figure 2.

Overall, TSUBAME's installation space is approximately $350m^2$ including the service area. There are approximately 80 compute/storage/network

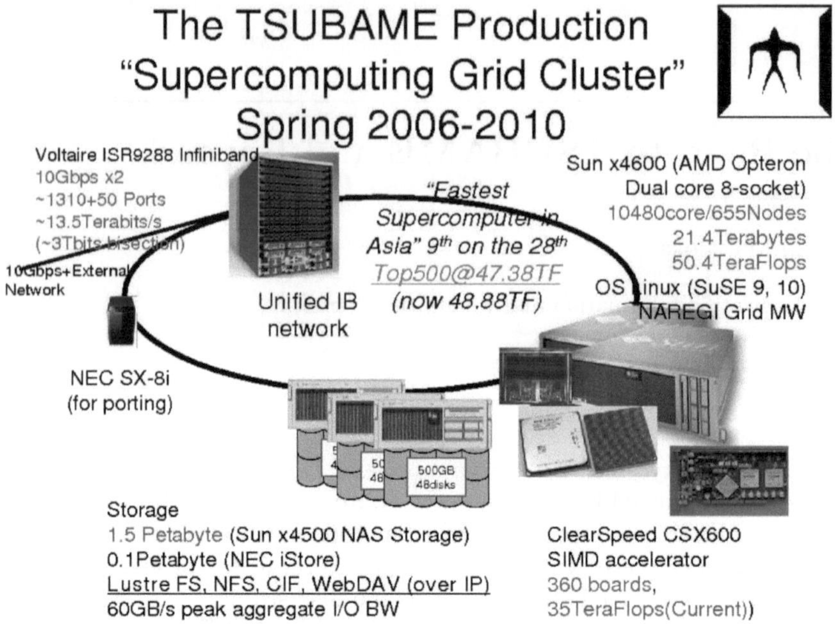

Fig. 1. Overview of TSUBAME Architecture

Fig. 2. Tsubame pictures

racks, as well as 32 CRC units for cooling, that are laid out in a customized fashion to maximize cooling efficiency, instead of the machine itself merely being placed as an afterthought. This allows for considerable density and much better cooling efficiency compared to other machines of similar performance. The total weight of TSUBAME exceeds 60 tons, requiring minor building reinforcements as the current building was designed for systems of much smaller scale. TSUBAME occupies three rooms, where room-to-room Infiniband connections are achieved via optical fiber connection, whereas uses the CX4 copper cable within a room. The total power consumption of TSUBAME is less than a Megawatt even at peak load, making it one of the most power-efficient supercomputer in the 100 Teraflops performance scale.

TSUBAME's lifetime was initially designed to be 4 years, until the spring of 2010. This could be extended up to a year with interim upgrades, such as upgrade to future quad-core processors or beyond. However, eventually the lifetime will expire, and we are already beginning the plans for designing "TSUBAME 2.0". One design consideration that is clear at the moment is that the success of "Everybody's Supercomputer" should be continued; however, simply waiting for processor improvements relying on CPU vendors would not be sufficient to meet the growing demands, as a result of success of "Everybody's Supercomputer", in growth of the supercomputing community itself, not just the individual needs.

Another requirement is not to increase the power or the footprint requirement of the current machine, resulting in a considerable challenge in supercomputer design we are researching at the current moment. One such research investment is in the area of acceleration technologies, which will provide vastly improved Megaflops/Watt ratio. In fact, even currently, two-fifth of TSUBAME's peak computing power is provided by the ClearSpeed Advanced Accelerator PCI-X board. However, acceleration technology is still narrowly scoped in terms of its applicability and user base; as such, we must generalize the use of acceleration via advances in algorithm and software technologies, as well as design a machine with right mix of various heterogeneous resources, including general-purpose processors, and various types of accelerators. Another factor is storage, where multi-Petabyte storage with high bandwidth must be accommodated. Challenges are in devising more efficient cooling, better power control, etc. etc.... there are various challenges abound, and it will require advances in multi-disciplinary fashion to meet this challenge. This is not a mere pursuit of FLOPS, but rather, "pursuit of FLOPS usable by everyone" — a challenge worthwhile taking for those of us who are computer scientists. And the challenge will continue beyond TSUBAME2.0 for many years to come.

References

1. Top500 Supercomputers Sites, http://www.top500.org/.

MIX
Papier aus verantwortungsvollen Quellen
Paper from responsible sources
FSC® C105338

If you have any concerns about our products,
you can contact us on
ProductSafety@springernature.com

In case Publisher is established outside the EU,
the EU authorized representative is:
Springer Nature Customer Service Center GmbH
Europaplatz 3, 69115 Heidelberg, Germany

Printed by Libri Plureos GmbH
in Hamburg, Germany